Engineering Applications of Finite Element Analysis

Engineering Applications of Finite Element Analysis

Edited by **Connie Mcguire**

NY RESEARCH PRESS
New York

Published by NY Research Press,
23 West, 55th Street, Suite 816,
New York, NY 10019, USA
www.nyresearchpress.com

Engineering Applications of Finite Element Analysis
Edited by Connie Mcguire

International Standard Book Number: 978-1-63238-178-1 (Hardback)

Printed in China.

Contents

	Preface	**IX**
Section 1	**Applications of FEA in "Fluid Mechanics and Heat Transfer"**	**1**
Chapter 1	**Control Volume Finite Element Methods for Flow in Porous Media: Resin Transfer Molding** Jamal Samir, Jamal Echaabi and Mohamed Hattabi	**3**
Chapter 2	**Electromagnetic and Fluid Analysis of Collisional Plasmas** Antonis P. Papadakis	**31**
Section 2	**Applications of FEA in "Structural Mechanics and Composite Materials"**	**63**
Chapter 3	**Finite Element Analysis of Functionally Graded Piezoelectric Spheres** A. Ghorbanpour Arani, R. Kolahchi, A. A. Mosalaei Barzoki, A. Loghman and F. Ebrahimi	**65**
Chapter 4	**Finite Element Analysis of Bias Extension Test of Dry Woven** Samia Dridi	**87**
Chapter 5	**Nonlinear Large Deflection Analysis of Stiffened Plates** Khosrow Ghavami and Mohammad Reza Khedmati	**109**
Chapter 6	**3D Nonlinear Finite Element Plastic Analysis of Cylindrical Vessels Under In-Plane Moment** B.H. Wu	**137**

Chapter 7 **Flexural Behavior of Functionally Graded Sandwich Composite** 153
Mrityunjay R. Doddamani and Satyabodh M Kulkarni

Section 3 Applications of FEA in "Wave Propagation and Failure-Analysis" 177

Chapter 8 **Modeling Dynamics Behaviour of Materials: Theoretical Framework and Applications** 179
Giovanni Leonardi and Michele Buonsanti

Chapter 9 **Perfectly Matched Layer for Finite Element Analysis of Elastic Waves in Solids** 197
Koji Hasegawa and Takao Shimada

Chapter 10 **Finite Element Analysis Applications in Failure Analysis: Case Studies** 217
Ahmad Ivan Karayan, Deni Ferdian, Sri Harjanto, Dwi Marta Nurjaya, Ahmad Ashari and Homero Castaneda

Section 4 Applications of FEA in "Machine Elements Analysis and Design" 235

Chapter 11 **Overview in the Application of FEM in Mining and the Study of Case: Stress Analysis in Pulleys of Stacker-Reclaimers: FEM vs. Analytical** 237
Jairo A. Martins and István Kövesdy

Chapter 12 **Finite Element Analysis Coupled with Feedback Control for Dynamics of Metal Pushing V-Belt CVT** 257
Toshihiro Saito

Chapter 13 **Development of Graphical Solution to Determine Optimum Hollowness of Hollow Cylindrical Roller Bearing Using Elastic Finite Element Analysis** 273
P.H. Darji and D.P. Vakharia

Chapter 14 **Optimization and Improvement of Throwing Performance in Baseball Pitching Machine Using Finite Element Analysis** 297
Shinobu Sakai and Hitoshi Nakayama

Section 5 **Applications of FEA in "Machining and Product Design"** **325**

Chapter 15 **Finite Element Analysis of Machining Thin-Wall Parts: Error Prediction and Stability Analysis** **327**
YongAn Huang, Xiaoming Zhang and Youlun Xiong

Chapter 16 **Concurrent Process-Product Design Optimization Using Coupled Nonlinear Finite-Element Simulations** **355**
Ali Najafi and Masoud Rais-Rohani

Permissions

List of Contributors

Preface

This book has been an outcome of determined endeavour from a group of educationists in the field. The primary objective was to involve a broad spectrum of professionals from diverse cultural background involved in the field for developing new researches. The book not only targets students but also scholars pursuing higher research for further enhancement of the theoretical and practical applications of the subject.

This book discusses the engineering applications of finite element analysis. In the past few years, Finite Element Analysis (FEA) has been refined into an important and necessary technology in the modeling and simulation of several engineering systems. This book includes contributions made by professionals from international scientific organizations. It compiles authentic and interesting research studies on the applications of FEA in some major areas of mechanical engineering like fluid mechanics and heat transfer, machine elements evaluation and design, machining and product design, wave propagation and failure evaluation, and structural mechanics and composite materials. This book presents an overview of contemporary research activities across the globe in this area and will be beneficial to a large number of researchers. The introductions, data and references provided in the book will help the readers learn more about this subject.

It was an honour to edit such a profound book and also a challenging task to compile and examine all the relevant data for accuracy and originality. I wish to acknowledge the efforts of the contributors for submitting such brilliant and diverse chapters in the field and for endlessly working for the completion of the book. Last, but not the least; I thank my family for being a constant source of support in all my research endeavours.

Editor

Section 1

Applications of FEA in "Fluid Mechanics and Heat Transfer"

Control Volume Finite Element Methods for Flow in Porous Media: Resin Transfer Molding

Jamal Samir, Jamal Echaabi and Mohamed Hattabi

Additional information is available at the end of the chapter

1. Introduction

The method of solution for the own problem is based on the control volume finite element approach which has been shown to be particularly well suited for a fast and efficient implementation of the Newton-Raphson linearization technique.

The basic idea of the control volume finite element approach is to obtain a discretized equation that mimics the governing mass conservation equation locally. A volume of influence, referred to as a control volume, is assigned to each node. The discretized equation for a given node then consists of a term describing the change in fluid mass storage for that volume which is balanced by the term representing the divergence of the fluid mass flux in the volume. The fluid mass flux will depend on the physical properties associated with the volume and the difference in the value of the primary variable between the node in question and its neighbors.

Discretization of the subsurface and the surface flow equations is identical except for the difference in dimensionality. For the sake of clarity, we present in this chapter, a detailed description of the control volume finite element method applied to discretize a simplified prototype continuity equation in Liquid composite molding (LCM) . The final discretized equations for all subsurface domains and for surface flow are then presented without providing the details of the derivation.

Liquid composite molding (LCM) processes are routinely considered as a viable option to manufacture composite parts. In this process, a fibrous preform is placed in a mold. The mold is sealed and a liquid thermoset resin is injected to impregnate the fibrous preform. All LCM processes involve impregnation of the resin into a bed of fibrous network. The goal is to saturate all the empty spaces between the fibers with the resin before the resin gels and then solidifies. In RTM, resin is injected slowly and little or no heat transfer and chemical reaction takes place until the mold is filled. Thus, the entire cycle can be viewed as two separate events, fill and subsequent cure.

The mould filling is considered as one of the most critical and complicated stages throughout the entire RTM process. It has a great influence on the performance and quality of the final parts. However, it is hard to understand effects of the filling parameters on the flow front pattern during mold filling. Therefore, it is necessary to understand interrelationship among filling parameters, flow behavior during RTM, and physical properties of the final parts.

The present study concerns the numerical modeling of resin transfer molding techniques (RTM). From a mechanical point of view, these processes can be treated in the same way as the problems of fluids in porous media. Some of authors use methods based on the systems of curvilinear coordinates adapted to a border. However, this approach becomes limited during divisions or fusion of the flow fronts [1-4]. The modeling methods currently elaborated are based on a control volume finite element method (CVFEM). This type of approach was first presented by Wang and others and was adopted in the case of thin shell injection molding [5]. Fracchia was the first to apply the CVFEM to simulate the RTM process [6] and other researchers also followed this approach [7-16]. The application of these methods generate several commercial software: RTM-FLOT (no longer supported), PAM-RTM, MyRTM and LIMS.

In the last years, despite the introduction of alternative methods for simulation of flow in porous media BEM (Trochu [17]) and SPH (Krawczak [18]), the CVFEM method has been, usually, used to simulate resin flow in the RTM process. However, this numerical approach has some inherent drawbacks. First, the flow front is difficult to define with the exact location because of using fixed mesh system. This problem in the resin flow front location limits the accuracy of CVFEM method [19]. Mass conservation problems have also been reported with the use of this numerical approach [20,21]. Researchers have addressed these numerical problems and put forward methods to improve the conventional CVFEM method [19, 22].

In this study, the simulation of the resin flow in the RTM process is developed by the control volume finite element method (CVFEM) coupled with the equation of the free surface location. The equation is solved at each time step using nonconforming linear finite elements on triangles, which allow the conservation of the resin flow rate along inter-element boundaries [23]. At each time step, the velocity and pressure in the saturated domain is calculated. The effective velocity is used to update the front position. The filling algorithm determines the time increment needed to fill up completely at least one new element, then the boundary condition is updated and the flow front is advanced for the next iteration. The flow front is refined in an adaptive manner at each time step by using our mesh generator to add new nodes, to get a smoother flow front and reduce the error in the pressure at the flow front of a CVFEM simulation of resin flow in a porous medium. Thus, the position of the flow front, the time-lapse and the rate of the unsaturated zone are calculated at every step. Our results will be compared with the experimental and analytical models in the literature.

On the whole, our study is concerned with the simulation of isothermal filling of moulds in RTM process while adopting the CVFEM and VOF method, taking into account the presence of obstacles and the thickness variation of the reinforcement. The elaborated code allows calculating the position of the injection points and vents, and injection pressure in order to optimize the process parameters. We present a mesh generator for 1D, 2D and 2.5 D geometries for one or several points of injection and vents, and a numerical method for the simulation of the RTM process. Numerical examples are used to validate and assess the applicability of the developed code in the case of anisotropic reinforcements, multilayered, several injection points and the existence of inserts.

2. Mathematical formalism

2.1. Continuity equations and Darcy law

In RTM process and when filling the mould, the resin flow passes through a bed of fibers, The process of injection in the mould is treated as part of the flows of fluids inside a porous medium.

On the basis of partial saturation concept, the mass balance at a point within the domain of an isothermal incompressible fluid flow inside a fiber preform can be expressed as [24] :

$$\phi \frac{\partial s}{\partial t} = -\nabla . q \tag{1}$$

Where q is the volumetric flow rate per unit area, ϕ is the porosity and the saturation level s is 1 for a fully saturated node and its value ranges between 0 and 1 for a partially saturated point. If the transient term on the left hand side of the above equation is removed (saturate case), the following equation for quasi-steady state situation is obtained:

$$\frac{\partial q}{\partial x_i} = 0 \tag{2}$$

As the fluid flows through the pores of the preform, the interstitial velocity of the resin is given by :

$$v_i = \frac{q_i}{\phi} \tag{3}$$

Where v_i the intrinsic phase average resin velocity within the pores and ϕ is the porosity of the solid.

Using the assumptions that the preform is a porous medium and that the flow is quasi-steady state, the momentum equation can be replaced by Darcy's law:

$$q_i = -\frac{K_{ij}}{\mu} \frac{\partial P}{\partial x_j} \tag{4}$$

where μ is the fluid viscosity, K_{ij} is the permeability tensor of the preform, and P is the fluid pressure.

Assuming that the resin is incompressible and substituting (4) into (2) gives the governing differential equation of the flow :

$$\frac{\partial}{\partial x_i}\left(\frac{K_{ij}}{\mu}\frac{\partial P}{\partial x_j}\right)=0 \tag{5}$$

This second order partial differential equation can be solved when the boundary conditions are prescribed. Two common boundary conditions for the inlet to the mould are either a prescribed pressure condition:

$$P_{inlet}=P_{inlet}(t) \tag{6}$$

Or a prescribed flow rate condition:

$$Q_n=n_i\frac{K_{ij}}{\mu}\frac{\partial P}{\partial x_j} \tag{7}$$

Where Q_n is the volumetric flow rate and n_i is the normal vector to the inlet.

The boundary conditions along the flow front are as follows:

$$P_{front}=0 \tag{8}$$

Since the resin cannot pass through the mould wall, the final boundary condition necessary to solve equation (5) is that the velocity normal to the wall at the boundary of the mould must be zero:

$$\overline{v}\,.\,\overline{n}=0 \tag{9}$$

Where n is the vector normal to the mould wall.

3. Discretization of the domain by CV /FEM -VOF method

3.1. Delaunay triangulations

In mathematics and computational geometry, a Delaunay triangulation for a set P of points in the plane is a triangulation DT (*P*) such that no point in *P* is inside the circumcircle (figure 1)of any triangle in DT (P). Delaunay triangulations maximize the minimum angle of all the angles of the triangles in the triangulation; they tend to avoid triangles with high aspect ratio [25].

Suppose $P = \{p_1,\ldots.., p_n\}$ is a point set in *d* dimensions. The convex hull of *d+1* affinely independent points from P forms a *Delaunay simplex* if the circumscribed ball of the simplex contains no point from P in its interior. The union of all Delaunay simplices forms the Delaunay diagram, DT (P). If the set P is not degenerate then the DT (P) is a simplex decomposition of the convex hull of P.

The geometric dual of Delaunay Diagram is the *Voronoi Diagram,* which consists of a set of polyhedra $V_1, \ldots, V_n$, one for each point in P, called the Voronoi Polyhedra. Geometrically, V_i is the set of points $p \in \Re^d$ whose Euclidean distance to p_i is less than or equal to that of any other point in P. We call p_i the center of *polyhedra* V_i. For more discussion, see [26, 27].

The DT has some very desired properties for mesh generation. For example, among all triangulations of a point set in 2D, the DT maximizes the smallest angle, it contains the nearest-neighbors graph, and the minimal spanning tree. Thus Delaunay triangulation is very useful for computer graphics and mesh generation in two dimensions. Moreover, discrete maximum principles will only exist for Delaunay triangulations. Chew [28] and Ruppert [29] have developed Delaunay refinement algorithms that generate provably good meshes for 2D domains.

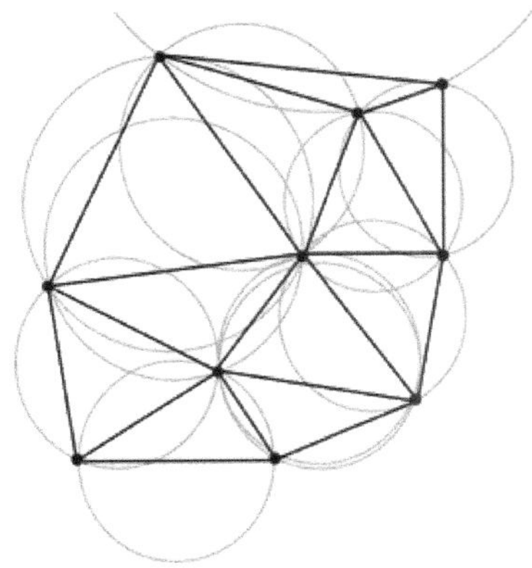

Figure 1. A Delaunay triangulation in the plane with circumcircles.

3.2. Discretization domain

In processes such as Resin Transfer Moulding (RTM), numerical simulations are usually performed on a fixed mesh, on which the numerical algorithm predict the displacement of the flow front. Error estimations can be used in the numerical algorithm to optimize the mesh for the finite element analysis. The mesh can also be adapted during mould filling to follow the shape of the moving boundary. In CVFEM, the calculation domain is first discretized using finite elements, and then each element is further divided into sub-volumes. For the discretization of the calculation domain in FEM, we developed a mesh generator (figure 2) allowing to generate 2 and 2.5 dimensional, unstructured Delaunay and constrained Delaunay triangulations in general domains.

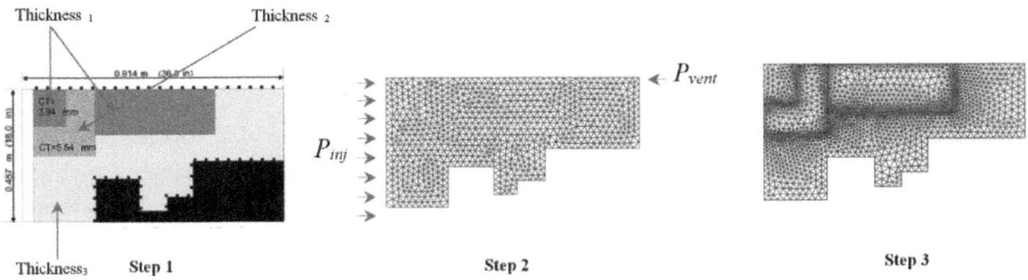

Figure 2. Discretization of calculation domain

3.3. Domain discretization CV/FEM

To use the method CV/FEM coupled with VOF, the mould is first divided into finite elements. Around each nodal location, a control volume is constructed by subdividing the elements into smaller volumes. These control volumes are used to track the location of the flow front.

The calculation domain is in a finite number of triangular elements. After connecting the centroides of the elements with the middles of the elements borders, the calculation domain another time being divided in a number of polygonal control volume, as indicated in figure 3. The borders of any element of the control volume constitute the control surface.

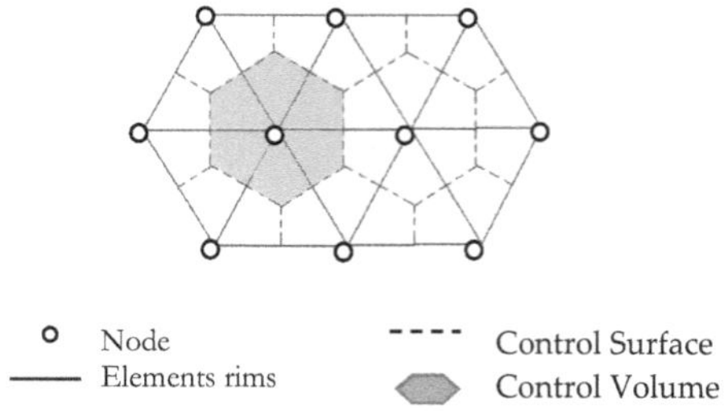

Figure 3. Discretization of the calculation domain during CV/FEM

3.4. Resin front tracking

The control volumes can be empty, partially filled, or completely filled. The amount of fluid in each control volume is monitored by a quantity called the fill factor. It is the ratio of the volume of fluid to the total volume of the control volume. The fill factor takes values from 0 to 1 where 0 represents totally empty and 1 represents totally full. The control volume method tracks the flow front by determining which control volumes are partially filled and connecting them to form the flow front. The numerical flow front is composed from the nodes with partially filled control volumes as shown in figure 4. The location of the fluid in the control volume cannot be identified, therefore the exact shape of the flow front is not known. Thus, the mesh density can affect seriously the accuracy of numerical solution of the flow front.

For any control volume and after integration equation 5, we obtained the following relationship:

$$\int_{cv} \frac{\partial}{\partial x_i}\left(\frac{K_{ij}}{\mu}\frac{\partial P}{\partial x_j}\right) d\Omega = \int_{cs} \frac{K_{ij}}{\mu}\frac{\partial P}{\partial x_i}\vec{n}\, ds = 0 \tag{10}$$

Where s,Ω, $\vec{n}$ are the control surface, the control volume and the normal vector of the control surface, respectively. CV and CS represent the control volume and the control surface domains respectively.

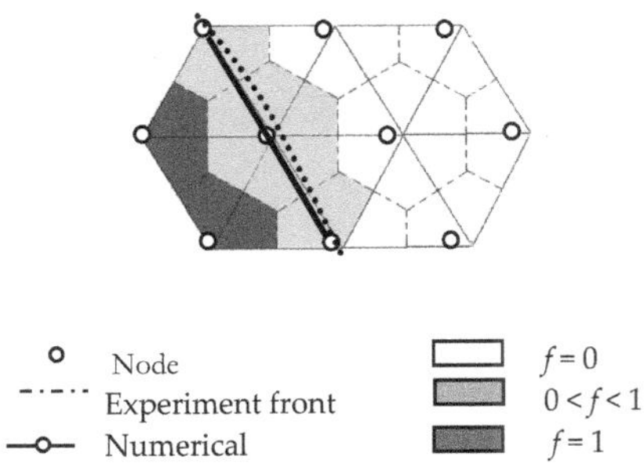

Figure 4. Treatment of the flow front during the fixed meshing method.

Moreover, in every iteration, the calculation matrix contains only elements that have at least one node with a filling ratio unity *f*=*1*. This approach requires a rigor during the development of code. However, the time of treatment of the problem is ameliorated.

4. Numerical simulation

4.1. Pressure

During the resolution of the pressure field, we adopted Galerkin's approximation to represent the distribution of the field of pressure.

Using the procedure outlined by Reddy [30], the finite element formulation of Equation (5) was found to be:

$$\left[K_{ij}^e\right]\left\{P_j^e\right\} = \left\{F_i^e\right\} \tag{11}$$

Where

$$K_{ij}^e = \int_{\Omega_e} \frac{K_{\alpha\beta}}{\mu}\frac{\partial \psi_i}{\partial x_\alpha}\frac{\partial \psi_j}{\partial x_\beta} d\Omega \tag{12}$$

And

$$F_i^e = \int_{\Omega_e} f\,\Psi_i d\Omega + \int_{\Gamma_e} Q_n \Psi_i d\Gamma \tag{13}$$

Here Ω_e is the domain of an element. Γ_e is the surface of an element, P_j^e is the pressure at each node, f is a volumetric source term, Q_n is a specified fluid flux through the face of the element, and Ψ_i is a linear interpolation function.

4.2. Velocity

After the pressures are calculated, the velocities are calculated at the centroid of each element using the volumetric flux equation 4 :

$$v_i = -\frac{K_{ij}}{\mu\phi}\frac{\partial P}{\partial x_j} \tag{14}$$

4.3. Calculation of the parameters of filling

The control volume method tracks the flow front by identifying the controls volume partially filled, and connecting them to form the flow front. The numerical flow front is made from the nodes with the partially filled control volumes.

4.3.1. Flow rate calculation

It is assumed that the velocity of the fluid is constant throughout each element (figure.5).

$$Q_{en} = \overline{v}_e \,\overline{a}_{en} = -\int_o^A \frac{K_{ij}}{\mu}\frac{\partial P}{\partial x_j}\cdot \vec{n}\, ds \tag{15}$$

Where Q_{en} is the volumetric flow rate in the control volume (n) from element (e), $\overline{v}_{en}$ is the fluid velocity in the element, and $\overline{a}_{en}$ is the area vector for the sub-volume.

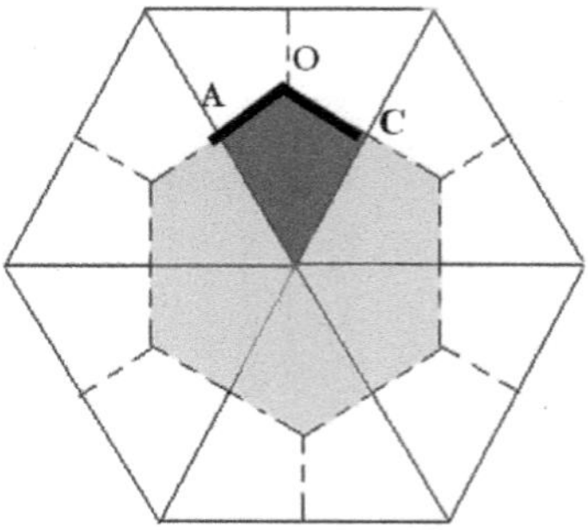

Figure 5. Calculation of the filling velocity (CV/FEM)

4.3.2. *Fill factor calculations*

After the flow rates in each control volume have been calculated, the fill factors can be updated. Given the current time step, the fill factors from the previous step, the calculated flow rate, and the volume of each CV, the new fill factors can be calculated with:

$$f_n^{i+1} = f_n^i + \frac{\Delta t \sum_e Q_{en}}{V_n} \tag{16}$$

where f_n is the fill factor, Δt is the time step, V_n is the volume of the control volume, and the superscripts indicate time level.

4.3.3. *Time step calculation*

The time step for the next iteration must be calculated before the solution can proceed. The optimal time step would be where the fluid just fills one control volume. If a larger step were chosen, the flow front would over-run the control volume and a loss of mass from the system would result. The time to fill the partially filled control volume "n" is calculated with the following relation:

$$\Delta t_n = \frac{(1 - f_n) V_n}{\sum_e Q_{en}} \tag{17}$$

Once Δt_n has been calculated for all the partially filled control volumes, the smallest Δt is chosen as the time step for the next iteration.

5. Adaptive mesh

The numerical schemes used in mould filling simulations are usually based on a time dependent resolution of an unsteady (free surface) boundary value problem. The boundary of the filled area in the mould cavity is constantly evolving, and it is difficult to generate a mesh suitable for all the successive calculation steps of a filling simulation. The fluid front cannot be approximated with a fine precision with an isotropic mesh. Such a mesh would have to be very fine everywhere in the geometrical domain in order to provide an accurate approximation. This would lead to time consuming calculations, although a fine mesh is required only in the vicinity of the flow front and near the inlet gates. For this reason, several researchers have proposed to construct a new mesh of the fluid saturated domain at each time step (Bechet et al. [31] for Eulerian scheme, Muttin et al. [32] for Lagrangian schemes). This approach is long in terms of computer time and rather complex, especially in the case of obstacles, merging flow fronts and for 3D problems Kang and Lee [19] proposed an algorithm, referred to as the Floating Imaginary Nodes and Elements (FINE) method, to get a smoother flow front and reduce the error in the pressure at the flow front of a CVFEM simulation of resin flow in a porous medium. With the FINE method, imaginary new nodes were added at the estimated flow front and the flow front elements were divided into two separate regions: the area of resin and the area of air. Thus, the flow front element was refined in an adaptive manner at each time step. In this study, the generation of the mesh is realized by a code developed by our team, included like a module of the code of numerical simulation. The mesh generator allows to discretize the field in unstructured triangular elements with possibility of local refinement (static and dynamic) and inclusion of inserts. The development of the mesh generator code and its use in our solution allows the refinement of the critical zone ($0<f<1$) in each iteration.

In the CVFEM process, the numerical flow front is composed from the nodes with partially filled control volumes. So, since the location of the fluid in the control volume is not known, the exact shape of the flow front cannot be identified. In our numerical code, for the zone positioned in the flow front, we have developed a technique of local refinement. This technique uses an iterative method to refine repeatedly an initial triangulation of Delaunay, by inserting new nodes in the triangulation until satisfying size criteria and the shape of the elements (figure.6). The number of calculation points necessary to characterize accurately the deformations of the flow front decreases, and thus, the numerical computing time is reduced.

The refinement of position and shape of the front flow consists in adding the new nodes to the initials meshes (triangulation of Delaunay) in the zone of the flow front. The integration of these refined nodes, in the computer code, is conditioned by the value of the filling rate. The algorithm adopted for the mesh generator makes possible to generate first the standard Delaunay elements and initial nodes for the calculation domain. Then, in a second time, these elements are re-meshed by a technique of addition of nodes *(figure.7)*. The criteria to be respected during refinement are:

The meshing coincides at the interfaces: the meshing technique used implies the coincidence of the grid at the interfaces between the neighbouring elements (of the first standard grid). The triangulation of Delaunay is based directly on the contour nodes discretization which are discretized only once.

An automatic identification of the nodes and the elements of refinement: During the refinement of the meshes created, the initial nodes and elements are identified by a traditional technique of programming called "the coloring". This technique consists in affecting a particular code to define the elements of the sub-domains. During the resolution of the equations of the linear system, the integration of the nodes and elements resulting from refinement, in the numerical code, is conditioned by the value of the filling rate. Only, the elements with partially filled nodes are taken into account.

Association of the sub-domains to the principal element: During the refinement of the initial elements, the numerical algorithm affects a code to each sub-domain realized by the mesh generator. This code corresponds to the principal element generating the sub-domains. Thus, the mesh code generates a structure of data that permits to associate the sub-domains, resulting from the refinement process, to their principal element.

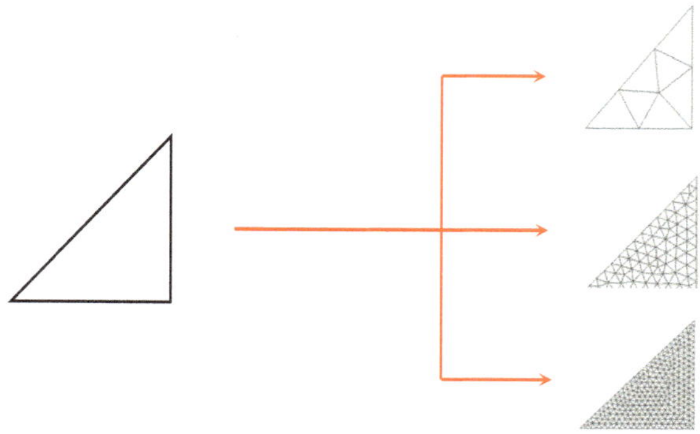

Figure 6. Refinement of the triangular element.

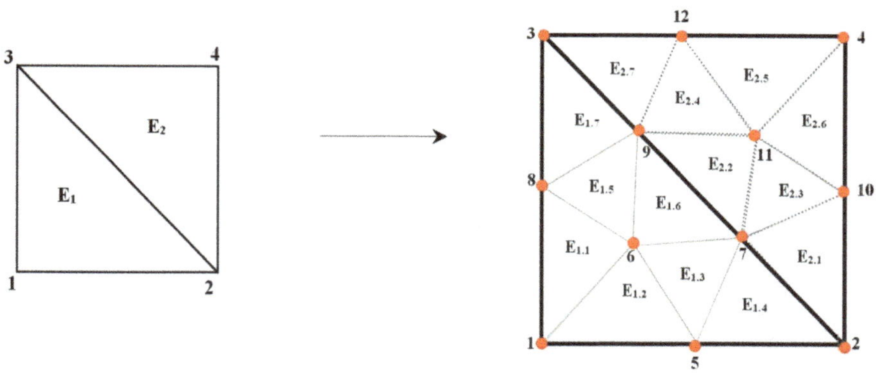

Figure 7. Refinement by inserting new nodes in the triangulation .

In the present approach, a local mesh generator module has been developed and integrated to the principal code of numerical simulation of the RTM process. This approach is useful to carry out the instruction of the meshing at each step time during the execution of the program, which improves calculation time. The meshing technique adopted is based on two concepts. The first one is the restricted triangulations of Delaunay and the second is the Delaunay refinement applied to elements of the flow front (figure.8).

Starting from the reference level of meshing G_0, we define for each element $E_{1,ps}$, the condition of refinement as :

$$Condition\ \Re : \begin{cases} \exists\, node_i \in E_{1,ps} \,/\, f_{node_i} = 1 \\ \exists\, node_j \in E_{1,ps} \,/\, f_{node_j} < 1 \end{cases}$$

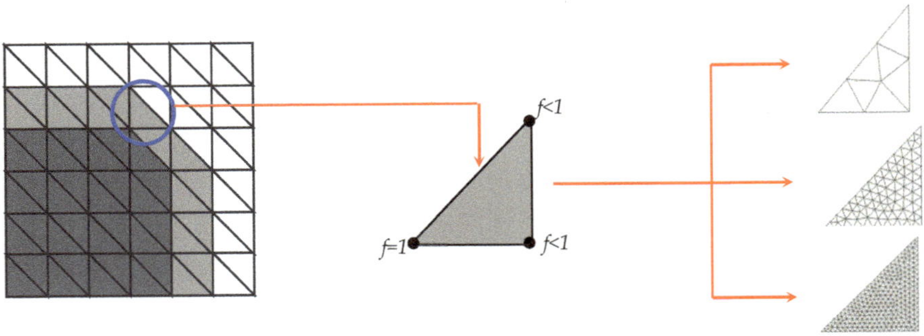

Figure 8. Identification of the elements for refinement

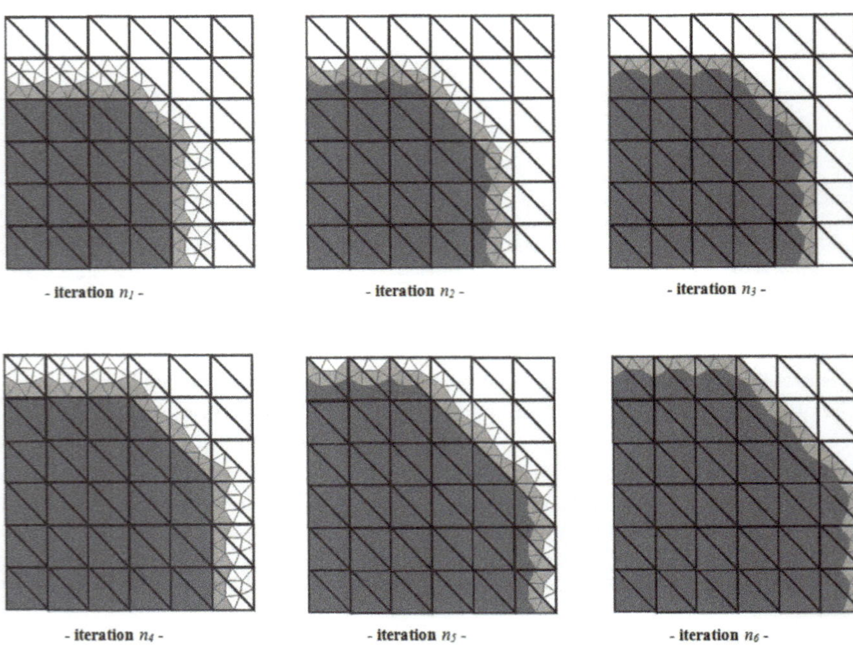

Figure 9. Example of numerical simulation in RTM process using refinement technique.

The operation for detecting which elements of G_0 must be refined, is repeated at each iteration of the resolution process. If the condition $\Re$ has value "true" for an element $E_{1,ps}$ of the level G_0, the numerical algorithm applies the refinement process and create sub-domains. Thus, the element $E_{1,ps}$ is divided into an number of sub-domains (7 in figure.9), in each direction to preserve good properties of connection between the domains, which creates a new meshing level G_l. The solutions at the refined G_l are initialized by a prolongation of the same numerical algorithm of the G_0 level. The system of equation is solved successively at each level of grid. Each refined element has a pressure boundary conditions of Dirichlet type. These operations are repeated at each iteration of the numerical resolution until the mould is completely filled.

The advantage of this technique is to increase the local precision of calculations while preserving the properties of a meshing. On the other hand, the resolution process at each meshing level can be carried out starting from the same system of equations, the same approximations and the same solver.

6. Numerical algorithm

Large complex structures need computational models that accurately capture both the geometric and physical phenomena. This may involve the use of flow simulations as warranted by geometry, thickness, and fiber preforms employed. There also exists a need for accuracy improvements by refining the discretization of the computational domain. All these have serious impact on the computational time and power requirements. Physical modeling and computational algorithms and methodologies play an important role in computational times. For large-scale computations, it becomes critical to have algorithms that are physically accurate and permit faster solution of the computational domain. It is not only essential to design efficient parallel algorithms, data structures, and communication strategies for highly scalable parallel computing, it is also very important to have improved computational algorithms and methodologies to further improve the computational performance of large-scale simulations.

In this study, the algorithm adopted uses techniques for the optimization of the execution time. For example, the management of the memory by the dynamic allocation of the tables and matrices allow the optimization of the resources machines. Also, the utilization of the pointers in the definition of the variables of the problems ensures the code the adaptation to the size of the data to be treated. In the same way, the adoption of algorithm of the sparse matrix for the inversion impacts the processing time seriously. Also in every iteration, the calculation matrix is dynamic and contains only elements that have a node with a filling ratio unity $f=1$. This approach imposes a rigor during the development of code, however, the time of treatment of the problem is ameliorated.

In this study, the numerical adopted is based on the various computational steps involved in adaptive meshing that accurately capture both the geometric and physical phenomena., CVFEM methodology for the update of the filled regions and refinement technique for flow front advancement are summarized below.

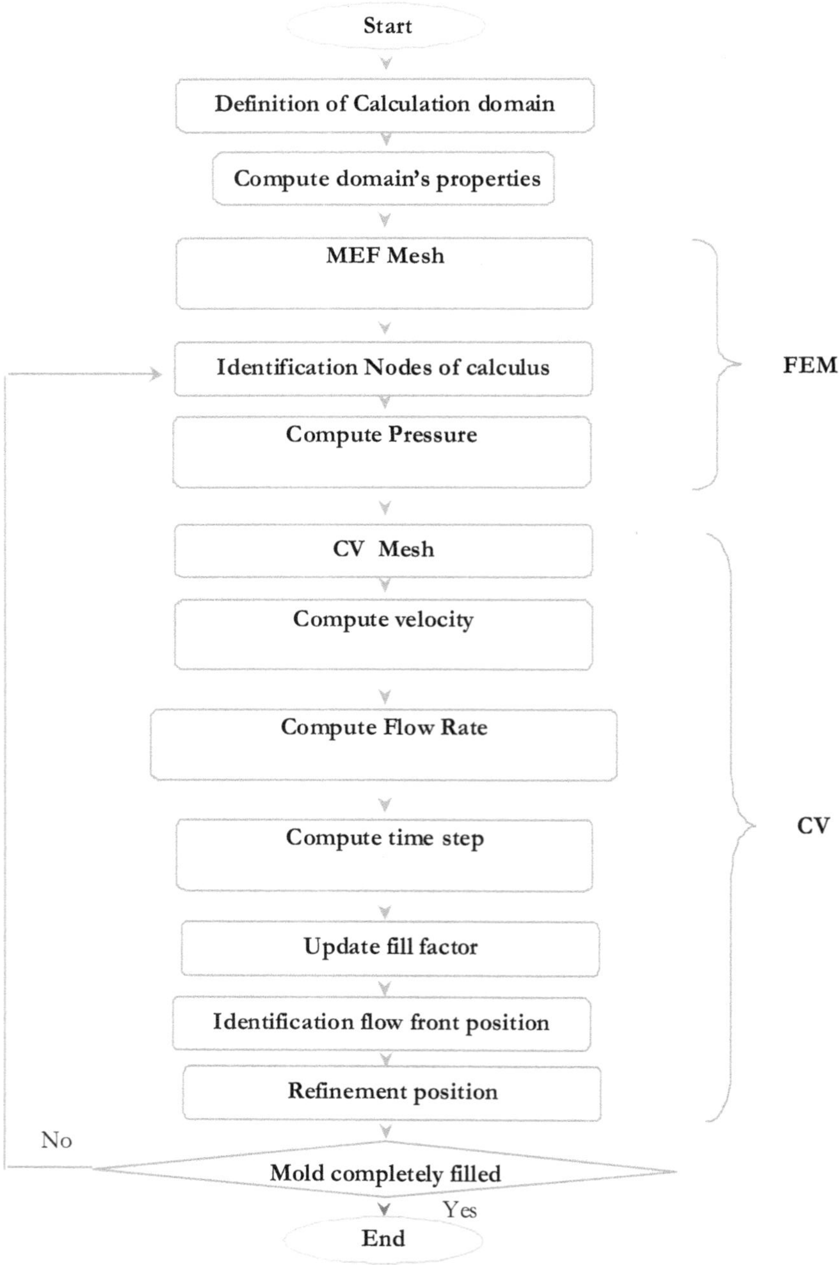

Figure 10. Numerical Algorithm of simulation of filling of moulds in RTM process

7. Results and discussion

7.1. Adaptive meshing

In the first case (figure.11), we present examples prove the accuracy provided by our adaptive meshing technique. These examples treat an injection mold for radial rectangular 400mmx400mm with a uniform thickness of 4mm. The injection is located at the center of the mold. The meshing generates the following data (table.1). The running time is related to a machine CPU 2Ghz Core Duo 2GB RAM

	Using Technique Adaptive Mesh	Number of principal elements	Number of secondary elements	Number of principal Nodes	Number of secondary Nodes	Time Consuming
Example A	No	160	0	97	0	12s
Example B	Yes	160	800	97	506	27s
Example C	Yes	160	3520	97	1972	97s

Table 1. Data of meshing

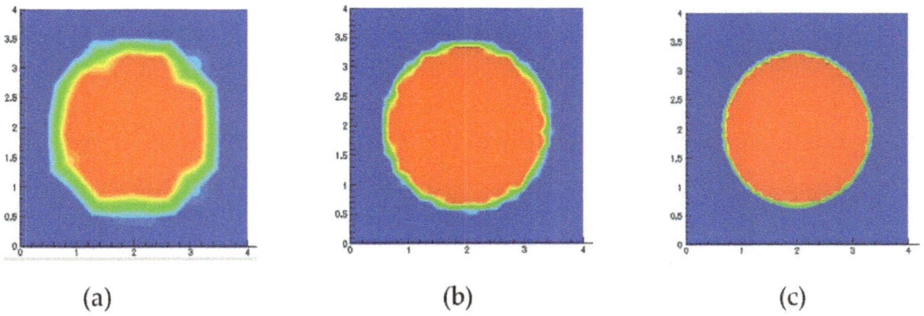

Figure 11. Numerical simulation with different type of meshing

In the second case (figure.12), we present two examples for comparison. These examples are related to the gain of the execution time for our adaptive meshing technique. These two examples treat an injection mold for radial rectangular 400mmx400mm with a uniform thickness of 4 mm. The first example is related to a discretization without adaptive meshing. This model has the same level of precision as the second example related to the adaptive mesh (same total number of nodes and elements). The meshing generates the following data (table.2). The running time is related to a machine CPU 2Ghz Core Duo 2GB RAM

	Using Technique Adaptive Mesh	Number of principal elements	Number of secondary elements	Number of principal Nodes	Number of secondary Nodes	Time Consuming
Example A	No	3680	0	2069	0	452s
Example B	Yes	160	3520	97	1972	97s

Table 2. Data of meshing

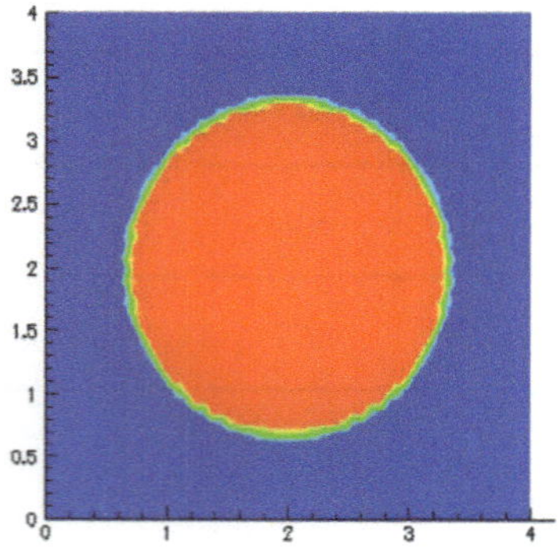

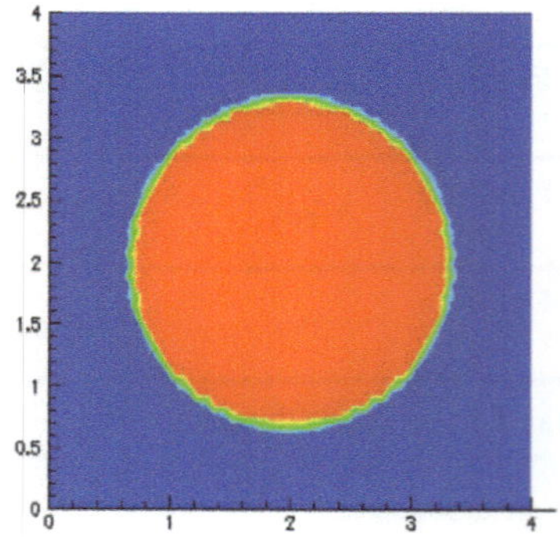

Figure 12. Numerical simulation with different type of meshing

7.2. Validation of the results relating to RTM flow with uniform thickness

In the validation examples presented, the moulds used, have a rectangular cavity, with dimensions: (1000 × 200) mm² and (400 × 400) mm², respectively. The first example is relative to the unidirectional validation (1D) of our numerical results, whereas the second is used in radial injection (2D). The fluid viscosity $\mu = 0{,}109$ Pa.s, the pressure injection $P_{inj} = 2 \times 10^5$ Pa, the permeability $K = 2{,}65 \times 10^{-10}$ m² and the porosity is $\phi = 0{,}696$.

7.2.1. Unidirectional validation

The mould used has a rectangular cavity of dimensions (1000 × 200) mm², the thickness is uniform and equal to 4 mm. The resin is injected from left side of the mould and the vents are placed on the right-side of the cavity. Under these conditions, the kinetics of the flow can be obtained analytically by the equations of table.3. The comparison of the three kinetics of the front flow obtained is presented in figure 13. It shows a good concordance with the solution obtained.

Pressure	Front's position	Filling time
$P(x)=P_{inlet}(1-\frac{x}{x_f(t)})$	$x_f=\sqrt{\frac{2KP_{inlet}\,t_f}{\varphi\,\mu}}$	$t_f=\frac{\varphi\,\mu L^2}{2KP_{inlet}}$

Table 3. Analytical solutions of unidirectional simulations in RTM

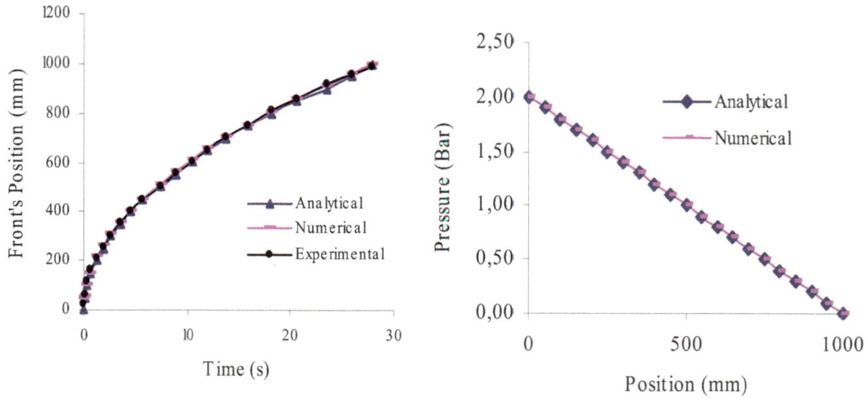

Figure 13. Unidirectional Validation

7.2.2. Bidirectional validation

The mould used has rectangular cavity of dimensions 400x400 mm^2, and the thickness is uniform 4mm. The resin is injected from the central point of the mould, the reinforcement is isotropic and the permeability is the same one in the various directions. Under these conditions, the form and the position of the flow front can be obtained analytically by the equation 18. The comparison of the two front's kinetics (analytical and numerical) is presented in figure.14. It shows a good concordance with the solution obtained.

$$\frac{r_f^2}{2}\ln\left(\frac{r_f}{r_0}\right)+\frac{1}{4}\left(r_0^2-r_f^2\right)=\frac{K}{\varphi\mu}(P_f-P_{inlet})t \tag{18}$$

With r_f is the radius of the front flow in a time t.

7.3. Validation of the results relating to RTM flow with variable thickness

7.3.1. Analytical validation

RTM process can be used to produce pieces with complex geometry. In the industry of the composite, the plates employed often consist of reinforcements with a variable number of plies and stacking sequences. A correct simulation of this process requires taking into account all these parameters. Lonné makes a modeling according to a formalism derived from the Thomson-Haskell method for the prediction of these geometrical variations on the ultrasound transmission [33].

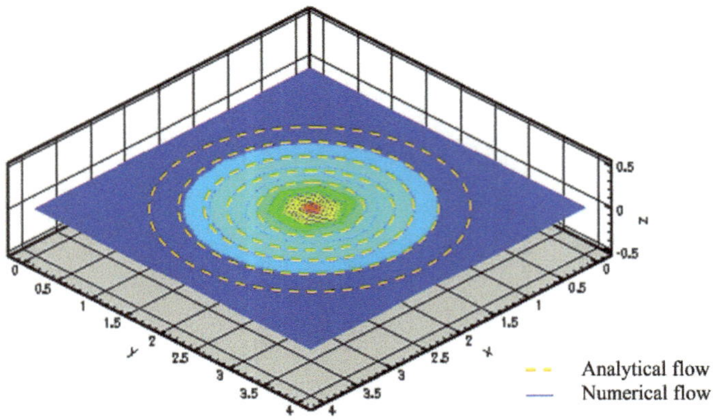

Figure 14. Bidirectional Validation with $K_{11}=K_{22}=K$

The reinforcement variation generates different pressures when closing the mold. Under the impact of the compressibility or the relaxation of the mold plates, a variation occurs in the volume and the pores distribution through the fabric and influences permeability and porosity. The mechanical performance of resin transfer molding depend on the fiber volume fraction [34], microstructure of the preform [35–36], void content [37], and impregnation parameters [38]. In most cases, mechanical properties of composite parts can be improved by increasing fiber volume fraction. Higher fiber volume fractions, however, require increased injection pressure and longer time to fill up the mould, which may significantly affect the properties of the final part. Patel et al. [39] molded composite parts containing glass fibers at constant injection pressure.

The study of Chen and al [40] showed that the initial compressibility of reinforcements is essentially related to that of the pores. This compressibility or « relaxation » effect directly influences the global volume and the distribution of the pores. During the mold closing, the variation of the reinforcement thickness generates, under the compressibility or the relaxation effects, a variation of the pores volume and their distribution through the fabric. The works of Buntain and Bickerton [41] were oriented to the way that compressibility affects permeability. Their results clearly showed that permeability (a property required to be perfectly controlled for a correct simulation of the flow front and the distribution of the pressure) was closely related to the pore volumetric fraction. Several models have been proposed to estimate the value of the permeability for various porous media. Capillary models such as those proposed by Carman [42] and Gutowski et al. [43] use the fiber radius and porosity to predict the permeability, but several discrepancies with experimental data have been reported [43–47]. Theoretical models have also been developed for different idealized media structures [44–47]. Most models may not give accurate prediction of permeabilities since fibrous mats used in RTM are often more complex than the idealized unit cell patterns used in theoretical derivations. Thus due to the lack of adequate predictive models, permeability of RTM preforms are usually determined experimentally.

A number of permeability measurement methods have been developed; however, there is no standardized measurement technique for RTM applications. Trevino et al. [48] and Young et al. [49] determined unidirectional permeability using two pressure transducers at each of the inlet and the exit in conjunction with an equation based on Darcy's law applied to their flow geometry. Calhoun et al. [50] presented a technique based on placing several pressure transducers at various locations inside the mold. Adams and Rebenfeld [51–53] developed a technique that quantifies the planar permeability using the position and shape of the advancing resin front as a function of time. A transparent mold was used to enable the monitoring of the advancing front. Other techniques based also on the observation of the moving resin front are common in the literature [43,51-59]. However, transparent mold walls may not have enough rigidity to avoid deflection, which has been shown to perturbate the measured permeability values [45,48,49,54]. In this context, the University of Plymouth radial flow permeability appartus was enhanced by the use of a laminate of two 25 mm toughened float glass sheets [60] as the upper mould tool to achieve a similar stiffness to the aluminium mould base [61].

In this study, the variation of the plies number (figure.15.a) and the stacking sequence are modeled by the variation of permeability and porosity. During the standard approach, these parameters are defined as an intrinsic property of the global discretized domain. In our approach, the permeability and porosity are defined at the level of the element. The comparison of the two kinetic fronts (analytical and numerical) is presented in figure.15.b ; it shows a good concordance with the solution obtained.

The analytical model for this type of reinforcement is indeed the prolongation of the linear model already adopted in the case of a medium with isotropic permeability. To ensure the accuracy of our numerical results, the elements of the initial meshing belong only to the one of the two zones. Also, in the static refinement of meshing, the creation of the new refined elements respects the condition of the uniformity of the permeability within the element (the element of refinement must belong only to the one of the two zones).

In addition, to illustrate the thickness effect, we present within the framework of the analytical validation, the case of a radial flow through a multi-thickness reinforcement.

In the setting of a bi-dimensional flow, we used the reinforcement with variation of the number of plies. The mould cavity had a uniform thickness.

1. Reinforcement with a constant thickness (10 plies) (Figure 16.a)
2. Reinforcement with half 10 plies and half 20 plies (Figure 16.b)
3. Reinforcement with ¼ 10 plies and ¾ 20 plies (Figure 16.c)

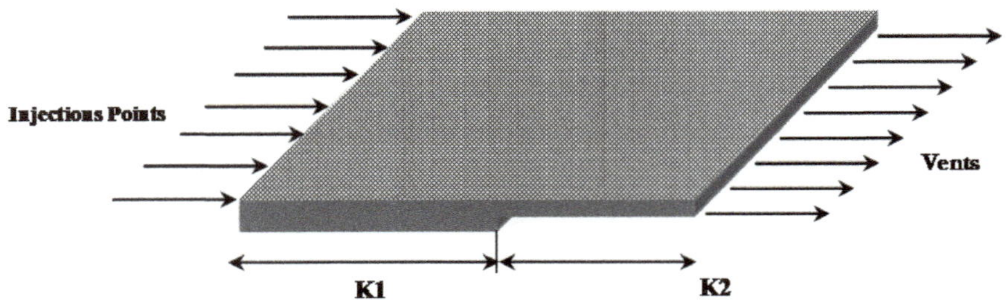

(a)

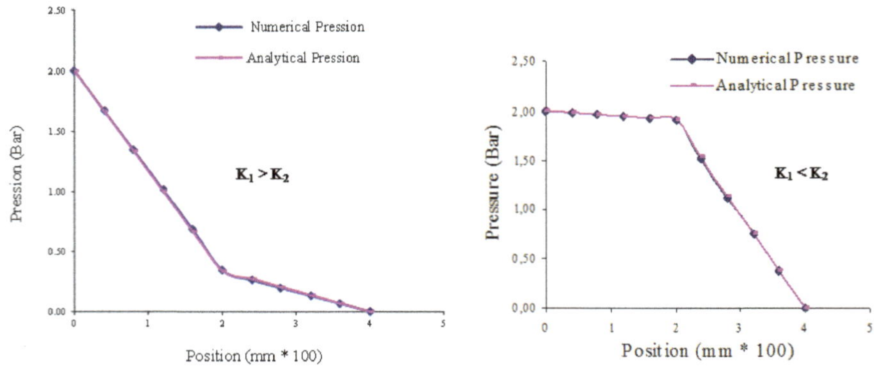

(b)

Figure 15. a) Reinforcement with multiple thickness, b) Evolution of the pressure-position for multiple thickness

For a reinforcement with uniform thickness (figure16.a), the permeability in the principal directions is constant. The shape of the flow front for a radial injection is a circle whose center is the point of injection. For the case of two different reinforcements thicknesses (figure 16.b, and 16.c), the position and the shape of the flow are variable according to the resistance presented by the fabric (permeability).

Thus, for the figure 16.b where each half of the reinforcement is characterized by a fixed value of the permeability, it is quite clear that the solution of the Darcy law concerning the shape of flow front, in each zone, has a form of half-circles spaced according to the value of the permeability. The connection between the two shapes respects the continuity of the flow front. Finally, and in order to ensure the required precision, we proceeded to each step of time, with a refinement in the zone of connection of the two parts. The same approach is adopted for the figure16 .c.

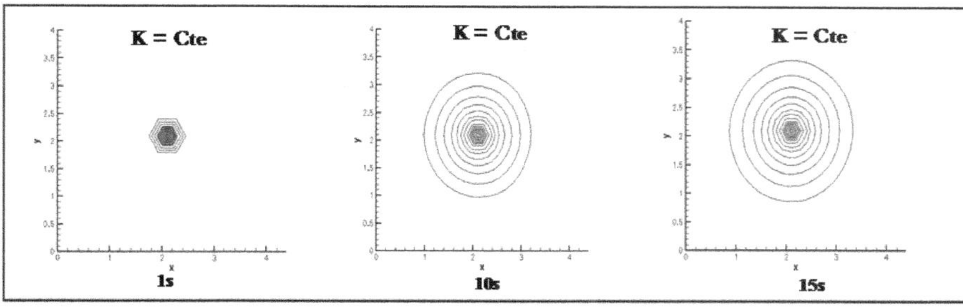

(a)

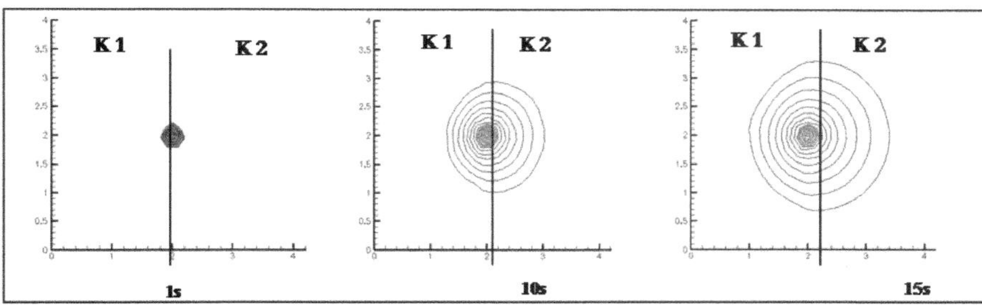

(b)

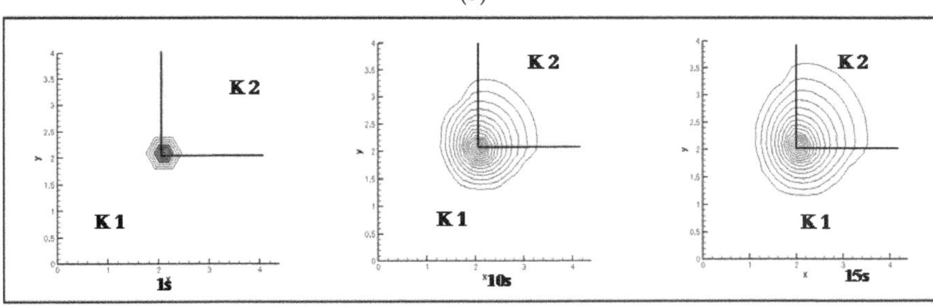

(c)

Figure 16. a) Flow Front for a bi-dimensional injection with Reinforcement with a constant thickness, b) Flow Front for a bi-dimensional injection with Reinforcement half 10 plies and half 20 plies, c) Flow Front for a bi-dimensional injection with Reinforcement ¼ 10 plies and ¾ 20 plies

7.4. Simulation of the pressure distribution for a multiple – Thickness reinforcement

The pressure distribution directly influences the injection pressure, the time needed to fill up the mould, the position of the injection points and vents and especially the mechanical properties of the final piece. The approach adopted by our team, use the permeability at the element's level. The thickness variation of the element directly influences its permeability and its porosity. The integration of these parameter's variation inside the code of resolution, gives more precision.

The model we treated is a wiring cover made of glass fibers and an Isophthalic Polyester resin. The model has the specificity of including a square insert inside the piece, with the possibility to vary the thickness around the insert (square zone, Figure 17).

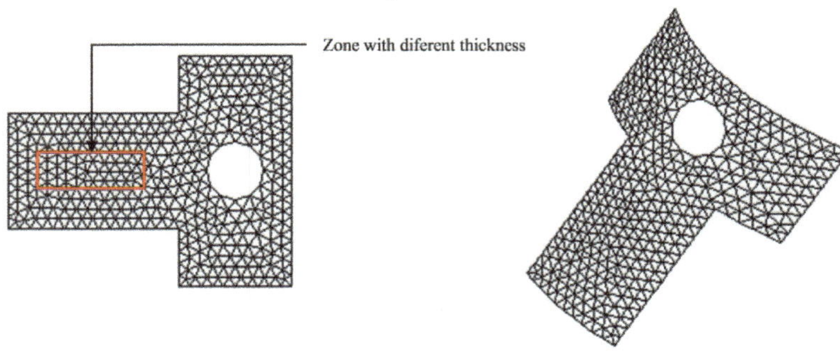

Figure 17. Piece with insert and reinforcement multiple thickness

During the moulding in RTM process, the resin injected infiltrates in empty spaces between fibers. However, a minor modification of the characteristics of the preform in specific places (around the insert for example), can cause significant deviations in the flow and the results on the final properties of the part can be disastrous generating the rejection of the whole process.

In order to simulate the filling process for 3D Dimension, logically a 3D model would be required. Since the thickness of composite parts is often much smaller than its length and width, thin part assumptions can be used for these simulation models [62]. For example, the resin flow in the thickness direction (here denoted as z) is neglected. Therefore, these models, although they describe 3D geometries, they are often called 2.5 D flow models [63].

The figures presented in 18.a, 18.b and 19 clearly show the impact of taking into account the variation thickness on the accuracy of the simulation of the flow front and the distribution of the pressure for 2.5 D models. These elements are paramount to control the parameters of the moulding and to obtain the required properties of the final part. the increase in thickness on a particular area by inserting new plies of reinforcement, leading to decreased permeability and increased porosity.

During numerical resolution, after the pressures are calculated, the velocities are calculated at the centroid of each element. Thus, given the reduced permeability and increased porosity, the velocity decreases while increased the thickness. In Figure 18.b we note the delaying of flow front in this area.

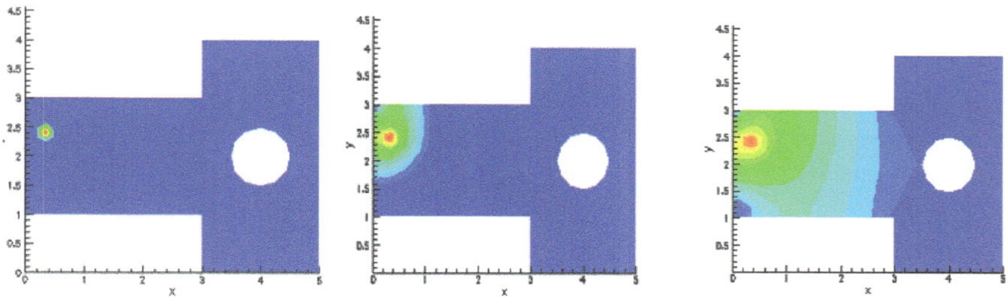

(a)

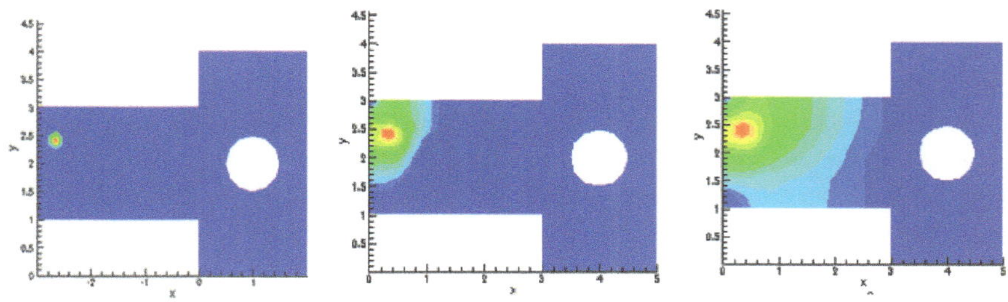

(b)

Figure 18. a) 2D Simulation without taking into account the effect of thickness variation, b) 2D Simulation when taking into account the effect of thickness variation

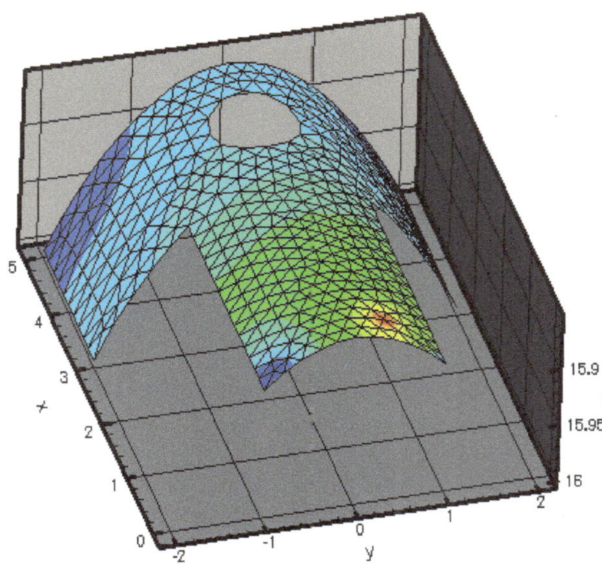

Figure 19. 2.5D simulation of pressure distribution with thickness variation

The approach adopted during this study ensures flexibility during the simulation of the heterogeneities of the problem. The properties of the medium are calculated at the level of the element of the mesh. Having developed an adaptive mesh generator specific to the team, we are able to ensure a better adaptation of the elements of grid to describe physical specificities generated in each problem. The results obtained on the figures of this paragraph show the relevance of the present approach.

8. Conclusion

During this study, we developed a mesh generator and a numerical code to simulate the filling of an isothermal mould in RTM process, by adopting the CV/FEM and VOF method. An adaptive static meshing to model the variation of the thickness, and dynamic to refine the flow front are used. This approach is useful to carry out the meshing at each step of time during the resolution process, which improves precision and calculation time. The algorithm adopted treats permeability and porosity at the level of the element of the mesh. The effects of the variations of the plies number and the stacking sequence, around the inserts for example, are modeled by the variation of permeability and porosity. The flows around obstacles and through the reinforced area near inserts are simulated by the present formalism with a local refinement of the meshing. The results obtained during the numerical simulation show a good concordance with the results: analytical, experimental and numerical. We can note the effectiveness of the numerical model developed in the predicted flow front and the distribution of pressure. An excellent reproduction of the form of the front and a good precision of its position are obtained. In our next studies, we will be interested in applying a similar approach to simulate saturation effects.

Author details

Jamal Samir, Jamal Echaabi and Mohamed Hattabi
Applied Research Team on Polymers, Department of Mechanical Engineering, ENSEM, Hassan II University, Casa Blanca, Oasis, Casablanca, Morocco

9. References

[1] Rigas EJ, Walsh SM, Spurgeon WA. "Development of a novel processing technique for vacuum assisted resin transfer molding (VARTM)". Int SAMPE Symp Exhibit (Proc) 2001; 46 I: 1086–94.

[2] Luo J, Liang Z, Zhang C, Wang B. "Optimum tooling design for resin transfer molding with virtual manufacturing and artificial intelligence". Compos Part A: Appl Sci Manuf, 2001;32(6):877–88.

[3] Loos AC, Fingerson JC, MacRae JD, "Verification of a three dimensional RTM/RFI flow model, technology transfer in a global community", Int SAMPE Tech Conf 1996; 28:393–403.

[4] Nielsen D, Pitchumani R. "Intelligent model-based control of preform permeation in liquid composite molding processes, with online optimization". Compos Part A: Appl Sci Manuf, 2001; 32(12):1789-803.

[5] Wang, V. W., Hieber, C.A. and Wang, K. K., "Simulation in Injection Molding of Three-Dimensional Thin Parts," In ANTEC 86, Boston, (Fairfield, Connecticut: Society of Plastics Engineer, pp. 97-102, 1986.

[6] Fracchia, C.A., Castro, J., and Tucker, Cl.L., "A Finite Element/Control Volume Simulation of Resin Transfer Mold Filling," In Proceedings of the American Society for Composites Fourth Technical Conference, Lancaster, PA, 1989, pp. 157-166.

[7] Bruschke, M.V., and Advani S.G., "A Finite Element Control Volume Approach to Mold Filling in Anisotropic Porous Media," Polymer Composites, Vol. 11, No. 6, 1990, pp. 398-405

[8] Li, S. and Gauvin, R., "Numerical Analysis of the Resin Flow in Resin Transfer Molding," Journal of Reinforced Plastics and Composites, Vol. 10, 1991, pp. 314- 327.

[9] Young, W.B. al., "Analysis of Resin Injection Molding in Molds with preplaced Fiber Mats: 2.Numerical Simulation and Experiments of Mold Filling," Polymer Composites, Vol. 12, N° 1, 1991, pp. 30-38

[10] Lin, R., Lee, L.J. and Liou, M., "Non-isothermal Mold Filing and Curing Simulation in Thin Cavities with Preplaced Fiber Mats," International Polymer Processing, Vol. 6, No. 4, 1991, pp. 356-369

[11] Trochu, F. and Gauvin, R., "Limitations of a Boundary-Fitted Finite Difference Method for the Simulation of the Resin Transfer Molding Process," Journal of Reinforced Plastics and Composites, 1992, Vol. 11, No. 7, pp. 772-786.

[12] Voller, V.R., and Chen, Y.F., "Prediction of Filling Time in Porous Cavities," International Journal for Numerical Methods in Fluids, Vol. 23, N° 7., 1996, pp. 661-672

[13] Spoerre J, Zhang C, Wang B, Parnas R. "Integrated product and process design for resin transfer molded parts". Journal Composite Mater 1998; 32(13):1244–72.

[14] Padmanabhan SK, Pitchumani R. Stochastic "modeling of non isothermal flow during resin transfer molding" Int J Heat Mass Transfer 1999; 42(16):3057–70.

[15] Moon Koo Kang, Woo Il Lee," A flow-front refinement technique for the numerical simulation of the resin-transfer molding process", Composites Science and Technology 59(11), 1999, pp. 1663-1674.

[16] Hattabi M., Snaike I., Echaabi J. and Bensaleh M.O., "Simulation du front d'écoulement dans les procédés de moulage des composites liquide », Comptes Rendus Mécanique, Volume 333, Issue 7, July 2005, pp. 585-591.

[17] Comas-Cardona S.; Groenenboom P.; Binetruy C. (1); Krawczak A generic mixed "FE-SPH method to address hydro-mechanical coupling in liquid composite moulding processes" Composites. Part A, Applied science and manufacturing ISSN 1359-835X 2005, vol. 36, no7, pp. 1004-1010

[18] Soukane S.; Trochu F.; "Application of the level set method to the simulation of resin transfer molding"Composites science and technology ISSN 0266-3538 CODEN CSTCEH 2006, vol. 66, no7-8, pp. 1067-1080

[19] Kang M. K. and W. Il Lee "A Flow Front Refinement Technique for the Numerical Simulation of the Resin Transfer Molding Process". Composites Science and Technology 59, 1663-1774. 1999
[20] Phelan R. F. JR " Simulation of the Injection Process in Resin Transfer Molding". Polymer Composites 18(4), 460-476.1997
[21] Bruschke M. V. and S. G. Advani "A Numerical Approach to Model Non- isothermal Viscous Flow through Fibrous Media with Free Surfaces". International Journal for Numerical Methods in Fluids 19, 575-603.1994
[22] Joshi S. C., Lam Y. C., and Liu X.-L. "Mass Conservation in Numerical Simulation of Resin Flow". Composites: Part A 31, 1061-1068.2000
[23] Tucker CL, Dessenberger RB. Governing equations for flow and heat transfer in stationary fiber beds. In: Advani SG, editor. Flow and rheology in polymer composites manufacturing. Amsterdam: Elsevier; p. 257–323. 1994.
[24] Lin M, Hahn HT, Huh H. A finite element simulation of resin transfer molding based on partial nodal saturation and implicit time integration. Composites Part A 1998;29:541–50.
[25] B. Delaunay: Sur la sphère vide, Izvestia Akademii Nauk SSSR, Otdelenie Matematicheskikh i Estestvennykh Nauk, 7:793–800, 1934
[26] F. P. Preparata and M. I. Shamos. Computational Geometry An Introduction. Texts and Monographs in Computer Science. Springer-Verlag, 1985.
[27] H. Edelsbrunner. Algorithms in Combinatorial Geometry, volume 10 of EATCS Monographs on Theoretical CS. Springer-Verlag, 1987.
[28] L. P. Chew. Guaranteed-quality triangular meshes. TR-89-983, Cornell, 1989.
[29] J. Ruppert. A new and simple algorithm for quality 2-dimensional mesh generation. Proc.4th ACM-SIAM Symp. Discrete Algorithms (1993) 83{92.
[30] J. P. Pontaza, H.C. Chen and J. N. Reddy, "A local-analytic-based discretization procedure for the numerical solution of incompressible flows", International Journal for Numerical Methods in Fluids, Vol. 49, No. 6, pp. 657-699, 2005.
[31] Bechet E, Ruiz E, Trochu F, Cuillière JC. Remeshing algorithms applied to mould filling simulations in resin transfer moulding. J Reinforced Plast Compos in press. 2003
[32] Muttin F, Coupez T; Bellet M.; Chenot J. L. "Lagrangian finite-element analysis of time-dependent viscous free-surface flow using an automatic remeshing technique" ISSN 0029-5981 CODEN IJNMBH vol. 36, no12, pp. 2001-2015 (26 ref.)1993
[33] S. Lonné, PhD thesis "Modélisation de la propagation ultrasonore dans les matériaux composites obtenus par le procédé de fabrication RTM (Resin Transfer Molding) Université Bordeaux 1, 2003.
[34] Naik, R.A. "Failure Analysis of Woven and Braided Fabric Reinforced Composites", Journal of Composite Materials, 29(17): 2334–2363. 1995.
[35] Wang, Y. "Effect of Consolidation Method on the Mechanical Properties of Nonwoven Fabric Reinforced Composites", Applied Composite Materials, 6(1): 19–34. 1999.
[36] Wang, Y. and Li, J. "Properties of Composites Reinforced with E-Glass Nonwoven Fabric", Journal of Advanced Materials, 26(3): 28–34. 1995.

[37] Goodwin, A.A., Howe, C.A. and Paton, R.J. "The Role of Voids in Reducing the Interlaminar Shear Strength in RTM Laminates", In: Scott, M.L. (ed.), Proceedings of ICCM-11, Vol. 4, pp. 11–19.1997.
[38] Lee, C.-L. and Wei, K.-H. "Effect of Material and Process Variables on the Performance of Resin-Transfer-Molded Epoxy Fabric Composites", Journal of Applied Polymer Science, 77(10): 2149–2155. 2000.
[39] Patel, N., Rohatgi, V. and Lee, L.J. "Influence of Processing and Material Variables in Resin-Fiber Interface in Liquid Composite Molding", Polymer Composites, 14(2): 161–172. 1993.
[40] B. Chen, AH-D Leng and T-W Chou, "A non linear compaction model for fibrous preforms", Composites Part A: Applied Science and Manufacturing, 32(5), pp. 701-707,. 2001,
[41] M.J. Buntain, S. Bickerton "Compression flow permeability measurement: a continuous technique" Composites Part A: Applied Science and Manufacturing, Vol. 34 (5), pp. 445-457 2003,
[42] Carman, P.C. "Fluid Flow Through a Granular Bed", Transaction of the Institution of Chemical Engineers, 15: 150–166. 1937.
[43] Gutowski, T.G., Morigaki, T. and Cai, Z. "The Consolidation of Laminate Composites", Journal of Composite Materials, 21(2): 172–188. 1987.
[44] Gebart, B.R. "Permeability of Unidirectional Reinforcements for RTM", Journal of Composite Materials, 26(8): 1100–1133. 1992.
[45] Bruschke, M.V. and Advani, S.G. A "Finite Element/Control Volume Approach to Mold Filling in Anisotropic Porous Media", Polymer Composites, 11(6): 398–405. 1990.
[46] Ngo, N.D. and Tamma, K.K. Microscale "Permeability Predictions of Porous Fibrous Media", International Journal of Heat and Mass Transfer, 44(16): 3135–3145. 2001.
[47] Williams, J.G., Morris, C.E.M. and Ennis, B.C. "Liquid Flow Through Aligned Fiber Beds", Polymer Engineering and Science, 14(6): 413–419. 1974.
[48] Trevino, L., Rupel, K., Young, W.B., Liou, M.J. and Lee, L.J. "Analysis of Resin Injection Molding in Molds with Preplaced Fiber Mats". I: "Permeability and Compressibility Measurements", Polymer Composites, 12(1): 20–29. 1991.
[49] Young, W.B., Rupel, K., Lee, L.J. and Liou, M.J. "Analysis of Resin Injection Molding in Molds with Preplaced Fiber Mats. II: Numerical Simulation and Experiments of Mold Filling", Polymer Composites, 12(1): 30–38. 1991.
[50] Calhoun, D.R., Yalvac¸, S., Wetters, D.G. and Raeck, C.A. "Critical Issues in Model Verification for the Resin Transfer Molding Process", Polymer Composites, 17(1): 11–22. 1996.
[51] Adams, K.L., Miller, B. and Rebenfeld, L. Forced In-Plane Flow of an Epoxy Resin in Fibrous Network, Polymer Engineering and Science, 26(20): 1434–1441. 1986.
[52] Adams, K.L. and Rebenfeld, L. "Permeability Characteristics of Multilayer Fiber Reinforcements. Part I: Experimental Observations", Polymer Composites, 12(3): 179–185. 1991.
[53] Adams, K.L. and Rebenfeld, L. "Permeability Characteristics of Multilayer Fiber Reinforcements. Part II: Theoretical Model", Polymer Composites, 12(3): 186–190. 1991.

[54] Bickerton, S., Sozer, E.M., Graham, P.J. and Advani, S.G "Fabric Structure and Mold Curvature Effects on Preform Permeability and Mold Filling in the RTM Process". Part I. Experiments, Composites Part A, 31(5): 423–438. . 2000.

[55] Bickerton, S., Sozer, E.M., Simacek, P. and Advani, S.G. "Fabric Structure and Mold Curvature Effects on Preform Permeability and Mold Filling in the RTM Process". Part II. Predictions and Comparisons with Experiments, Composites Part A, 31(5): 439–458. (2000).

[56] Gauvin, R., Trochu, F., Lemenn, Y. and Diallo, L. "Permeability Measurement and Flow Simulation Through Fiber Reinforcement", Polymer Composites, 17(1): 34–42. 1996.

[57] Sheard, J., Senft, V., Mantell, S.C. and Vogel, J.H. "Determination of Corner and Edge Permeability in Resin Transfer Molding", Polymer Composites, 19(1): 96–105. 1998.

[58] Ferland, P., Guittard, D. and Torchu, F. "Concurrent Methods for Permeability Measurement in Resin Transfer Molding", Polymer Composites, 17(1): 149–158. 1996.

[59] Dungan, F.D., Senoguz, M.T., Sastry, A.M. and Faillaci, D.A. "Simulation and Experiments on Low Pressure" Permeation of Fabrics: Part I-3D Modeling of Unbalanced Fabric, Journal of Composite Materials, 35(14): 1250–1284. 2001.

[60] Pearce N.R.L., Summerscales J. and Guild F.J., Improving the resin transfer moulding process for fabric-reinforced composites by modification of the fibre architecture, Composites Part A, 31A(12); 1433-1441. 2000.

[61] Carter E.J., Fell A.W., Griffin P.R. and Summerscales J., Data validation procedures for the automated determination of the two-dimensional permeability tensor of a fabric reinforcement, Composites Part A, A27(4); 255-261. 1996.

[62] M.J. Buntain, S. Bickerton "Compression flow permeability measurement: a continuous technique" Composites Part A: Applied Science and Manufacturing, Vol. 34 (5), 2003, pp. 445-457

[63] Josef F.A. Kessels, Attie S. Jonker, Remko Akkerman, "Fully 2.1/2D flow modeling of resin infusion under flexible tooling using unstructured meshes and wet and dry compaction properties" Composites Part A: Applied Science and Manufacturing, Vol. 38 (1), 2007, pp. 51-60

Chapter 2

Electromagnetic and Fluid Analysis of Collisional Plasmas

Antonis P. Papadakis

Additional information is available at the end of the chapter

1. Introduction

For the analysis of collisional plasmas one needs to analyze the different constituent particles which characterize these discharges, mainly the charged particles such as electrons, positive and negative ions as well as the neutral gas particles. Due to the presence of the charged particles, one needs to calculate also the electromagnetic fields within the plasma which are characterized by the Maxwell's equations. The general Maxwell equations can be characterized by substituting a scalar and vector potential equations using Lorentz Gauge transformation to calculate the electric and magnetic fields. In the case of electrostatic fields, Maxwell equations reduce to Poisson equation that characterizes the electric field in the absence of a magnetic field, whereas in the magnetostatic case, a steady current exists invariant in time with the magnetic field related quantities considered constant. At high frequencies, the electromagnetic wavelength is small and if one is outside this single wavelength, the electric and magnetic field are directly coupled to each other such that if one of the two parameters is calculated, then the other is known. In the case of low frequencies, where the wavelength is high, if you are within the near field region, the electric and magnetic fields are completely independent, therefore the solution of both electric and magnetic field equations is necessary to calculate the electric and magnetic field distribution within the collisional plasma.

The charged particles behaviour is characterized using the continuity conservation equations of mass, momentum and energy for electrons, positive and negative ions including convective, diffusive and source term phenomena. Regarding the behaviour of the neutral gas particles within collisional plasmas, the Navier-Stokes equations, which are the conservation equations for mass, momentum and energy for the neutral gas particles need to be solved including convective, diffusive as well as source terms phenomena such as shear stresses, momentum transfer by elastic collisions, Lorentz forces, Joule heating and thermal conduction. Having established the necessary equations, the implementation of the solution procedure for solving Poisson, charged particle continuity and Navier-Stokes equations is presented. Thereafter, the Finite Element-Flux Corrected Transport method (FE-FCT) formulation follows in two-dimensional Cartesian, two-dimensional cylindrical axisymmetric and three-dimensional Cartesian coordinates, comprising of the predictor-corrector step based on the Taylor-Galerkin finite element method to calculate the high and low order schemes. The validation of the fluid flow equations using the FE-FCT is performed using the shock tube type problem, the shock wave incident on a wedge test case and the energy source term that result in sound and shock wave generation. Having validated the fluid model thoroughly, an adaptive mesh generation technique is discussed which reduces computational needs significantly and at the same time guaranteeing the stability and accuracy of the results. Finally, different collisional plasma configurations are analyzed including avalanche, primary and secondary streamer propagations and finally heating effects in constant voltage Dielectric Barrier Discharges and normal and abnormal glow discharges in ambient atmospheric air.

2. Model description

The complete plasma model in its multidimensional form consists of the Maxwell equations to account for the electromagnetic field, the continuity equations for charges to account for the charged particles (electrons, positive and negative ions) and the Navier-Stokes equations to account for the neutral gas charges.

2.1. Maxwell's equations

The Maxwell equations consist of the following four differential equations in macroscopic form:

$$\nabla \cdot \boldsymbol{D} = \rho_c \tag{1}$$

$$\nabla \cdot \boldsymbol{B} = 0 \tag{2}$$

$$\nabla \times \boldsymbol{E} = -\frac{\partial \boldsymbol{B}}{\partial t} \tag{3}$$

$$\nabla \times \boldsymbol{H} = J + \frac{\partial \boldsymbol{D}}{\partial t} \tag{4}$$

where **D** is the electric field strength, **B** is the magnetic flux density, **E** is the electric field strength, **H** is the magnetic field strength, ϱ_c is the net charge density, **J** is the current density and t is the time.

Equation (1) is known as the Gauss's law equation for electricity, equation (2) the Gauss's law for magnetism, equation (3) as the Michael-Faraday's or Faraday's law of induction and finally equation (4) shows the Ampere's circuital law with Maxwell's correction.

where:

$$\boldsymbol{J} = \sigma(\boldsymbol{E} + \boldsymbol{u}x\boldsymbol{B}) \tag{5}$$

$$\boldsymbol{D} = \varepsilon_c \cdot \boldsymbol{E} \tag{6}$$

$$\boldsymbol{B} = \mu_c \cdot \boldsymbol{H} \tag{7}$$

where ε_c and μ_c are the dielectric and magnetic permeability constants. One can define a vector potential **A** that **B** is the curl of:

$$\boldsymbol{B} = \nabla x \boldsymbol{A} \tag{8}$$

which satisfies equation (2) above of Maxwell equations which basically states that no magnetic monopoles exist such that:

$$\nabla.\boldsymbol{B} = \nabla.(\nabla x \boldsymbol{A}) = 0 \tag{9}$$

since the divergence of a curl is zero satisfying the magnetic monopole constraint.

Now by substituting equation (8) into Faraday's equation (3), one gets:

$$\frac{\partial(\nabla x \boldsymbol{A})}{\partial t} = -\nabla x \boldsymbol{E} \tag{10}$$

By rewriting equation (10) above, one gets:

$$\nabla x(\boldsymbol{E} + \frac{\partial \boldsymbol{A}}{\partial t}) = 0 \tag{11}$$

The term in the parenthesis of equation (11) above has no curl present implying that a potential V exists such that:

$$\boldsymbol{E} + \frac{\partial \boldsymbol{A}}{\partial t} = -\nabla V \tag{12}$$

which gives the electric field to be:

$$\boldsymbol{E} = -\frac{\partial \boldsymbol{A}}{\partial t} - \nabla V \tag{13}$$

Substituting equation (13) into Amperes law of electromagnetism equation (4) gives:

$$\frac{\partial \boldsymbol{E}}{\partial t} - c^2(\nabla x \boldsymbol{B}) - \frac{\boldsymbol{J}}{\varepsilon_c} = 0 \tag{14}$$

where :

$$c^2 = \frac{1}{\mu_c \varepsilon_c} \tag{15}$$

with c being the speed of propagation within the medium. From vector calculus, one has that:

$$\nabla \times \boldsymbol{B} = \nabla \times \nabla \times \boldsymbol{A} = \nabla(\nabla \cdot \boldsymbol{A}) - \nabla^2 \boldsymbol{A} \tag{16}$$

Substituting equation (16) above and equation (13) into equation (14) gives:

$$\frac{\partial(-\frac{\partial \boldsymbol{A}}{\partial t} - \nabla V)}{\partial t} - c^2(\nabla(\nabla \cdot \boldsymbol{A}) - \nabla^2 \boldsymbol{A}) - \frac{\boldsymbol{J}}{\varepsilon_c} = 0 \tag{17}$$

which gives:

$$\frac{\partial^2 \boldsymbol{A}}{\partial t^2} - \nabla(\frac{\partial V}{\partial t}) + c^2 \nabla^2 \boldsymbol{A} - c^2(\nabla(\nabla \cdot \boldsymbol{A}) - \frac{\boldsymbol{J}}{\varepsilon_c} = 0 \tag{18}$$

Rearranging equation (18) above gives:

$$\frac{\partial^2 \boldsymbol{A}}{\partial t^2} - c^2 \nabla^2 \boldsymbol{A} + c^2 \nabla(\nabla \cdot \boldsymbol{A} + \frac{1}{c^2}\frac{\partial V}{\partial t}) - \frac{\boldsymbol{J}}{\varepsilon_c} = 0 \tag{19}$$

In order to make the choice of vector potential **A** not arbitrary, one imposes the Lorentz gauge which sets the parenthesis of equation (19) equal to zero:

$$(\nabla \cdot \boldsymbol{A} + \frac{1}{c^2}\frac{\partial V}{\partial t}) = 0 \tag{20}$$

with equation (20) above becoming:

$$\frac{\partial^2 \boldsymbol{A}}{\partial t^2} - c^2 \nabla^2 \boldsymbol{A} - \frac{\boldsymbol{J}}{\varepsilon_c} = 0 \tag{21}$$

Equation (21) above is Maxwell's equation for the vector potential. On another note, by combining equations (1) and (13), one gets:

$$\nabla \cdot \left(-\nabla V - \frac{\partial \boldsymbol{A}}{\partial t}\right) = \frac{\rho_c}{\varepsilon_c} \tag{22}$$

which gives:

$$-\nabla^2 V - \frac{\partial(\nabla . \boldsymbol{A})}{\partial t} = \frac{\rho_c}{\varepsilon_c} \tag{23}$$

Now using the Lorentz Gauge tranformation from equation (20) into equation (23) above gives:

$$-\nabla^2 V + \frac{\partial\left(\frac{1}{c^2}\frac{\partial V}{\partial t}\right)}{\partial t} = \frac{\rho_c}{\varepsilon_c} \tag{24}$$

$$\frac{\partial^2 V}{\partial t^2} - c^2\nabla^2 V - \frac{c^2}{\varepsilon}\rho_c = 0 \tag{25}$$

Finally, equations (21) and (25) are respectively the vector and scalar potential equations of the Maxwell's equations which when solved by using the equations (26) and (27):

$$\boldsymbol{B} = \nabla x \boldsymbol{A} \tag{26}$$

and:

$$\boldsymbol{E} = -\nabla V - \frac{\partial \boldsymbol{A}}{\partial t} \tag{27}$$

one can calculate the electric and magnetic field of time-dependent Maxwell's equations.

2.1.1. *Electrostatic case*

In the electrostatic case, one assumes that $\partial/\partial t$, **J**, **H** and **B** are all zero, and equation (25) becomes:

$$-c^2\nabla^2 V - \frac{c^2}{\varepsilon}\rho_c = 0 \tag{28}$$

$$\nabla^2 V = -\frac{\rho_c}{\varepsilon_c} \tag{29}$$

which is the Poisson equation. In the presence of no free charges, $\varrho = 0$ and the above equation (29) becomes the Laplace equation:

$$\nabla^2 V = 0 \tag{30}$$

2.1.2. *Magnetostatic case*

In the magnetostatic case, there are steady currents in the system under consideration, which generate magnetic fields (ferromagnetic media are ignored), therefore ones sets $\partial/\partial t=0$ and assumes that **J**, **H** and **B** are constant vectors. Substituting these relations into the scalar and vector potential equations (21) and (25) of Maxwell's gives:

$$-c^2\nabla^2 V - \frac{c^2}{\varepsilon}\rho_c = 0 \tag{31}$$

which is again the Poisson equation:

$$\nabla^2 V = -\frac{\rho_c}{\varepsilon_c} \tag{32}$$

and:

$$-c^2\nabla^2 \boldsymbol{A} - \frac{\boldsymbol{J}}{\varepsilon_c} = 0 \tag{33}$$

$$\nabla^2 \boldsymbol{A} = \frac{\boldsymbol{J}}{c^2\varepsilon_c} \tag{34}$$

$$\nabla^2 \boldsymbol{A} = -\mu \boldsymbol{J} \tag{35}$$

which is the magnetic equivalent of the electrostatic Poisson equation.

2.2. Continuity equations for charges

The continuity equations which comprise of the conservation of mass for electrons, positive and negative ions are described in equations (36) to (38) below:

$$\frac{\partial N_e}{\partial t} + \nabla.(N_e \mathbf{W_e}) = S_e \tag{36}$$

$$\frac{\partial N_p}{\partial t} + \nabla.(N_p \mathbf{W_p}) = S_p \tag{37}$$

$$\frac{\partial N_n}{\partial t} + \nabla.(N_n \mathbf{W_n}) = S_n \tag{38}$$

where N_e, N_p, N_n are respectively the electron, positive and negative ion densities, $\mathbf{W_e}$, $\mathbf{W_p}$ and $\mathbf{W_n}$ the corresponding velocity vectors and S_e, S_p, S_n are the source terms for the electrons, positive and negative ions respectively, which are calculated according to equations (39) to (41) below:

$$S_e = \alpha N_e |\mathbf{W_e}| - \eta N_e |\mathbf{W_e}| - \beta_{ep} N_e N_p + \nabla.(D_e \nabla N_e) \tag{39}$$

$$S_p = \alpha N_e |\mathbf{W_e}| - \beta_{ep} N_e N_p - \beta_{pn} N_p N_n \tag{40}$$

$$S_n = \eta N_e |\mathbf{W_e}| - \beta_{pn} N_p N_n \tag{41}$$

where α the ionisation coefficient, η the attachment coefficient, β_{ep} the recombination coefficient between electrons and positive ions, β_{pn} the recombination coefficient between positive and negative ions and D_e the electron diffusion coefficient. The positive and negative ion diffusion coefficients are omitted since their effect is negligible compared to electron diffusion for the time scales considered here.

The conservation of momentum and energy density for electrons, positive and negative ions equations are described below in equations (42) and (43) and have the general form:

$$\frac{\partial(\rho_s \mathbf{W_s})}{\partial t} + \nabla.(\rho_s \mathbf{W_s W_s}) = -\nabla P_s - \nabla.\boldsymbol{\tau}_s - \sum_s \mathbf{MT_s} \quad (42)$$

$$\frac{\partial \varepsilon_s}{\partial t} + \nabla.(\varepsilon_s W_s) = -\nabla.(Q_s) - \nabla.(W_s(P_s I + \tau_s)) - \sum_s MT_s.v - \sum_s f_{ts} J_s.E \quad (43)$$

Subscripts s, p and n represent the electron, positive and negative ions, and the various terms are similar to the ones used for the neutral gas particle conservation equations that are explained below.

2.3. Navier-Stokes equations

2.3.1. Navier-Stokes general form

The Navier-Stokes equations which account for the neutral gas particles are described by equations (44) to (46) below:

$$\frac{\partial \rho}{\partial t} + \nabla.(\rho \mathbf{v}) = S \quad (44)$$

$$\frac{\partial(\rho \mathbf{v})}{\partial t} + \nabla.(\rho \mathbf{vv}) = -\nabla P - \nabla.\boldsymbol{\tau} + \sum_s \mathbf{MT_s} + \mathbf{J}x\mathbf{B} \quad (45)$$

$$\frac{\partial \varepsilon}{\partial t} + \nabla.(\varepsilon \mathbf{v}) = -\nabla.(\mathbf{Q}) - \nabla.(\mathbf{v}(P\mathbf{I} + \boldsymbol{\tau})) + \sum_s \mathbf{MT_s}.\mathbf{v} + \sum_s f_{ts} \mathbf{J}.\mathbf{E} + (\mathbf{J}x\mathbf{B}).\mathbf{v} \quad (46)$$

where ϱ the neutral gas density, **v** the neutral gas velocity vector, S the neutral gas density source term, P the pressure, $\boldsymbol{\tau}$ the shear stress tensor which comprises of the τ_{ii} and τ_{jj} components, $\mathbf{MT_s}$ the momentum transfer of charged to the neutral particles due to elastic collisions, ε the neutral gas thermal energy density, **Q** is the thermal conductivity term, f_{ts} the percentage of energy density of charged particle with subscript s that is transferred as thermal energy to the neutral particles due to inelastic collisions and **J** the current density, **JxB** is the Lorentz force and **(JxB).v** is the magnetic Lorentz force acting on the energy of the flow.

The neutral gas density source term is calculated by:

$$S = m(-\alpha N_e |\mathbf{W_e}| - \eta N_e |\mathbf{W_e}| + 2\beta_{pn} N_p N_n + \beta_{ep} N_e N_p) \quad (47)$$

where m is the neutral gas particle mass which is constant.

The thermal conductivity, current density and pressure are calculated respectively by equations (48) to (50):

$$\mathbf{Q} = -k\nabla T \quad (48)$$

$$\mathbf{J} = q_s N_s \mathbf{W_s} \quad (49)$$

$$P = NKT \quad (50)$$

where k is the thermal conductivity, T is the neutral gas temperature, q_s, N_s, $\mathbf{W_s}$ the charge, density per unit volume and velocity vector of charged particle s, N the number of neutral gas particles per unit volume and K the Boltzmann constant.

The shear stress tensors are calculated by:

$$\tau_{ii} = -\mu(2\frac{\partial v_i}{\partial x_i} - \frac{2}{3}\nabla.\mathbf{v}) \quad (51)$$

$$\tau_{ij} = -\mu(\frac{\partial v_i}{\partial x_j} + \frac{\partial v_j}{\partial x_i}) \quad (52)$$

where μ the viscosity coefficient, and subscript i and j represent the different degrees of freedom in the different space directions.

The momentum transfer of charged to the neutral particles due to elastic collisions $\mathbf{MT_s}$ is calculated by:

$$\mathbf{MT_s} = N_s \frac{m}{m + m_s} \frac{q_s}{\mu_s}(\mathbf{W_s} - \mathbf{v}) \quad (53)$$

where μ_s is the mobility of charged particle s.

The numerical algorithm and its implementation are presented below in two-dimensional Cartesian, two-dimensional axisymmetric cylindrical and three-dimensional Cartesian coordinates in the following sections.

2.3.2. Two-dimensional Cartesian coordinates of Navier-Stokes

The Navier-Stokes equations in two-dimensional Cartesian (x, y) coordinates are formulated as follows:

$$\frac{\partial Q}{\partial t} + \frac{\partial K_x}{\partial x} + \frac{\partial K_y}{\partial y} + \frac{\partial L_x}{\partial x} + \frac{\partial L_y}{\partial y} = M \quad (54)$$

$$Q = \begin{bmatrix} \rho \\ \rho v_x \\ \rho v_y \\ \varepsilon \end{bmatrix} \tag{55}$$

$$K_x = \begin{bmatrix} \rho v_x \\ \rho v_x v_x + P \\ \rho v_y v_x \\ (\varepsilon + P) v_x \end{bmatrix} \quad K_y = \begin{bmatrix} \rho v_y \\ \rho v_x v_y \\ \rho v_y v_y + P \\ (\varepsilon + P) v_y \end{bmatrix} \tag{56}$$

$$L_x = \begin{bmatrix} 0 \\ \tau_{xx} \\ \tau_{yx} \\ -K\dfrac{\partial T}{\partial x} + \tau_{xx} v_x + \tau_{xy} v_y \end{bmatrix} \quad L_y = \begin{bmatrix} 0 \\ \tau_{xy} \\ \tau_{yy} \\ -K\dfrac{\partial T}{\partial y} + \tau_{yy} v_y + \tau_{yx} v_x \end{bmatrix} \tag{57}$$

$$M = \begin{bmatrix} m(-\alpha N_e \,|\, \mathbf{W_e} \,|\, - \eta N_e \,|\, \mathbf{W_e} \,|\, + \beta_{ep} N_e N_p + 2\beta_{pn} N_p N_n) \\ MT_s\{x\} + (\mathbf{J}x\mathbf{B})\{x\} \\ MT_s\{y\} + (\mathbf{J}x\mathbf{B})\{y\} \\ \sum_s f_{ts} \mathbf{J.E} + \sum_s MT_s\{x\} v_x + \sum_s MT_s\{y\} v_y + (\mathbf{J}x\mathbf{B})\{x\} v_x + (\mathbf{J}x\mathbf{B})\{y.\} v_y \end{bmatrix} \tag{58}$$

2.3.3. Two-dimensional cylindrical axisymmetric coordinates of Navier-Stokes

The Navier-Stokes equations in two-dimensional cylindrical (r, z) axisymmetric coordinates are formulated as follows:

$$\frac{\partial Q}{\partial t} + \frac{1}{r}\frac{\partial (K_r r)}{\partial r} + \frac{\partial K_z}{\partial z} + \frac{1}{r}\frac{\partial (L_r r)}{\partial r} + \frac{\partial L_z}{\partial z} = M \tag{59}$$

$$Q = \begin{bmatrix} \rho \\ \rho v_r \\ \rho v z \\ \varepsilon \end{bmatrix} \tag{60}$$

$$K_r = \begin{bmatrix} \rho v_r \\ \rho v_r v_r + P \\ \rho v_z v_r \\ (\varepsilon + P) v_r \end{bmatrix} \quad K_z = \begin{bmatrix} \rho v_z \\ \rho v_r v_z \\ \rho v_z v_z + P \\ (\varepsilon + P) v_z \end{bmatrix} \tag{61}$$

$$L_r = \begin{bmatrix} 0 \\ \tau_{rr} \\ \tau_{zr} \\ -K\dfrac{\partial T}{\partial r} + \tau_{rr}v_r + \tau_{rz}v_z \end{bmatrix} \quad L_z = \begin{bmatrix} 0 \\ \tau_{rz} \\ \tau_{zz} \\ -K\dfrac{\partial T}{\partial z} + \tau_{zz}v_z + \tau_{zr}v_r \end{bmatrix} \tag{62}$$

$$M = \begin{bmatrix} m(-\alpha N_e \,|\, \mathbf{W_e} \,| - \eta N_e \,|\, \mathbf{W_e} \,| + \beta_{ep}N_eN_p + 2\beta_{pn}N_pN_n) \\ MT_s\{r\} + \dfrac{P}{r} + (\mathbf{J}x\mathbf{B})\{r\}v_r \\ MT_s\{z\} + (\mathbf{J}x\mathbf{B})\{z.\}v_z \\ \sum_s f_{ts}\mathbf{J.E} + \sum_s MT_s\{r\}v_r + \sum_s MT_s\{z\}v_z + (\mathbf{J}x\mathbf{B})\{r\}v_r + (\mathbf{J}x\mathbf{B})\{z.\}v_z \end{bmatrix} \tag{63}$$

2.3.4. Three-dimensional Cartesian coordinates of Navier-Stokes

The Navier-Stokes equations in three-dimensional Cartesian (x, y, z) coordinates are formulated as follows:

$$\frac{\partial Q}{\partial t} + \frac{\partial K_x}{\partial x} + \frac{\partial K_y}{\partial y} + \frac{\partial Kz}{\partial z} + \frac{\partial L_x}{\partial x} + \frac{\partial L_y}{\partial y} + \frac{\partial L_z}{\partial z} = M \tag{64}$$

$$Q = \begin{bmatrix} \rho \\ \rho v_x \\ \rho v_y \\ \rho v_z \\ \varepsilon \end{bmatrix} \tag{65}$$

$$K_x = \begin{bmatrix} \rho v_x \\ \rho v_x v_x + P \\ \rho v_x v_y \\ \rho v_x v_z \\ (\varepsilon + P)v_x \end{bmatrix} \quad K_y = \begin{bmatrix} \rho v_y \\ \rho v_y v_x \\ \rho v_y v_y + P \\ \rho v_y v_z \\ (\varepsilon + P)v_y \end{bmatrix} \quad Kz = \begin{bmatrix} \rho v_z \\ \rho v_z v_x \\ \rho v_z v_y \\ \rho v_z v_z + P \\ (\varepsilon + P)v_z \end{bmatrix} \tag{66}$$

The two-step Lax-Wendroff technique comprising of the predictor-corrector steps is used for time stepping. The charged particle continuity and Navier-Stokes are discretised using Taylor-Galerkin Finite Elements [1], whereas Poisson's equation using Galerkin Finite Elements. Accuracy and efficiency for these long calculations are a crucial factor. The Flux Corrected Transport method (FCT) [2-4] ensures that accurate and efficient results are obtained free from inaccuracies building up and from non-physical oscillations.

$$L_x = \begin{bmatrix} 0 \\ \tau_{xx} \\ \tau_{yx} \\ \tau_{zx} \\ -K\dfrac{\partial T}{\partial x} + \tau_{xx}v_x + \tau_{xy}v_y + \tau_{xz}v_z \end{bmatrix}$$

$$L_y = \begin{bmatrix} 0 \\ \tau_{xy} \\ \tau_{yy} \\ \tau_{zy} \\ -K\dfrac{\partial T}{\partial y} + \tau_{yx}v_x + \tau_{yy}v_y + \tau_{yz}v_z \end{bmatrix} \quad (67)$$

$$L_z = \begin{bmatrix} 0 \\ \tau_{xz} \\ \tau_{yz} \\ \tau_{zz} \\ -K\dfrac{\partial T}{\partial z} + \tau_{zx}v_x + \tau_{zy}v_y + \tau_{zz}v_z \end{bmatrix}$$

$$M = \begin{bmatrix} m(-\alpha N_e \mid \mathbf{W_e} \mid -\eta N_e \mid \mathbf{W_e} \mid + \beta_{ep} N_e N_p + 2\beta_{pn} N_p N_n) \\ MT_s\{x\} + (\mathbf{J}x\mathbf{B})\{x\}v_x \\ MT_s\{y\} + (\mathbf{J}x\mathbf{B})\{y.\}v_y \\ MT_s\{z\} + (\mathbf{J}x\mathbf{B})\{z\}v_z \\ \sum_s f_{ts}\mathbf{J}.\mathbf{E} + \sum_s MT_s\{x\}v_x + \sum_s MT_s\{y\}v_y + \sum_s MT_s\{z\}v_z + (\mathbf{J}x\mathbf{B})\{x\}v_x + (\mathbf{J}x\mathbf{B})\{y.\}v_y + (\mathbf{J}x\mathbf{B})\{z\}v_z \end{bmatrix} \quad (68)$$

3. Implementation of the solution procedure

The solution procedure follows the pattern shown in Figure 1 for a single time step. It assumes that an initial distribution of electrons and positive ions at a given neutral gas temperature, velocity, density and applied voltage exists. The charged particle densities at time n (ϱ^n) are fed to Poisson's equation (PO) to obtain the electric field distribution at time n (E^n). The electric field and neutral gas parameters (N^n) are used to calculate the transport parameters (TP) at time n (TP^n). Then, the transport parameters are passed to the charge particle continuity equation (CON) to calculate the charge densities at time n+1/2 ($\varrho^{n+1/2}$). Next, the charge densities (ϱ^n), electric field (E^n) and transport properties (TP^n) at time n are passed to the Navier-Stokes solver (NS) to calculate the neutral gas properties at time n+1/2

($N^{n+1/2}$). This completes the half time step solution. Subsequently, the charge particle densities at time n+1/2 are passed to the Poisson solver to calculate the electric field at time n+1/2 ($E^{n+1/2}$). The electric field and neutral gas parameters at time n+1/2 are used to calculate the transport properties at time n+1/2 ($TP^{n+1/2}$). Consequently the transport properties are fed to the charged particle continuity solver to calculate the charge densities at time n+1 (ϱ^{n+1}). Finally, the Navier-Stokes solver uses the transport parameters, the electric field and the charge densities at time n+1/2 to calculate the neutral gas parameters at time n+1 (N^{n+1}) and this process is repeated continuously to proceed forward in time.

In fluid analysis, the collision coefficients and drift velocities of the charged particles can be calculated either using experimental or numerical techniques. Experimentally they are calculated as a function of the ratio of the electric field to the number of neutral gas particles per unit volume (E/N), depending also on the density and pressure of the neutral gas that has a direct effect on the frequency of collisions between the charged and neutral gas particles.

Numerically, commercial software packages calculate the transport properties by solving the electron distribution Boltzmann equation by utilizing the collisional cross sectional data and such software examples are ELENDIF [5], BOLSIG [6], and BOLSIG+ [7] which are used to calculate the above transport parameters [8].

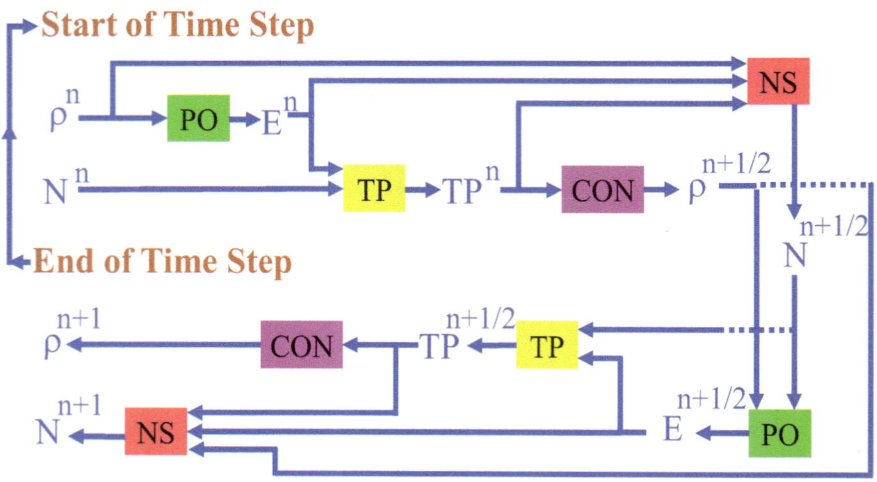

Figure 1. Solution procedure of the complete model

The charged particle continuity equations are coupled to Poisson's equation via the electric field strength and to the Navier-Stokes equations via the number of neutral gas particles per unit volume and their pressure. Poisson's equation is coupled to the continuity equations via the net charge density. As far as the Navier-Stokes equations are concerned, they are coupled to Poisson's equation via the electric field strength and to the charged particle continuity equations via production and loss processes, and via momentum and kinetic energy exchange between charged and neutral particles.

4. Finite element formulation of continuity and Navier-Stokes fluid equations

4.1. Two-dimensional Cartesian coordinates

The fluid transport equations take the general form in the two-dimensional Cartesian coordinates as:

$$\frac{\partial \bar{Q}}{\partial t}+\sum_{j=1}^{2}\frac{\partial \bar{K}_j}{\partial x_j}+\sum_{j=1}^{2}\frac{\partial \bar{L}_j}{\partial x_j}=\bar{M} \tag{69}$$

where $\bar{Q}$ is the unknown or independent variable, $\bar{K}$ and $\bar{L}$ are respectively the convective and diffusive fluxes, $\bar{M}$ is the source term.

The Taylor Galerkin scheme is used to develop the FE-FCT scheme, with the time discretization preceding the space discretization. The time stepping is performed in two steps, the advective predictor and the corrector step.

4.1.1. *Advective predictor step*

The advective predictor step is formulated using the Taylor series expansion for the unknown variable $\bar{Q}$ at a first order approximation, thereby resulting in the half time values of the independent variable $\bar{Q}^{n+1/2}$:

$$\bar{Q}^{n+1/2}=\frac{A_e}{3}\sum_{i=1}^{3}\bar{Q}_i^{n}+\frac{\Delta t}{4}\sum_{i=1}^{3}\bar{K}_i^{n}b_i+\frac{\Delta t}{4}\sum_{i=1}^{3}\bar{K}_i^{n}c_i-\frac{\Delta t}{6}A_e\sum_{i=1}^{3}\bar{M}_i^{n} \tag{70}$$

where A_e is the area of a triangular element, Δt is the time step, b_i and c_i are the linearly interpolated piecewise shape functions. The velocities at time (n+1/2) are assumed to be the average value of the nodal velocities at time n.

4.1.2. *Corrector step*

The corrector step utilizes the Taylor series expansion for the full time step to the second order approximation to get:

$$D_c\Delta\bar{Q}^{n}=-\frac{\Delta t b_i^e}{2}K_x^{n+1/2}-\frac{\Delta t c_i^e}{2}K_y^{n+1/2}-\frac{\Delta t b_i^e}{2}L_x^{n+1/2}-\frac{\Delta t c_i^e}{2}L_y^{n+1/2}$$

$$+\Delta t M^{n+1/2}\frac{A_e}{3}+\Delta t\int_{\Gamma_e}(K_n^{n+1/2}+L_n^n)N_i^e d\Gamma \quad (71)$$

where Dc is the consistent mass matrix, $\bar{K}_n^{n+1/2}$ and $\bar{L}_n^{\,n}$ are the convective term at time n+1/2 and diffusive term at time n fluxes outward in the normal direction of the boundary, Γ_e is the boundary surrounding element e and N_i^e is the interpolation function. Writing the above equation in the general form results in:

$$D_c \Delta \bar{Q}^{n} = B^n \quad (72)$$

where $\Delta \bar{Q}^{n}$ is the vector of nodal increments, B^n is the vector of added element contributions to the nodes. Instead of using a consistent mass matrix, a lumped mass matrix (Dl) is used for computational purposes, resulting in:

$$D_l \Delta \bar{Q}^{a,n} - (D_c - D_l)\Delta \bar{Q}^{a-1,n} = B^n \quad (73)$$

where α is an integer representing the iteration number.

To calculate the low order scheme, mass diffusion is added of the form:

$$c_d (D_c - D_l)\bar{Q}^{n} \quad (74)$$

to get the low order scheme:

$$D_l \Delta \bar{Q}^{a,n} - (D_c - D_l)\Delta \bar{Q}^{a-1,n} - c_d (D_c - D_l)\bar{Q}^{n} = B^n \quad (75)$$

where cd is the variable diffusion coefficient and is dependent on the mesh size, time step and speed of an element. A detailed analysis of the above formulation procedure is found in [9].

4.2. Two-dimensional cylindrical axisymmetric coordinates

The fluid transport equations take the general form in the two-dimensional cylindrical axisymmetric coordinates as:

$$\frac{\partial \bar{Q}}{\partial t}+\frac{1}{r}\frac{\partial(\bar{K}_r r)}{\partial r}+\frac{\partial(\bar{K}_z)}{\partial z}+\frac{1}{r}\frac{\partial(\bar{L}_r r)}{\partial r}+\frac{\partial(\bar{L}_z)}{\partial z}=\bar{M} \quad (76)$$

where $\bar{K}_r$ $\bar{K}_z$ $\bar{L}_r$ $\bar{L}_r$ are respectively the radial and axial convective and diffusive fluxes, and r is the radial coordinate in cylindrical axisymmetric coordinates.

The Taylor Galerkin scheme is used to develop the FE-FCT scheme, with the time discretization preceding the space discretization. The time stepping is performed in two steps, the advective predictor and the corrector step.

4.2.1 *Advective predictor step*

The advective predictor step is formulated using the Taylor series expansion for the unknown variable $\bar{Q}$ at a first order approximation, thereby resulting in the half time values of the independent variable $\bar{Q}^{n+1/2}$:

$$\bar{Q}^{n+1/2} = \frac{1}{12}\sum_{i=1}^{3}(3+r_i^{'})(\bar{Q}_i^{n} + \frac{\Delta t}{2}\bar{M}^{n}) - \frac{\Delta t}{4A_e}\sum_{i=1}^{3}((\bar{K}_z)_i^{n} c_i + (\bar{K}_r)_i^{n} b_i) - \frac{\Delta t}{6\bar{r}_e}\sum_{i=1}^{3}(\bar{K}_r)_i^{n} \quad (77)$$

where A_e is the area of a triangular element, Δt is the time step, b_i and c_i are the linearly interpolated piecewise shape functions.

where $r_i^{'}$ is given by:

$$r_i^{'} = \frac{r_i}{\bar{r}_e} \quad (78)$$

where $\bar{r}_e$ is the r-coordinate of the centroid of the element e given by:

$$\bar{r}_e = \sum_{i=1}^{3}\frac{r_i}{3} \quad (79)$$

The velocities at time (n+1/2) are assumed to be the average value of the nodal velocities at time n.

4.2.2. *Corrector step*

The corrector step utilizes the Taylor series expansion for the full time step to the second order approximation to get:

$$\bar{r}_e D_c \Delta\bar{Q}^{n} = -\Delta t\bar{r}_e\frac{b_i^e}{2}\bar{K}_r^{n+1/2} - \Delta t\bar{r}_e\frac{c_i^e}{2}\bar{K}_z^{n+1/2} - \Delta t\bar{r}_e\frac{b_i^e}{2}\bar{L}_r^{n}$$

$$-\Delta t\bar{r}_e\frac{c_i^e}{2}\bar{L}_z^{n} + \Delta t\bar{r}_e\frac{A_e}{3}\bar{M}^{n+1/2} + \Delta t\bar{r}_e\int_{A_e}(\bar{K}_n^{n+1/2} + \bar{L}_n^{n})N_i^e dA \quad (80)$$

The above discretization results in the following equations:

$$D'_c \Delta \bar{Q}^n = B'^n \tag{81}$$

where:

$$D'_c = \bar{r} D_c \tag{82}$$

and

$$B'^n = \bar{r} B^n \tag{83}$$

where $\bar{r}$ denoting a matrix of $\bar{r}_e$ entries everywhere, D'_c is the consistent mass matrix and B'^n is the vector of added element contributions to the nodes in the cylindrical axisymmetric case. In order to implement the Cartesian and axisymmetric cases together, the half time step values $\bar{Q}^{n+1/2}$ have to be calculated slightly differently, and the consistent mass matrix D_c and lumped matrix D_l need to be multiplied by $\bar{r}_e$ in the cylindrical axisymmetric case, Finally, during the limiting procedure, the antidiffusive element contribution, the consistent and lumped matrices need to be multiplied by $\bar{r}_e$ for each element.

4.3. Three-dimensional Cartesian coordinates

The fluid transport equations take the general form in the three-dimensional Cartesian coordinates as:

$$\frac{\partial \bar{Q}}{\partial t} + \sum_{j=1}^{3} \frac{\partial \bar{K}_j}{\partial x_j} + \sum_{j=1}^{3} \frac{\partial \bar{L}_j}{\partial x_j} = \bar{M} \tag{84}$$

The Taylor Galerkin scheme is used to develop the FE-FCT scheme, with the time discretization preceding the space discretization. The time stepping is performed in two steps, the advective predictor and the corrector step.

4.3.1. Advective predictor step

The advective predictor step is formulated using the Taylor series expansion for the unknown variable $\bar{Q}$ at a first order approximation, thereby resulting in the half time values of the independent variable $\bar{Q}^{n+1/2}$:

$$\bar{Q}^{n+1/2} = \frac{V_e}{4}\sum_{i=1}^{4}\bar{Q}_i^{n} + \frac{\Delta t}{12}\sum_{i=1}^{4}\bar{K}_i^{n} b_i + \frac{\Delta t}{12}\sum_{i=1}^{4}\bar{K}_i^{n} c_i + \frac{\Delta t}{12}\sum_{i=1}^{4}\bar{K}_i^{n} d_i - \frac{\Delta t}{8}V_e\sum_{i=1}^{4}\bar{M}_i^{n} \tag{85}$$

where V_e is the volume of a tetrahedral element, b_i, c_i and d_i are the linearly interpolated piecewise shape functions. The velocities of the charged species at time (n+1/2) are assumed to be the average value of the nodal velocities at time n.

4.3.2. Corrector step

The corrector step utilizes the Taylor series expansion for the full time step to the second order approximation to get:

$$D_c\Delta\bar{Q}^{n} = -\frac{\Delta t b_i^e}{3}K_x^{n+1/2} - \frac{\Delta t c_i^e}{3}K_y^{n+1/2} - \frac{\Delta t d_i^e}{3}K_z^{n+1/2} - \frac{\Delta t b_i^e}{3}L_x^{n+1/2} - \frac{\Delta t c_i^e}{3}L_y^{n+1/2} - \frac{\Delta t d_i^e}{3}L_z^{n+1/2}$$

$$+\Delta t M^{n+1/2}\frac{V_e}{4} + \Delta t\int_{A_e}(K_n^{n+1/2} + L_n^{n})N_i^e dA \tag{86}$$

where A_e is the boundary surrounding element e. Writing the above equation in the general form results in:

$$D_c\Delta\bar{Q}^{n} = B^n \tag{87}$$

Instead of using a consistent mass matrix, a lumped mass matrix is used for computational purposes, resulting in:

$$D_l\Delta\bar{Q}^{a,n} - (D_c - D_l)\Delta\bar{Q}^{a-1,n} = B^n \tag{88}$$

To calculate the low order scheme, mass diffusion is added of the form:

$$c_d(D_c - D_l)\bar{Q}^{n} \tag{89}$$

to get the low order scheme:

$$D_l\Delta\bar{Q}^{a,n} - (D_c - D_l)\Delta\bar{Q}^{a-1,n} - c_d(D_c - D_l)\bar{Q}^{n} = B^n \tag{90}$$

5. Fluid flow validation of the above FE-FCT formulation

In order to validate the fluid flow equations and the methodology developed, the Euler equations which are the Navier-Stokes equations without diffusion and source terms were tested using the shock tube type problem, where different parameters, such as density, diffusion coefficient and mesh size were considered. Furthermore, a shock wave incident on a wedge was also tested in the Cartesian case. Finally, energy source terms are introduced into the two-dimensional Cartesian and axisymmetric cylindrical geometries, which result in sound and shock wave generation, resembling the heating expected during the development of a spark and an arc plasma gas discharge.

5.1. Riemann test case

The Euler equations are validated in one-dimensional Cartesian coordinates using the shock tube or Riemann test case in air with the ratio of the specific heats, γ, equal to 5/3, where analytical results are available. Figure 2 shows a one-dimensional comparison between the analytical and numerical solution of density for three different meshes at times t = 0.1 and 0.5 s. It is shown that in the case of the finer mesh, the overshoot observed is reduced. Figure 3 shows the analytical and numerical solution for density for the finest mesh (Mesh 3) at three different instants in time of t = 0.1, 0.3 and 0.5. A direct comparison shows clearly that the Euler solver developed is capable of simulating the propagation of shock waves and that the results are of adequate accuracy. This is also verified by calculating the maximum percentage error for the three different meshes as shown in Table 1 below.

Mesh number	Maximum density error
1	1.6177 %
2	1.0895 %
3	0.2345 %

Table 1. Maximum density error for three different meshes

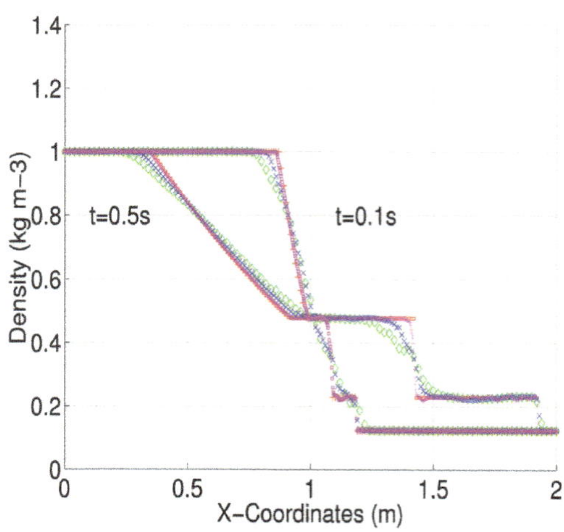

Figure 2. Comparison of analytical and numerical results using three different meshes at two different instants in time t = 0.1 and 0.5 s. □ line: Analytical solution, ◊ line: Mesh 1, 500 nodes, x line: Mesh 2, 1000 nodes and + line: Mesh 3, 5000 nodes

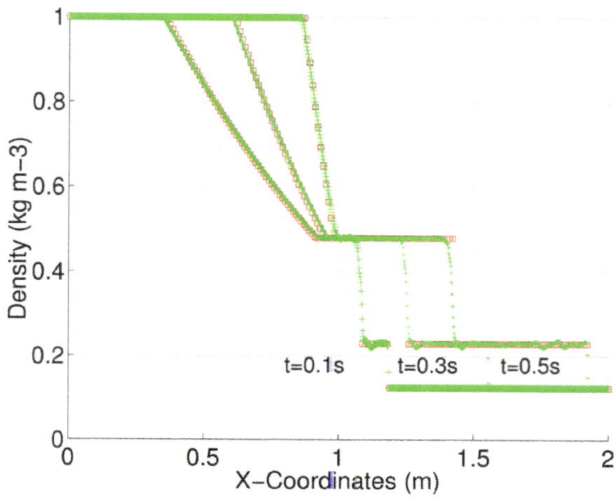

Figure 3. Comparison of analytical and experimental results using the finer mesh at three different instants in time t = 0.1, 0.3 and 0.5 s. □ line: Analytical solution, + line: Mesh 3, 5000 nodes

5.2. Heating source term test case

The two-dimensional Cartesian Euler equations in air with γ equal to 1.4 are tested by introducing a heating source term of magnitude $1\text{x}10^{12}$ J/m^2 for a duration of $3\text{x}10^{-8}$ s within a square block of size 0.002 m x 0.002 m, with the simulation being run for 5000 steps at a constant time step increments of $3\text{x}10^{-8}$ s. The contour plot of the momentum in the x direction is shown in Figure 4, where at the exterior of the wave front, the wave is moving outwards in the x direction with maximum momentum. At the interior of the wave front, there is a sudden decrease and a change in the direction of the momentum. This is due to the fact that the inertial effects of the sudden explosion cause an overexpansion of the wave, therefore a rarefraction wave is created moving inwardly. Figure 5 shows a contour plot of the energy at a time of $9\text{x}10^{-5}$ s. High energies occur at the shock front of the wave with low energies forming behind the shock front, again due to the overexpansion of the wave, which are in agreement with blast wave theory Baker (1973) [10].

5.3. Shock wave incident on a wedge test case

A shock wave of Mach number 2 in air with γ equal to 1.4 is incident on a wedge angle of 46° degrees. The neutral gas is at a temperature of 300 $^{\circ}$K and a pressure of 30 kPa. The simulation is run for 13000 time steps of $1\text{x}10^{-7}$ s duration. The simulation is terminated 0.1 m before the shock wave hits the boundary on the right. Figure 6 shows the Cartesian contour plots of the density at a time of $1.3\text{x}10^{-3}$ s with the results being compared with the benchmark test case of a shock wave impinging on a wedge Takayama (1997) [11]. The results are found to be in good agreement capturing the shock wave created at (0.9 m, 0.7 m) coordinates.

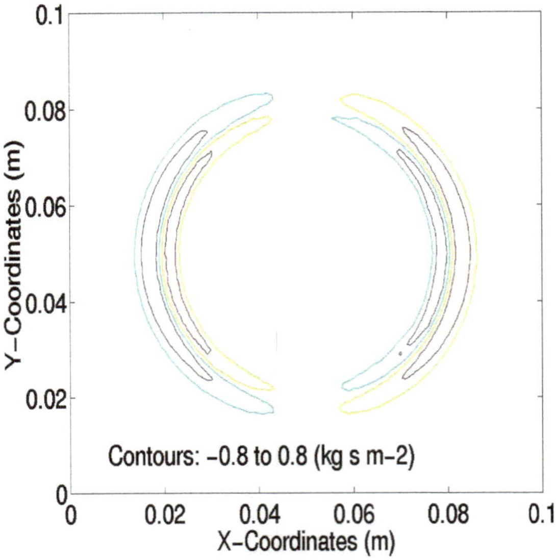

Figure 4. Contour plot of the x-momentum at time t = 9x10^{-5} s

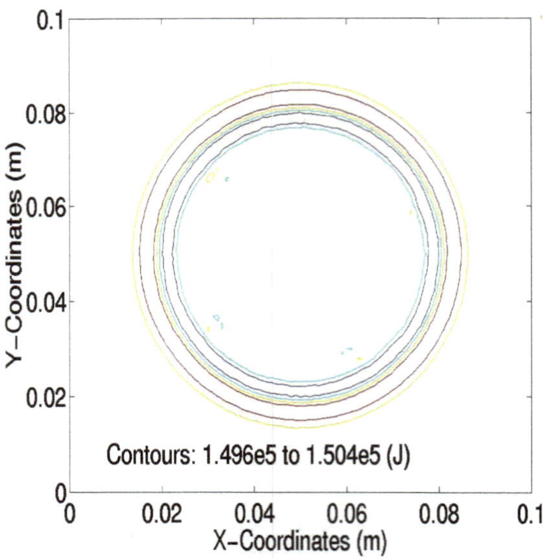

Figure 5. Contour plot of the energy at time t = 9x10^{-5}s

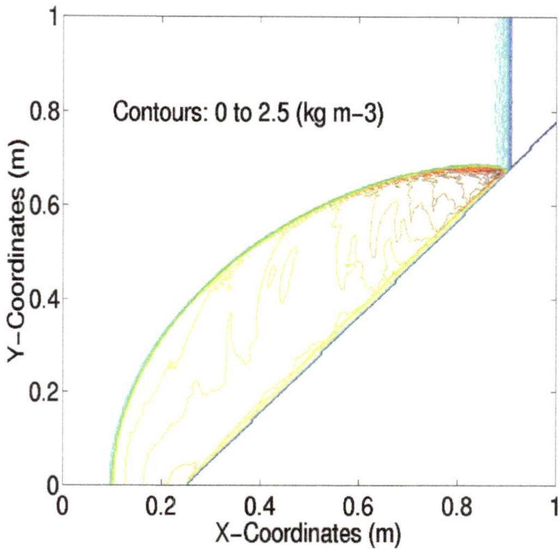

Figure 6. Contour plot of the density at a time of $1.3x10^{-3}$ s

5.4. Micro-blast waves test case

The two-dimensional Euler solver is validated in the cylindrical axisymmetric case using numerical blast wave tests that involve the release of sudden burst of energy by focusing a laser beam within small volumes in air with γ equal to 1.4, leading to the development of shock waves. Jiang et al (1998) [12] has analyzed both experimentally and numerically the propagation of micro-blast waves in ambient air. The duration of the pulse was of the order of 18 ns and the total amount of energy released was measured to be 1.38 J. This sudden release of energy raises instantaneously the pressure, density and temperature to values many times higher than those of the ambient conditions, thereby creating highly compressive acoustic waves that will in turn generate strong shock waves.

Figures 7 and 8 compare respectively one-dimensional plots of the density against time along the axis of symmetry obtained by Jiang and the authors for the same mesh and different diffusion coefficients and for different mesh and the same diffusion coefficient. Given the diversities of the two numerical schemes, the results show to be in good agreement.

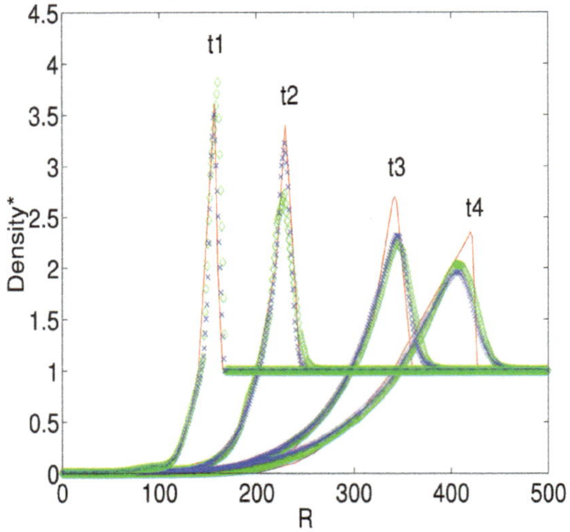

Figure 7. Non-dimensionalized density (ϱ^*) in radial direction at t_1 = 5.36 μs, t_2 = 6.65 μs, t_3 = 10.05 μs, t_4 = 12.58 μs. - line: Jiang, ◊ line : Mesh 2 : 27570 nodes, Diff. Coeff., 0.0012, x line: Mesh 2:, 27570 nodes, Diff. Coeff., 0.002

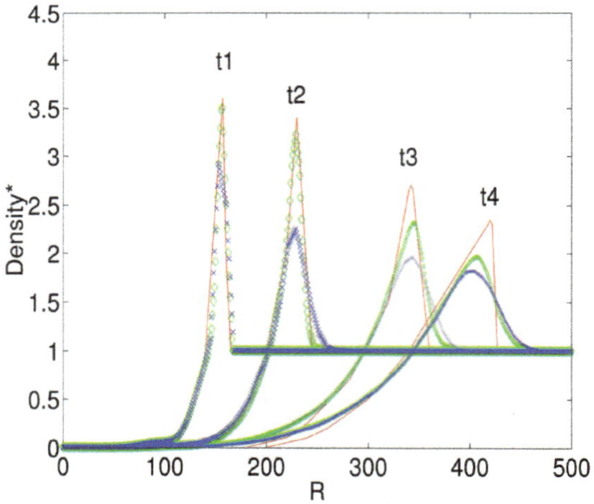

Figure 8. Non-dimensionalized density (ϱ^*) in the radial direction at t_1 = 5.36 μs, t_2 = 6.65 μs, t_3 = 10.05 μs, t_4 = 12.58 μs. - line: Jiang, ◊ line: Mesh 2, Diff. Coeff., 0.002, x line: Mesh 1, Diff. Coeff., 0.002

6. Adaptive mesh algorithm

In order to reduce computational needs considerably, an adaptive mesh generator has been developed, rendering possible the analysis of heating effects both in short and long gaps.

For an adaptive mesh generator, an error indicator is necessary to decide on the amount of refinement. The error indicator used by the author is the one used by Lohner [13], which in multidimensional form is calculated as 999999follows:

$$E^i = \sqrt{\frac{\sum_{k,l}(\int_\Omega N_k^i N_l^j d\Omega . U_j)^2}{\sum_{k,l}(\int_\Omega |N_k^i|[|N_l^j U_j| + \varepsilon(|N_l^j||U_j|)]d\Omega)^2}} \tag{91}$$

where N^i_k is the shape function of node i in element k, N^j_l is the shape function of node j in element l, U_j is the value of the variable chosen to be used as error indicator at node j, E^i is the error indicator value at node i, and ε is a factor varying from 0 to 1. The value of ε is added as noise filter, so that any loss of monotonicity such as wiggles or ripples are not refined. The above error indicator is dimensionless, fast to calculate, varies from 0 to 1 such that prefixed tolerances and many variables as error indicators can be used at the same time.

In this case, an adaptive mesh algorithm in two-dimensional Cartesian, and two-dimensional cylindrical axisymmetric coordinates has been developed by the author in order to analyze plasmas much faster and more accurately. This is achieved by constructing a computer resource efficient algorithm, which is automated in providing the necessary results. Generally, the numerical solution of time dependent differential equations is classified into the categories of static and dynamic. For static methods, any addition of nodes, and edge swap and topological movement of nodes is performed at a fixed time. In the dynamic case, or moving mesh methods, a mesh equation is introduced that involves node velocities such that a fixed number of nodes are moved in such a way that the nodes are always concentrated near regions of rapid variation of the solution. Thereby the simultaneous solution of the differential and mesh equations is necessary, having the advantage that no interpolation between existing and future meshes is necessary [14]. The author of this paper has adopted the static method, which is more widely used and tested in the literature.

The algorithm incorporates an innovative element quality improvement procedure that has the ability to guarantee the generation of new meshes, as well as the treatment of existing bad element quality meshes, to nearly ideal standards, guaranteeing a minimum element quality of 0.85, and above 0.90 of average element quality in uniform and non-uniform geometric domains. Furthermore, another novelty of the paper is also the utilisation of an interpolation method between meshes which is very fast, making the adaptive meshing more attractive. The re-meshing times are decided optimally according to the maximum speeds of the fastest particle within the simulation to guarantee that the desired mesh resolution is in place at all times, whereas the numerical diffusion error due to the interpolation between meshes is minimized by interpolating between meshes of similar size.

Figure 9 shows a flow chart of the implementation procedure of the adaptive mesh developed by the author. The first step is to create a reference coarse mesh that will be used as a reference mesh for future refinement and coarsening of the meshes. The second step is to apply the initial conditions on the reference coarse mesh, and calculate the amount of refinement by multiplying the error by a constant factor in each element. The refinement factor is decided on a basis of experience, and trial and error. Then a newly adapted initial mesh is developed using a freely available two-dimensional software package which uses the divide and conquer method to refine elements [15]. Since the elements that are created from this package are of bad element equality, they are first treated using adaptive mesh techniques. Specifically, the h-refinement/coarsening technique is firstly used, which includes the edge swap and node additions/removals to improve the interconnectivity between adjacent nodes. Secondly, the r-refinement technique, which involves jiggling of the mesh, follows, i.e. the movement of nodes around the geometry in a controlled way, such that the overall element quality is improved.

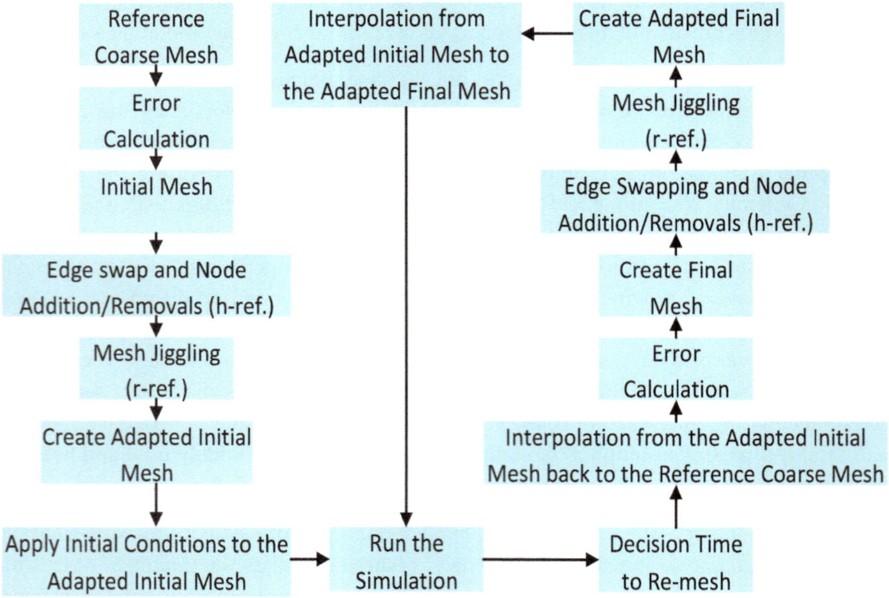

Figure 9. Flow chart of the implementation procedure of the adaptive mesh generator

Once the adapted initial mesh is created that can be used for the solution of the differential equations, an interpolation of the results from the reference coarse mesh to the adapted initial mesh is performed. Having defined the mesh to be used and calculated the corresponding values that all the variables have at the nodes of this mesh, the simulation proceeds forward in time, until an indicator signifies that the solution has reached the outer boundary of the refined region, and a re-meshing operation needs to be performed. The indicator calculates the maximum distance that the fastest particle travels, and ensures that it does not exceed the geometric tolerance of the initial coarse mesh, thereby it is guaranteed that the correct resolution is provided at all times. Then in order to decide on the appropriate mesh to run the simulations, the results are interpolated from the adapted initial mesh back to the reference coarse mesh and then the error is calculated to create a final mesh. Then this mesh is treated as above, using edge swap and node additions/removals operations to improve the interconnectivity between adjacent nodes, and finally mesh jiggling operations to improve the overall element quality, creating the adapted final mesh. Then one needs not to interpolate from the reference coarse mesh to the created adapted final mesh, but instead from the created adapted initial mesh to the created adapted final mesh. This operation is performed due to the fact that interpolation is a source of numerical diffusion on the results, and by interpolating just once, instead of twice, makes the results during the interpolation processes more accurate. Then having decided on the adapted final mesh and having interpolated the results, then one runs the simulation forward in time and the procedure is repeated many times. This completes the implementation of the adaptive mesh algorithm. For the implementation of the above adaptive mesh algorithm, three tools are necessary, which are (a) the error calculation, (b) the element quality improvement algorithm, and (c) the algorithm for interpolation between meshes tools, and are thoroughly explained below.

7. Application of electromagnetic and fluid code on microplasmas

7.1. RF applicator at 40 MHz

In this section, results are presented showing the avalanche and streamer propagation in a gap of 1 cm in an RF applicator at 40 MHz. Figures 10 and 11 refer to times $t_1 = 50$, $t_2 = 55$, $t_3 = 62.5$, $t_4 = 63.5$ ns. Figure 10 shows the radial field along the symmetry axis from times t_1 to t_4. It is shown that during the time interval t_1 to t_2, its magnitude increases to a value of approximately 0.5×10^5 V/m due to the net charge that exists and that it stays fairly constant at the bottom electrode end, but gradually increases at the upper electrode end, with both radial fields always extending closer towards the electrodes as time progresses. It is interesting to note that the radial field is positive during this time at the upper end and negative towards the bottom end. This initiates the propagation of streamers towards the electrodes. At time t_3, the radial field reverses at both ends, when compared with the time interval of t_1 to t_2, with the radial field magnitude being larger than before and this is due to the streamer impinging on the two electrodes. At time t_4, the radial field increases even further at the bottom electrode side, but changes sign at the upper electrode side. This is because the streamer that hits the upper electrode forms earlier than the one that hits the lower electrode, as the discharge is overall exposed to a larger negative voltage cycle. The change of sign of the radial field at the upper electrode is due to the absorption of the electrons into the electrode and the start of the formation of a cathode fall region of net positive charge near it. Figure 11 shows the axial electric field along the symmetry axis from times t_1 to t_4. At time t_1, at the middle of the gap, the axial electric field gets distorted from the Laplacian field which initiates the streamer propagation. During the time interval t_1 to t_2, the distortion increases in magnitude and spreads out further towards the two electrodes. At time t_3, there is a sudden decrease of the axial electric field at the two electrodes, leaving a nearly constant plateau in the middle of the gap due to the streamer impinging on the two electrodes. At a later stage of t_4,, the ripple profile reverses shape and starts to increase again at an ever higher rate, which corresponds to the times when the streamers charge is accumulated on the two electrodes.

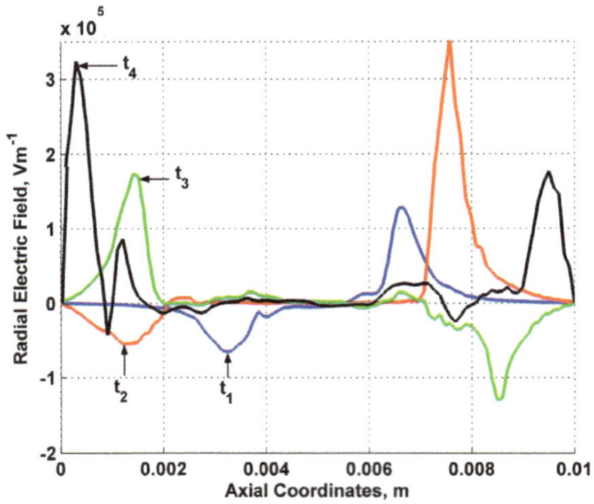

Figure 10. One-dimensional plots of the radial field at times t_1 to t_4 along the symmetry axis

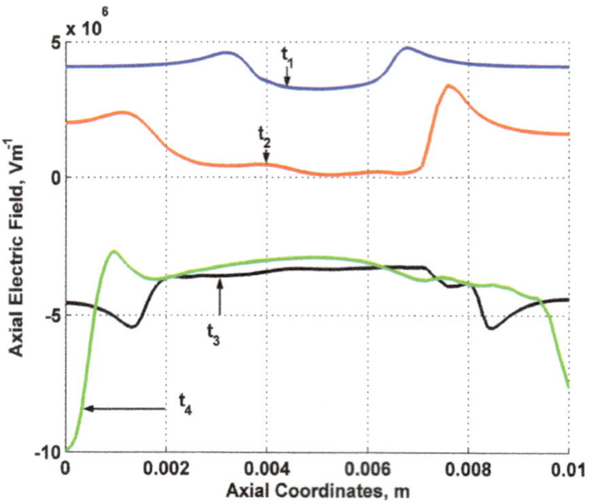

Figure 11. One-dimensional plots of the axial field at times t_1 to t_4 along the symmetry axis

7.2. Heating effects of normal and abnormal glow discharge

The development of a glow discharge in atmospheric pressure air and its associated Joule heating are analyzed by solving the Poisson, charged particle continuity and Navier-Stokes equations using the FE-FCT method in two-dimensional cylindrical axisymmetric coordinates. An applied direct current voltage of 20% above the breakdown voltage is applied at the anode in a 1 mm air gap between two parallel plate electrodes and a normal glow discharge is shown to consist of the cathode fall, negative glow, positive column and anode regions. Neutral gas heating occurs with the initiation of the glow discharge, with the temperature at the anode shown to increase by only a few degrees °K as shown in Figure 12 due to the low electron densities and axial fields at the adjacency of the anode, whereas at the proximity of the cathode the temperature increases by approximately 180 °K as shown in Figure 13, with the temperature maximum at the cathode fall region, reducing abruptly at almost ambient temperature just outside it, where it remains almost constant.

The numerical results for the development of a glow discharge and its associated Joule heating by modeling its transition from a single electron, in a uniform applied electric field in atmospheric air, to a fully fledged state which consists of the cathode fall, negative glow, positive column and anode regions are presented. It is shown that the positive column travels towards the anode in the form of a return wave and maximum heating by approximately 180 °K occurs at the cathode fall region during the glow discharge development.

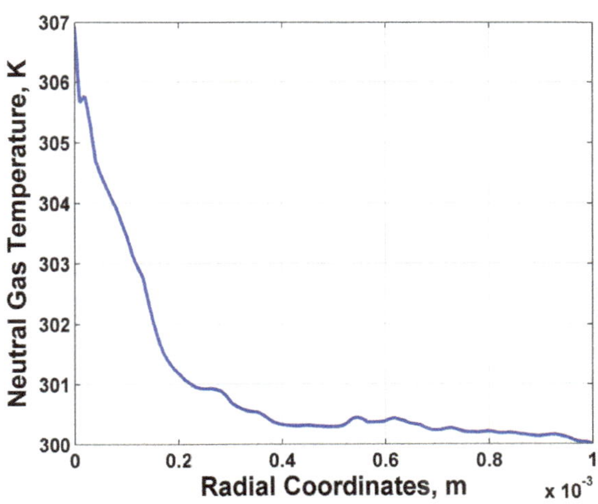

Figure 12. One-dimensional plots of the temperature along the symmetry axis close to the cathode at t = 6.75 ns at V= 5600 V

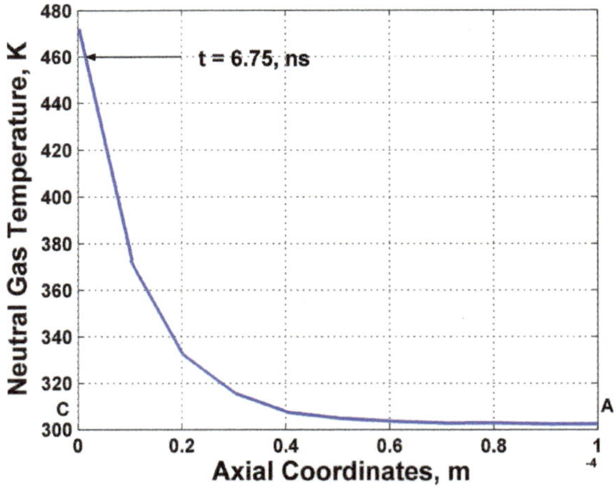

Figure 13. One-dimensional plots of the temperature along the anode at four instances in time at V = 5600 V

7.3. Secondary streamers

A constant voltage of 130 kV is applied on a dielectric barrier discharge configuration with the two metallic electrodes placed 4 mm apart, and with each one covered by alumina dielectric of 1 mm thickness, leaving an air gap of 2 mm for the discharge to develop. The two-dimensional neutral gas temperature distribution at time t = 0.3263 μs is shown in Figure 14, with most of the heating within the microplasma occurring along the symmetry axis where most of the activity takes place [8]. The column of heated air extends to a radial distance of 0.2 mm, and the temperature is raised approximately 120 °K. It is also observed that striations occur between the anode and the cathode. Along the dielectric electrodes, the temperature is larger than in the inter-electrode gap, but smaller than that at the symmetry axis. On average, higher neutral gas temperatures are observed at the cathode due to the arrival of the primary and secondary streamer charges on the dielectric cathode [16]. The two-dimensional axial electric field distribution at time t = 11.8 ns is shown below in Figure 15. Along the symmetry axis, there exists a cathode fall, negative glow, positive column and anode regions, similar to those of a normal glow discharge. A secondary streamer is shown to form at a radial distance of 1.5 mm that travels from the anode towards the cathode, extending radially up to the outer boundaries of the discharge and at later stages, when it hits the cathode, the sheath region spreads throughout the cathode dielectric surface.

7.4. Adaptive mesh results on DC avalanche and streamers in ambient air

The adaptive mesh generator has been tested in a geometric configuration that comprises of two metallic parallel-plates that are placed 1 mm apart in ambient atmospheric air. With a single electron released at the cathode as initial condition, the development of the avalanche and the primary streamer are analyzed as shown in Figures 16a-f, which show the mesh plots and the corresponding electron densities along the symmetry axis at three different instances in time. Figures 16a and 16b show respectively the initial mesh refinement and electron density distribution along the symmetry axis, whereas Figures 16c and 16d show the mesh and electron density during cathode directed primary streamer propagation, and Figures 16e and 16f show the mesh and electron density distributions just before the streamer hits the cathode. In the simulations, the photoemission effect is included in the calculation using the model developed by the authors [17], whereas photoionization and impact ionization phenomena are excluded. It has been shown that the avalanche and streamer propagation are captured by the adaptive mesh generator in an optimum way.

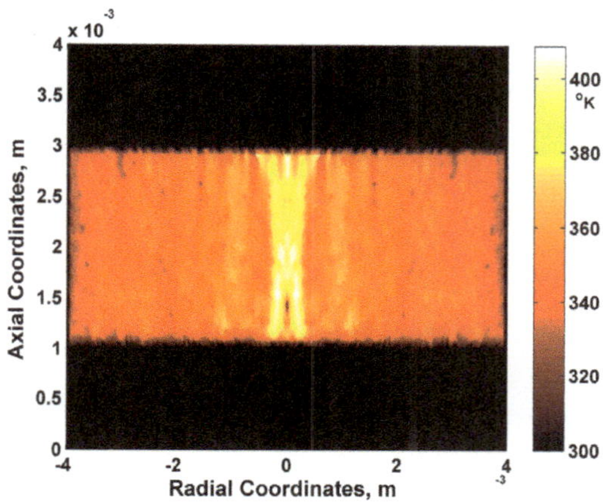

Figure 14. Plot of the two-dimensional cylindrical axisymmetric neutral gas temperature at time t = 0.3263 μs

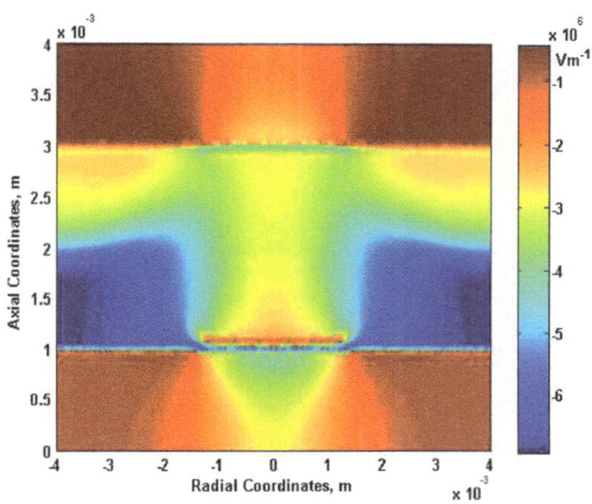

Figure 15. Plot of the two-dimensional axial electric field distribution at a time t = 11.8 ns

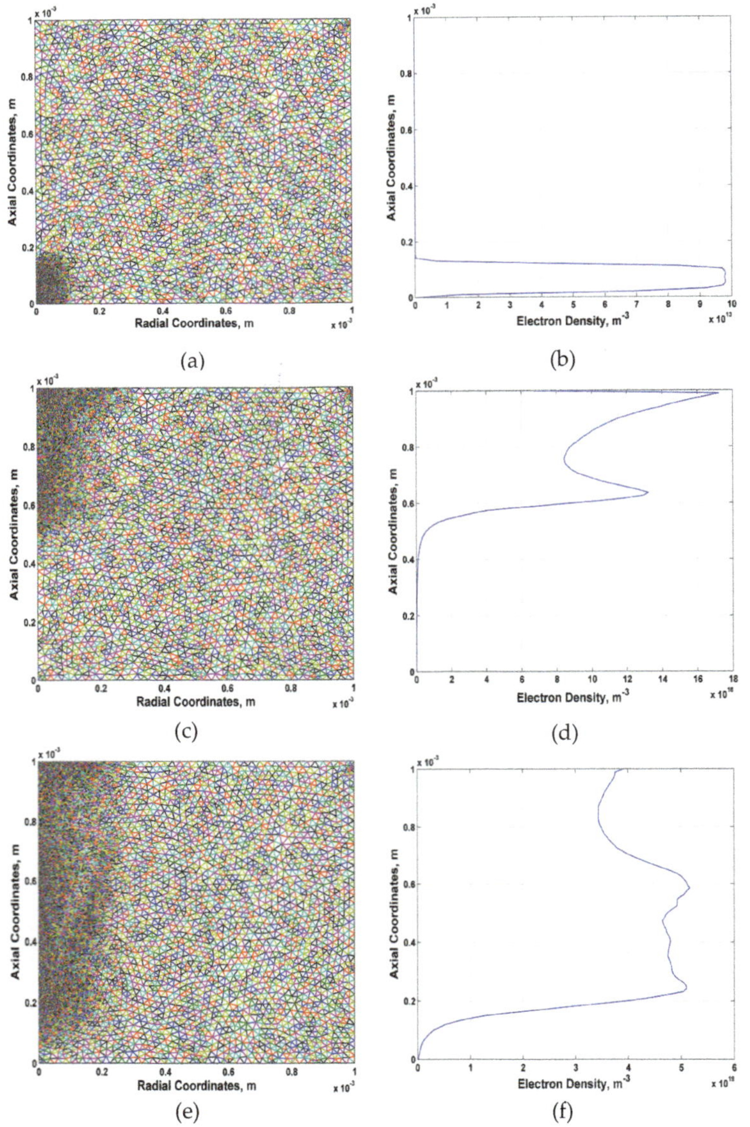

Figure 16. (a) Two-dimensional cylindrical mesh at time t = 0 s at a voltage of 5600 V, (b) One-dimensional electron distribution along the symmetry axis at time t = 0 s at a voltage of 5600 V, (c) Two-dimensional cylindrical mesh at time t = 3.59 ns at a voltage of 5600 V, (d) One-dimensional electron distribution along the symmetry axis at time t = 3.59 ns at a voltage of 5600 V, (e) Two-dimensional cylindrical mesh at time t = 4.54 ns at a voltage of 5600 V, (f) One-dimensional electron distribution along the symmetry axis at time t = 4.54 ns at a voltage of 5600 V

8. Conclusions

The electromagnetic and fluid analysis of collisional plasmas has been thoroughly discussed including the necessary differential conservation equations to characterize such plasmas. The finite-element formulations were presented for the solution of these equations and the implementation procedure to couple the above set of equations was discussed. Thereafter, the FE-FCT algorithm was validated against theoretical and experimental fluid flow results and it was then used in a variety of collisional plasma configurations to study different plasma phenomena.

Author details

Antonis P. Papadakis
Department of Electrical Engineering, Frederick University, Nicosia, Cyprus

9. References

[1] Georghiou GE, Morrow R, Metaxas AC (1999) An Improved Finite Element Flux-Corrected Transport Algorithm. J. Comput. Phys. 148: 605-620.
[2] Zalesak S (1979) Fully Multidimensional Flux-Corrected Transport Algorithms For Fluids. J. Comput. Phys. 31: 335-362.
[3] Lohner R (1987) Finite Element flux-corrected transport (fem-fct) for the euler navier stokes equations. Int. J. Numer. Meth. Fl. 7: 1093-1109.
[4] Georghiou GE, Morrow R, Metaxas AC (2000) Two-dimensional simulation of streamers using the FE-FCT algorithm. J. Phys. D: Appl. Phys. 33: 27-32.
[5] Morgan WL, Penetrante BMA (1990) ELENDIF: A time-dependent Boltzmann solver for partially ionized plasmas. Comput. Phys. Commun. 58: 127-152.
[6] BOLSIG 2011 CPAT (2011) Available: http://www.siglo-kinema.com/bolsig.htm. Accessed 2011 Mar. 10.
[7] BOLSIG+ 2011 CPAT (2011) Available: http://www.bolsig.laplace.univ-tlse.fr/. Accessed 2011 Mar. 10.
[8] Papadakis AP, Rossides S, Metaxas AC (2011) Microplasmas: A Review. Open Applied Physics Journal 4: 45-63.
[9] Papadakis AP (2004) Modelling of Gas Discharges using Finite Elements: Incorporation of Navier-Stokes equations. Ph.D. dissertation, Department of Electrical Engineering, University of Cambridge, Trinity College.
[10] Baker WE (1973) Explosions in Air. University of Texas Press, Austin and London.
[11] Takayama K, Jiang Z (1997) Shock wave reflection over wedges: a benchmark test for cfd and experiments. Shock Waves, 7: 191-203.
[12] Jiang Z, Moosad KPB, Takayama K, Onodera O (1998) Numerical and experimental study of a micro-blast wave generated by pulsed laser beam. Shock waves 8: 337-349.
[13] Lohner R (1987) An adaptive finite element scheme for transient problems in CFD. Comput. Method. Appl. M. 61: 323-338, North Holland.

[14] Huang W, Ren Y, Russell RD (1994) Moving mesh methods based on moving mesh partial differential equations. J. Comput. Phys. 113: 279-290.
[15] Dwyer R (1989) A faster divide and conquer algorithm for constructing Delaunay triangulations. Algorithmica 2: 137-151.
[16] Papadakis AP (2012) Numerical analysis of the heating effects of an atmospheric air dielectric barrier discharge. IEEE Trans. Plasma Sci. 40: 811 - 820.
[17] Georghiou GE, Morrow R, Metaxas AC (2001) The effect of photoemission on the streamer development and propagation in short uniform gaps. J. Phys. D: Appl. Phys. 34: 200-208.

Section 2

Applications of FEA in "Structural Mechanics and Composite Materials"

Chapter 3

Finite Element Analysis of Functionally Graded Piezoelectric Spheres

A. Ghorbanpour Arani, R. Kolahchi, A. A. Mosalaei Barzoki, A. Loghman and F. Ebrahimi

Additional information is available at the end of the chapter

1. Introduction

A smart structure typically comprises of one or more active (or functional) materials. These active materials act in a unique way in which couple at least two of the following fields to provide the required functionality: mechanical, electrical, magnetic, thermal, chemical and optical. Through this coupling, these materials have the ability to change their shape, respond to external stimuli and vary their physical, geometrical and rheological properties. In modern technologies there has been an intense interest in FGPMs which are used in smart structures. It is well known that piezoelectric materials produce an electric field when deformed, and undergo deformation when subjected to an electric field. The coupling nature of piezoelectric materials has conducted wide applications in electro-mechanical and electric devices, such as electro-mechanical actuators, sensors and transducers. For example, piezoelectric actuators can be used to modify the shape of an airfoil, thereby reducing transverse vortices [1], or to maintain proper tension with overhead electrical wires on a locomotive pantograph [2].

For homogeneous piezoelectric media, problems of radially-polarized piezoelectric bodies were considered and solved analytically by Chen [3]. Sinha [4] obtained the solution of the problem of static radial deformation of a piezoelectric spherical shell and under a given voltage difference between these surfaces, coupled with a radial distribution of temperature from the inner to the outer surface. Ghorbanpour et al. [5] investigated the stress and electric potential fields in piezoelectric hollow spheres. Stress in piezoelectric hollow sphere under thermal environment was developed by Saadatfar and Rastgoo [6]. Dai and Wang [7] presented the thermo-electro-elastic transient responses in piezoelectric hollow structures. Dai and Fu [8] studied the electromagneto transient stress and perturbation of magnetic field vector in transversely isotropic piezoelectric solid spheres.

In-homogenity was considered in a number of studies. Elastic analysis of internally pressurized thick-walled spherical pressure vessels of functionally graded materials (FGMs) investigated by You et al. [9]. Analytical solution for a non-homogeneous isotropic piezoelectric hollow sphere was presented by Ding et al. [10]. Effect of material in-homogeneity on electro-thermo-mechanical behaviors of functionally graded piezoelectric rotating cylinder was considered by Ghorbanpour et al. [11]. Wang and Xu [12] studied the effect of material inhomogeneity on electromechanical behaviors of functionally graded piezoelectric spherical structures. Magnetothermoelastic problems of FGM spheres are studied by Ghorbanpour et al. [13].

Sladek et al. [14] derived Local integral equations for numerical solution of 3-D problems in linear elasticity of FGMs viewed as 2-D axisymmetric problems while the meshless local Petrov-Galerkin method was applied to transient dynamic problems in 3D axisymmetric piezoelectric solids with continuously non-homogeneous material properties subjected to mechanical and thermal loads by Sladek et al.[15]. They concluded that this method is promising for numerical analysis of multi-field problems like piezoelectric or thermoelastic problems, which cannot be solved efficiently by the conventional boundary element method.

Motivated by these ideas, new applications of piezoelectric sensors and actuators are being introduced and expanded for a number of geometric configurations. In this chapter, a hollow sphere composed of a radially polarized transversely isotropic piezoelectric material, e.g., PZT-4, which is subjected to mechanical and thermal loads, together with a potential difference induced by electrodes attached to the inner and outer surfaces of the annular sphere is considered. All mechanical, thermal and piezoelectric properties of the FGPM hollow sphere, except for the Poisson's ratio, are assumed to depend on the radius r and expressed in terms of its power functions. Hence, the equation of equilibrium in the radially polarized form is reduced to a system of second–order ordinary differential equation and is solved analytically for four different sets of boundary conditions. Finally, the thermal stresses, electric potential and displacement distributions are shown for different material in-homogeneity Also, a three-dimensional finite element analysis of asymmetric closed and open spheres with different boundary conditions subjected to an internal pressure and a uniform temperature field has also been carried out using ANSYS software.

2. Electromechanical coupling

The subsequent characterization of electromechanical coupling covers the various classes of piezoelectric materials. Details with respect to definition and determination of the constants describing these materials have been standardized by the Institute of Electrical and Electronics Engineers [16]. Stresses σ and strains ε on the mechanical side, as well as flux density D and field strength E on the electrostatic side, may be arbitrarily combined as follows [17,18]

$$\begin{Bmatrix} \sigma \\ D \end{Bmatrix} = \begin{bmatrix} C^E & -e \\ e^T & \epsilon^{\varepsilon} \end{bmatrix} \begin{Bmatrix} \varepsilon \\ E \end{Bmatrix}, \tag{1}$$

where C^E, $\in^\varepsilon$, e and e^T are the fourth-order elasticity tensor, the dielectric permittivity tensor, third order tensor of piezoelectric coefficient and transpose of it, respectively.

Assuming total strain tensor to be the sum of mechanical (M) and thermal (T) strains [19, 20]

$$\varepsilon = \varepsilon^M + \varepsilon^T, \tag{2}$$

where

$$\varepsilon = \begin{Bmatrix} \varepsilon_{rr} \\ \varepsilon_{\theta\theta} \\ \varepsilon_{\zeta\zeta} \\ \varepsilon_{r\zeta} \\ \varepsilon_{\zeta\theta} \\ \varepsilon_{r\theta} \end{Bmatrix}, \tag{3}$$

$$\varepsilon^M = \begin{pmatrix} u_{,r} \\ \dfrac{w_{,\theta}}{r\sin\zeta} + \dfrac{\cot\zeta}{r} v + \dfrac{u}{r} \\ \dfrac{v_{,\zeta}}{r} + \dfrac{u}{r} \\ \dfrac{u_{,\zeta}}{r} + v_{,r} - \dfrac{v}{r} \\ \dfrac{v_{,\theta}}{r\sin\zeta} + \dfrac{w_{,\zeta}}{r} - \dfrac{\cot\zeta}{r} w \\ \dfrac{u_{,\theta}}{r\sin\zeta} + w_{,r} - \dfrac{w}{r} \end{pmatrix}, \qquad \varepsilon^T = \begin{pmatrix} -\alpha_r T \\ -\alpha_\theta T \\ -\alpha_\zeta T \\ 0 \\ 0 \\ 0 \end{pmatrix}, \tag{4}$$

It is also noted that the electric field tensor E can be written in terms of electric potential ϕ as [21]

$$\mathrm{E} = -grad\,\phi. \tag{5}$$

3. Formulation for electrothermoelastic FGPM spheres

A hollow FGPM sphere with an inner radius r_i and outer radius r_o is considered. The sphere is subjected to an internal and external pressures P_i and P_o, an electric potential φ and a distributed temperature field $T(r)$ (Fig. 1). It is assumed that, only the radial displacement U_r is nonzero and electric potential is the functions of radial coordinate r, Thus

$$U_r = u(r),\ U_\zeta = U_\theta = 0,\ \phi = \phi(r). \tag{6}$$

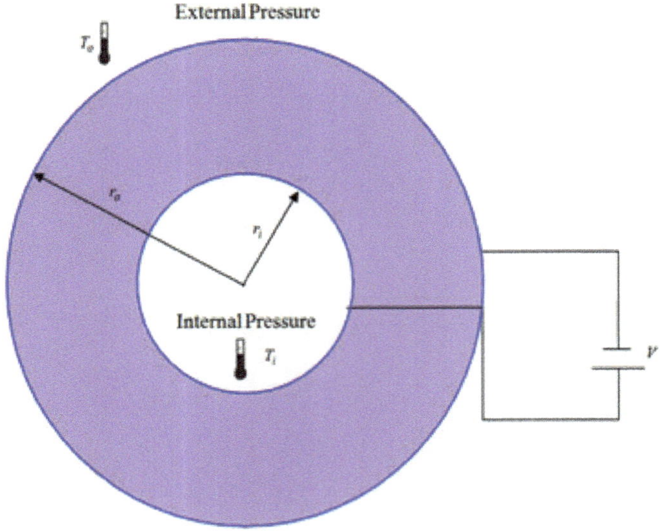

Figure 1. Hollow FGPM sphere subject to uniform temperature field, uniform internal pressure, uniform external pressure and applied voltage V.

The equilibrium equation of the FGPM sphere in the absence of body force and the Maxwell's equation for free electric charge density are [18, 22]

$$\sigma_{rr,r} + \frac{2(\sigma_{rr} - \sigma_{\theta\theta})}{r} = 0, \tag{7}$$

$$D_{rr,r} + \frac{2}{r} D_{rr} = 0, \tag{8}$$

where $\sigma_{ii}\left(i = r, \theta\right)$ is the stress tensor and D_{rr} is the radial electric displacement.

Also, the radial and circumferential strain and the relation between electric field and electric potential are reduced to

$$\varepsilon_{rr} = u_{,r}, \tag{9}$$

$$\varepsilon_{\theta\theta} = \varepsilon_{\zeta\zeta} = \frac{u}{r}, \tag{10}$$

$$E_{rr} = -\phi_{,r}. \tag{11}$$

The constitutive relations of spherically radially polarized piezoelectric media and the component of radial electric displacement vector also can be written as [23, 24]

$$\begin{Bmatrix} \sigma_{rr} \\ \sigma_{\theta\theta} \\ \sigma_{\zeta\zeta} \\ D_{rr} \end{Bmatrix} = \begin{bmatrix} C_{11} & C_{12} & C_{13} & e_{11} \\ C_{21} & C_{22} & C_{23} & e_{12} \\ C_{31} & C_{32} & C_{33} & e_{13} \\ e_{11} & e_{12} & e_{13} & -\in_{11} \end{bmatrix} \begin{Bmatrix} \varepsilon_{rr} - \alpha_r T(r) \\ \varepsilon_{\theta\theta} - \alpha_\theta T(r) \\ \varepsilon_{\zeta\zeta} - \alpha_\zeta T(r) \\ E_{rr} \end{Bmatrix} \tag{12}$$

For transversely isotropic properties, when the concerned axis of rotation is oriented in the radial direction, the elasticity and piezoelectric coefficient tensors are summarized to [25]

$$C_{12} = C_{13} = C_{21}, \quad C_{22} = C_{33}, \quad C_{32} = C_{23}, \qquad e_{12} = e_{13}. \tag{13}$$

It is appropriate to introduce the following dimensionless quantities as

$$\begin{gathered} \sigma_i = \frac{\sigma_{ii}}{C_{22}}(i = r,\theta), \; c_i = \frac{C_{1i}}{C_{22}}(i = 1,2,3), \quad E_i = \frac{e_{1i}}{E_0}(i = 1,2,3), \qquad E_0 = \sqrt{C_{22} \in_{11}}, \\ U = \frac{u_r}{r_i}, \; \xi = \frac{r}{r_i}, \quad \eta = \frac{r_o}{r_i}, \quad \beta = \frac{\beta_1}{E_0}, \; \Phi = \frac{\phi}{\phi_0}, \; \phi_0 = r_i\sqrt{\frac{C_{22}}{\in_{11}}}, \qquad D_r = \frac{D_{rr}}{E_0}. \end{gathered} \tag{14}$$

Using the above dimensionless variables, Eqs. (7) and (8) can be expressed as

$$\sigma_{r,\xi} + \frac{2(\sigma_r - \sigma_\theta)}{\xi} = 0, \tag{15}$$

$$D_{r,\xi} + \frac{2D_r}{\xi} = 0. \tag{16}$$

Before substituting the component of the electric field in Maxwell's equation, appropriate power functions for all properties are assumed as [26]

$$\Gamma_r = \Gamma_0(\xi)^\gamma, \tag{17}$$

in which Γ_r represents the general properties of the sphere such as the elastic, piezoelectric, dielectric coefficients and thermal conductivity, and Γ_0 corresponds to the value of the coefficients at the outer surface. Substituting Eqs. (14) and (17) into Eq. (12), yields

$$\begin{Bmatrix} \sigma_r \\ \sigma_\theta \\ \sigma_\zeta \\ D_r \end{Bmatrix} = \xi^\gamma \left(\begin{bmatrix} C_1 & C_2 & C_2 & E_1 \\ C_2 & 1 & C_3 & E_2 \\ C_2 & C_3 & 1 & E_2 \\ E_1 & E_2 & E_2 & -1 \end{bmatrix} \begin{Bmatrix} U_{,\xi} - \xi^\gamma \alpha_r T(\xi) \\ \frac{U}{\xi} - \xi^\gamma \alpha_\theta T(\xi) \\ \frac{U}{\xi} - \xi^\gamma \alpha_\theta T(\xi) \\ \Phi_{,\xi} \end{Bmatrix} \right) \tag{18}$$

4. Electrothermoelastic analysis of FGPM spheres

The solution of Eq. (16) is

$$D_r = \frac{A_1}{\xi^2}, \tag{19}$$

where A_1 is a constant. Substituting Eq. (19) into Eq. (18), we obtain

$$\begin{Bmatrix} \sigma_r \\ \sigma_\theta \end{Bmatrix} = \xi^\gamma \left(\left(\begin{bmatrix} C_1 & C_2 & C_2 \\ C_2 & 1 & C_3 \end{bmatrix} + \begin{bmatrix} E_1^2 & E_1E_2 & E_1E_2 \\ E_1E_2 & E_2^2 & E_2^2 \end{bmatrix} \right) \left(\begin{Bmatrix} U_{,\xi} - \xi^\gamma \alpha_r T(\xi) \\ \frac{U}{\xi} - \xi^\gamma \alpha_\theta T(\xi) \\ \frac{U}{\xi} - \xi^\gamma \alpha_\theta T(\xi) \end{Bmatrix} \right) - \begin{bmatrix} E_1 \\ E_2 \end{bmatrix} \left(A_1 \xi^{-\gamma-2} \right) \right) \tag{20}$$

In this study a distributed temperature field due to steady-state heat conduction has been considered. Using Eq. (17) for the thermal conductivity property, the heat conduction equation without any heat source is written in spherical coordinate as [22, 27]

$$\frac{1}{\xi^2}\left(K_0 \, \xi^{\gamma+2} \, T(\xi)_{,\xi} \right)_{,\xi} = 0, \tag{21}$$

$$\begin{aligned} &at\ \xi = 1 && T(\xi) = T_a, \\ &at\ \xi = \eta && T(\xi)_{,\xi} + hT(\xi) = 0, \end{aligned} \tag{22}$$

where h is the ratio of the convective heat-transfer coefficient and K_0 is the nominal heat conductivity coefficient. Integrating Eq. (21) twice yields

$$T(\xi) = -\frac{B_1}{\gamma+1}\xi^{-\gamma-1} + B_2, \tag{23}$$

Constants B_1 and B_2 are obtained using thermal boundary conditions which shown in Eq. (22).

Finally, substituting Eq. (20) and (24) into Eq. (15) yields the following non-homogeneous Cauchy differential equation

$$\xi^2 \frac{\partial^2 U}{\partial \xi^2} + D_1 \xi \frac{\partial U}{\partial \xi} + D_2 U = D_4 B_1 + D_5 B_2 \xi^{(1+\gamma)} + D_3 A_1 \xi^{-(1+\gamma)} \tag{24}$$

where $D_i \left(i = 1,...8 \right)$ are defined in Appendix A.

The exact solution for Eq. (24) is written as follows

$$u_g = \underbrace{K_1 \exp(q_1 \xi)}_{u_{g1}} + \underbrace{K_2 \exp(q_2 \xi)}_{u_{g2}}, \tag{25}$$

$$q_1, q_2 = \frac{(1 - D_1) \pm \sqrt{\left(D_1 - 1\right)^2 - 4D_2}}{2}. \tag{26}$$

The particular solution of the differential Eq. (24) may be obtained as

$$u_p = \xi^{q_1} u_1 + \xi^{q_2} u_2, \tag{27}$$

where

$$u_1 = -\int \frac{\xi^{q_2} R(\xi)}{W(q_1, q_2)}, \qquad u_2 = \int \frac{\xi^{q_1} R(\xi)}{W(q_1, q_2)}, \tag{28}$$

in which $R(\xi)$ is the expression on the right hand side of Eq. (24) and $W(\xi)$ is defined as

$$W(q_1, q_2) = \begin{vmatrix} u_{g1} & u_{g2} \\ (u_{g1})' & (u_{g2})' \end{vmatrix}. \tag{29}$$

Combining Eqs. (25)-(29) one can obtain the particular solution as

$$u_p = \frac{D_3 \xi^{1-\gamma}}{(q_2+\gamma-1)(q_1+\gamma-1)} A_1 + \frac{D_5 \xi^{3+\gamma}}{(3-q_2+\gamma)(3-q_1+\gamma)} B_2 + \frac{D_4 \xi^3}{(q_2-3)(q_1-3)} B_1. \tag{30}$$

The complete solution for U^m in terms of the non-dimensional radial coordinate is written as

$$U = u_g + u_p, \tag{31}$$

where K_1, K_2 and A_1 are unknown constants. Substituting the displacement from Eq. (31) into Eq. (20) the radial and circumferential stresses are obtained as

$$\begin{aligned}
\sigma_r(\xi) = {} & \left(\left(2C_2 + C_1 q_1 + 2E_1 E_2 + E_1^2 q_1\right)\xi^{q_1 - 1 + \gamma}\right) K_1 + \left(\left(2C_2 + C_1 q_2 + 2E_1 E_2 + E_1^2 q_2\right)\xi^{q_2 - 1 + \gamma}\right) K_2 \\
& + \left(\frac{\left(2C_2 + 2E_1 E_2 + (3+\gamma)(E_1^2 + C_1)\right) D_4 \xi^2}{(q_2-3)(q_1-3)} + \frac{\left(2(C_2 + E_1 E_2)\alpha_\theta + (E_1^2 + C_1)\alpha_r\right)\xi^{\gamma-1}}{\gamma+1}\right) B_1 \\
& + \left(\frac{\left(2C_2 + 2E_1 E_2 + (3+\gamma)(E_1^2 + C_1)\right) D_5 \xi^{2+2\gamma}}{(3-q_2+\gamma)(3-q_1+\gamma)} - \left(2(C_2 + E_1 E_2)\alpha_\theta + (E_1^2 + C_1)\alpha_r\right)\xi^{2\gamma}\right) B_2 \\
& + \left(\frac{\left(2C_2 + 2E_1 E_2 + (1-\gamma)(E_1^2 + C_1)\right) D_3}{(q_2+\gamma-1)(q_1+\gamma-1)} - 2\left(E_1\right)\xi^{-2}\right) A_1
\end{aligned} \tag{32}$$

$$\sigma_\theta(\xi)=\Big(\big(C_3+C_2q_1+E_1E_2q_1+2E_2^2+1\big)\xi^{q_1-1+\gamma}\Big)K_1+\Big(\big(C_3+2E_2^2+E_1E_2q_2+C_2q_2+1\big)\xi^{q_2-1+\gamma}\Big)K_2$$
$$+\left(\frac{\big(C_3+3C_2+2E_2^2+3E_1E_2+1\big)D_4\xi^{2+\gamma}}{(q_2-3)(q_1-3)}+\frac{\big((C_3+2E_2^2+1)\alpha_\theta+(C_2+E_1E_2)\alpha_r\big)\xi^{\gamma-1}}{\gamma+1}\right)B_1$$
$$+\left(\frac{\big(C_3+2E_2^2+1+(3+\gamma)(E_1E_2+C_2)\big)D_5\xi^{2+2\gamma}}{(3-q_2^m+\gamma)(3-q_1^m+\gamma)}-\big((1+C_3+2E_2^2)\alpha_\theta+(C_2+E_1E_2)\alpha_r\big)\xi^{2\gamma}\right)B_2 \quad (33)$$
$$+\left(\frac{\big(C_3+2E_1E_2+2E_2^2+1+(1-\gamma)(E_1E_2+C_2)\big)D_3}{(q_2+\gamma-1)(q_1+\gamma-1)}-E_2\xi^{-2}\right)A_1$$

Substituting U from above into the last term of Eq. (18), ($\Phi_{,\xi}$) and combining with Eq. (19) and performing the integrating, electric potential is obtained as

$$\Phi(\xi)=\left(\left(\frac{2E_2}{q_1}+E_1\right)\xi^{q_1}\right)K_1+\left(\left(\frac{2E_2}{q_2}+E_1\right)\xi^{q_2}\right)K_2$$
$$+\left(\frac{\big(\gamma E_1-E_1-2E_2\big)D_3\xi^{1-\gamma}}{(q_2+\gamma-1)(q_1+\gamma-1)(1-\gamma)}+\frac{\xi^{-\gamma-1}}{(1+\gamma)}\right)A_1$$
$$+\left(\frac{(E_1+2/3\,E_2)D_4\xi^3}{(q_2-3)(q_1-3)}+\frac{(2E_2\alpha_\theta+E_1\alpha_r)\ln(\xi)}{\gamma+1}\right)B_1 \quad (34)$$
$$+\left(\frac{\big(3E_1+\gamma E_1+2E_2\big)D_5\xi^{\gamma+3}}{(q_2+\gamma-1)(q_1+\gamma-1)(3+\gamma)}-\frac{(2E_2\alpha_\theta+E_1\alpha_r)\xi^{\gamma+1}}{\gamma+1}\right)B_2+A_2,$$

where σ_r, σ_θ and Φ are radial stress, hoop stress and electric potential, respectively. Two sets of mechanical and electrical loading boundary conditions are considered in this investigation which in normalized form are written as

$$\text{case 1}: \sigma_r(1)=-1,\ \ \sigma_r(\eta)=0,\ \ \Phi(1)=0,\ \ \Phi(\eta)=0 \quad (35)$$

$$\text{case 2}: \sigma_r(1)=0,\ \ \sigma_r(\eta)=0,\ \ \Phi(1)=1,\ \ \Phi(\eta)=0 \quad (36)$$

In case 1, the FGPM hollow sphere is subjected to an internal uniform pressure without any imposed electric potential and external pressure. However in this case the induced electric potential existed across the thickness. In this case, the sphere acts as an sensor. In the second case, an electrical potential difference is applied between the inner and outer surfaces of the sphere without any internal and external pressures. In this case, the sphere acts as an actuator.

For the above mentioned cases 1, and 2 the system of linear algebraic equations for the constants K_1, K_2, A_1 and A_2 of the Eqs. (32), (33) and (34) can be written in the following from

$$\begin{bmatrix} K_1 \\ K_2 \\ A_1 \\ A_2 \end{bmatrix} = inv\left(\begin{bmatrix} m_{11} & m_{12} & m_{13} & m_{14} \\ m_{21} & m_{22} & m_{23} & m_{24} \\ m_{31} & m_{32} & m_{33} & m_{34} \\ m_{41} & m_{42} & m_{43} & m_{44} \end{bmatrix}\right) \times \begin{bmatrix} b_1 \\ b_2 \\ b_3 \\ b_4 \end{bmatrix}, \tag{37}$$

where the m_{ij} and b_i $(i,j=1,...4)$ are defined in Appendix B

5. Numerical results and discussion

5.1. Analytical solution

The numerical results are showing the variation of stresses, electric potential and displacement across the thickness of the FGPM sphere for different material inhomogenity parameter γ. Presented results are for the two cases of different boundary conditions with aspect ratio $\eta = 1.3$. The plots in these figures correspond to $T_i = 323K$ and $T_o = 298K$. The piezoelectric material PZT-4 has been selected because of its technical applications. Mechanical and electrical properties of piezoelectric material, PZT-4 are tabulated in Table 1[28].

Property	C_{11}	C_{12}	C_{22}	C_{23}	e_{11}	e_{12}	e_{11}		α_{r0}	ρ
PZT-4	115 *Gpa*	74.3 *Gpa*	139 *Gpa*	77.8 *Gpa*	15.1 C/m^2	-5.2 C/m^2	3.87e-9 F/m	2 e - 5 $1/K$	2 e -6 $1/K$	*7500* kg/m^3

Table 1. Mechanical, electrical and thermal properties for PZT-4

5.1.1. Case 1

Results of the first case are illustrated in Figs. 2 to 5. Radial stresses for different material in-homogeneity parameters γ are shown in Fig. 2. Radial stresses satisfy the mechanical boundary conditions at the inner and outer surfaces of the FGPM sphere. The maximum absolute values of radial stresses belong to a material identified by in-homogeneity parameter $\gamma = 1.5$ the minimum absolute values of which belong to $\gamma = -1.5$. In this case there is no imposed electric potential however, the induced electric potentials for different material in-homogeneity parameters γ are shown in Fig. 3. Electric potentials satisfy the electrical boundary conditions at the inner and outer surfaces of the FGPM sphere. It is also obvious that higher induced electric potentials belong to higher absolute values of compressive radial stresses. In this case circumferential stresses shown in Fig. 4 are highly tensile throughout thickness and except for the material $\gamma = 1.5$ their maximum values located at the inner and their minimum values located at the outer surfaces of the sphere. Displacements are illustrated in Fig. 5 for all material properties. Displacements are positive throughout the thickness and they smoothly decrease from their maximum value at the inner surface to their minimum value at the outer surface of the FGPM sphere. Maximum values of displacements belong to $\gamma = -1.5$ and minimum values belong to $\gamma = 1.5$

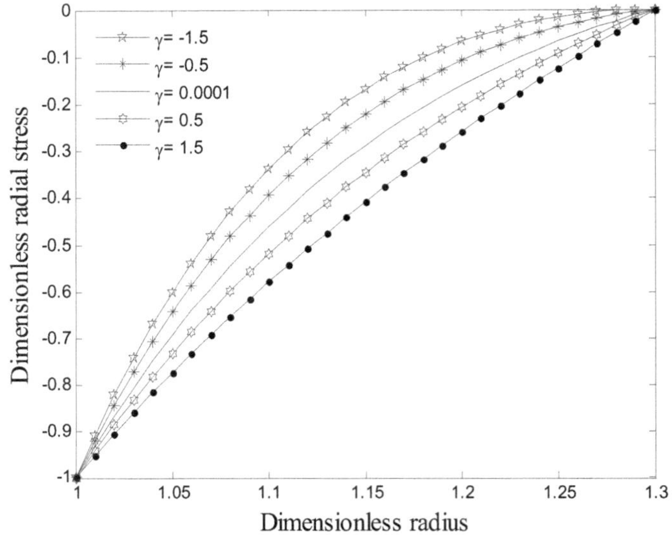

Figure 2. Case 1: Distributions of the radial stress for different values of γ.

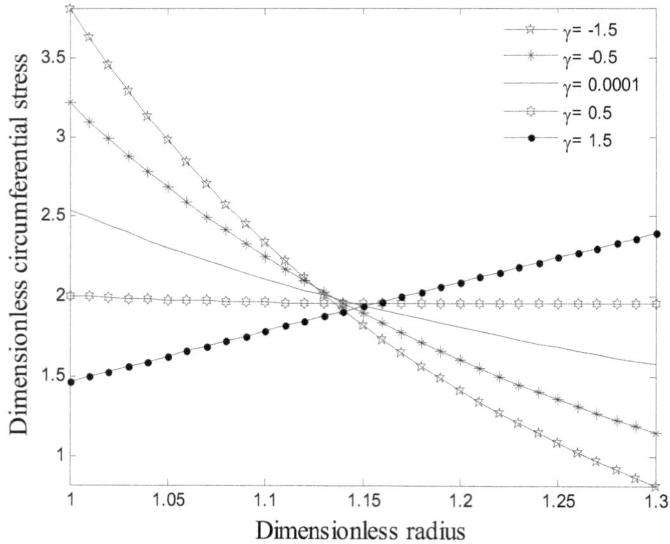

Figure 3. Case 1: Distributions of the circumferential stress for different values of γ.

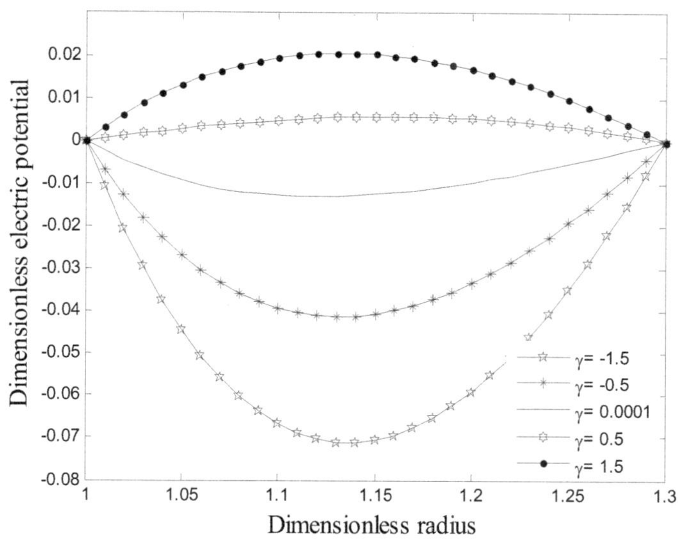

Figure 4. Case 1: Distributions of the electric potential for different values of γ.

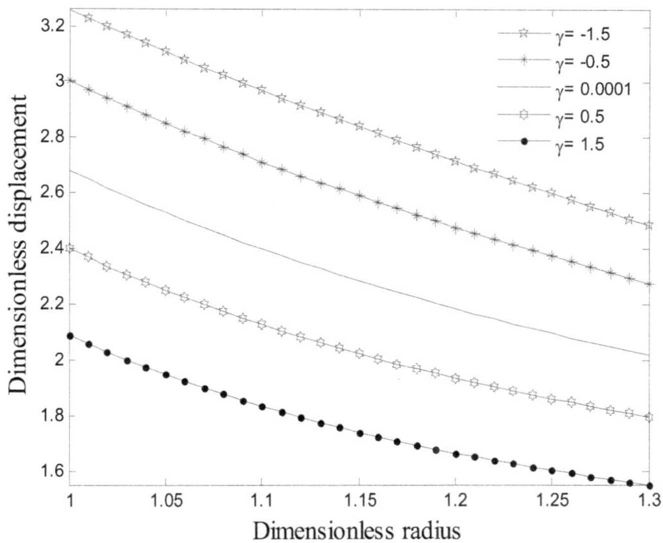

Figure 5. Case 1: Distributions of the radial displacement for different values of γ.

5.1.2. *Case 2*

Results of this case are illustrated in Figs. 6 to 9. In this case the imposed electric potential satisfies the electrical boundary conditions at the inner and outer surfaces of the sphere. The maximum electric potentials belong to $\gamma = -1.5$ the minimum values of which belong to $\gamma = 1.5$. In this case there is no applied pressure at the inner and outer surfaces of the sphere however the induced compressive radial stresses satisfy the free mechanical boundary conditions. The maximum absolute values of the induced compressive radial stresses belong to the same maximum value of electric potential. Circumferential induced stresses are compressive throughout thickness for different material in-homogeneity parameters γ . However, for negative parameters γ the minimum values of circumferential stresses located at the inner surface while for positive parameters γ their minimum values located at the outer surface of the FGPM sphere. The induced displacement is negative across the thickness for all material parameters. Their minimum values located at the inner and their maximum values located at the outer surfaces of the FGPM sphere. It is interesting to compare the induced radial and circumferential stresses in this case with the residual stresses locked in the sphere during the autofrettage process of spheres made of uniform material. One might conclude that by easily imposing an electric potential there is no need to autofrettage these vessels.

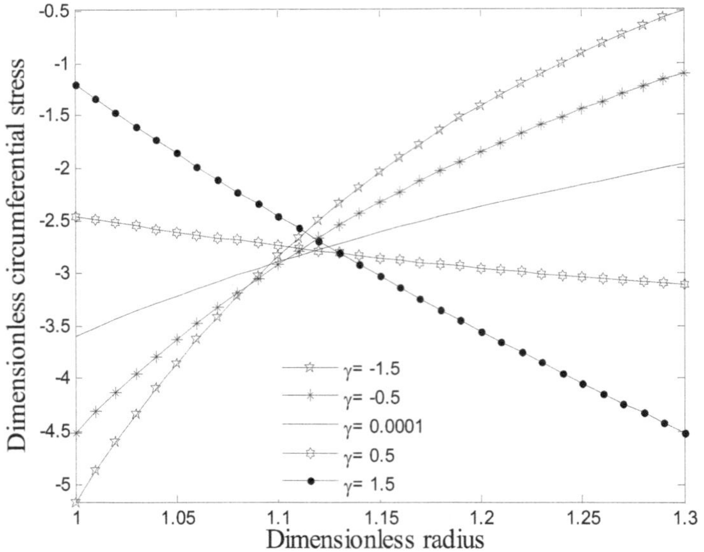

Figure 6. Case 2: Distributions of the radial stress for different values of γ.

5.2. Validation

The results of this investigation are validated with the recently published paper by Wang and Xu [12] which is shown in Figs. 10 and 11. There are a very good agreement among the results and the only small differences are due to thermal stresses which are not considered by Wang and Xu .

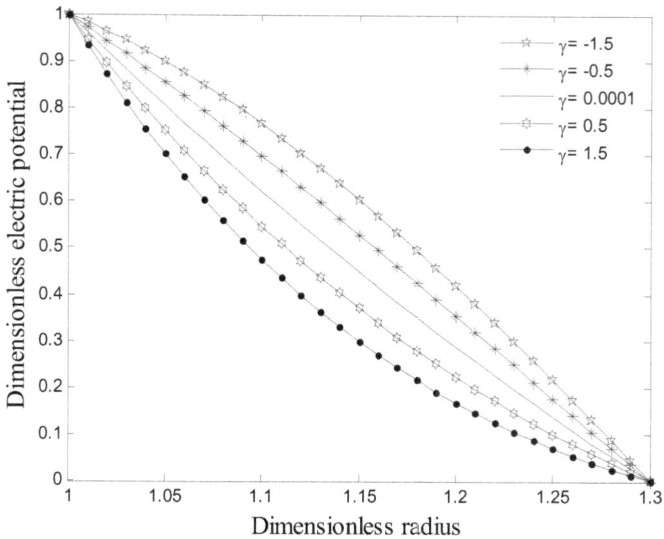

Figure 7. Case 2: Distributions of the circumferential stress for different values of γ.

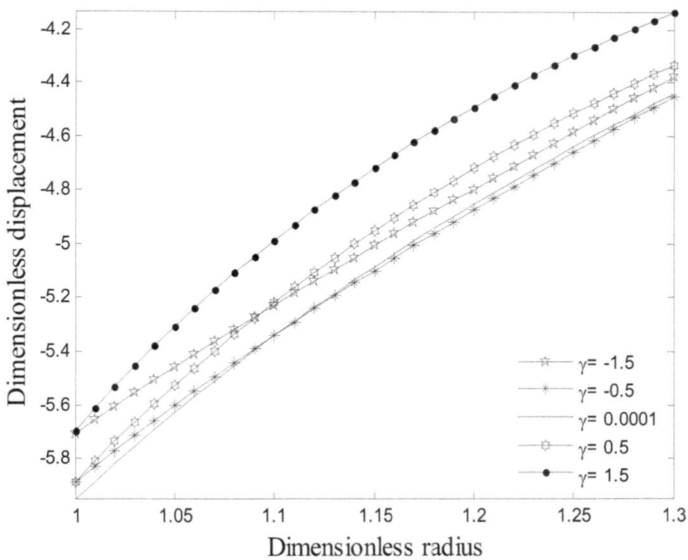

Figure 8. Case 2: Distributions of the electric potential for different values of γ.

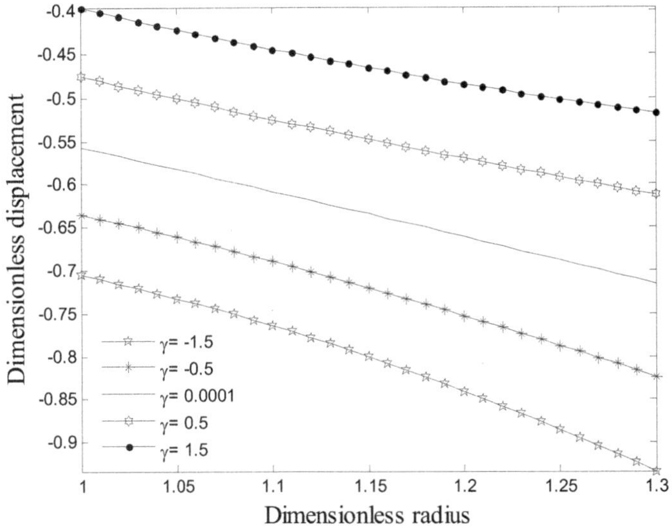

Figure 9. Case 2: Distributions of the radial displacement for different values of γ.

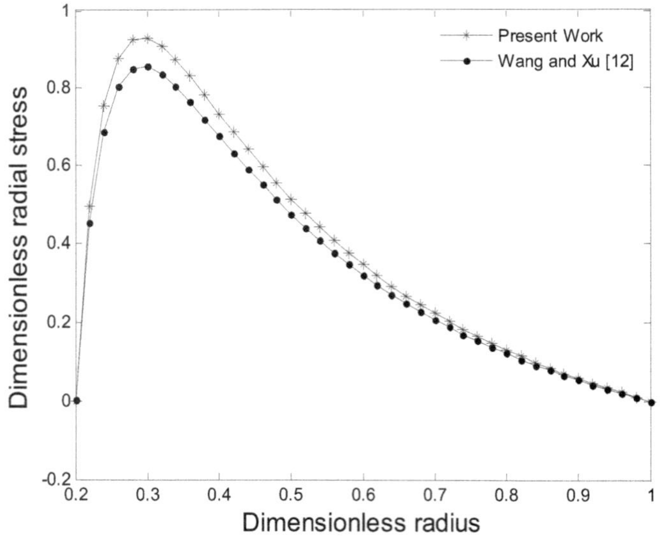

Figure 10. Case 2: Comparison of the radial stress distributions with Ref. [12] for homogeneous piezoelectric hollow sphere.

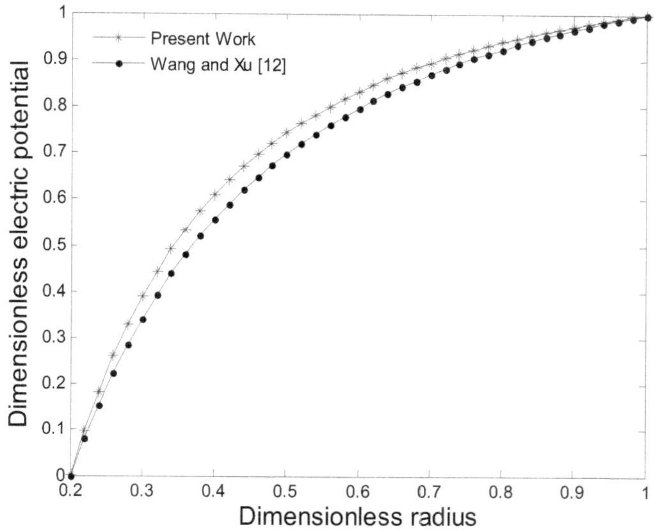

Figure 11. Case 2: Comparison of the electric potential distributions with Ref. [12] for homogeneous piezoelectric hollow sphere.

5.3. Finite element solution

In order to develop the one-dimensional solution to a three-dimensional approach finite element analysis of a sphere subjected to an internal pressure and a uniform temperature field has been carried out using ANSYS finite element software. A three-dimensional element identified by solid 191 is selected because it is an appropriate element for the FGPM structures. Sphere has been divide into eight layers by a controlled mesh system along radius and the mechanical, electrical and thermal properties are functionally defined according to power law Eq. (17) for $\gamma = 1.5$. A controlled mesh in which very fine elements are located at the supports where stress concentration existed is employed in this method. However farther from the supports a coarser mesh is dominated. In this work, two cases for sphere are considered as follows:

5.3.1. Three- dimensional sphere

In this case, consider a sphere with two asymmetric simply supported boundary conditions on the outer surface of the sphere as shown in Fig. 12. For this boundary condition dimensionless effective stresses versus normalized radius at two cross sections (i.e. A-A and B-B) are depicted in Fig. 13. Section A-A is selected to pass through supported point on the outer surface of the sphere and section B-B is an arbitrary section as shown in Fig. 12. It can be seen from this figure that the maximum effective stress for the above mentioned sections occur at the inner surface of the sphere and the effective stresses are decreasing with increasing radius for $\eta = 1.3$. Total dimensionless displacement versus dimensionless radius for $\eta = 1.3$ at two cross sections of A-A and B-B are demonstrated in Fig. 14. As can be seen from this figure the maximum displacement occur at the inner surface of the sphere and displacement value is decreasing with increasing dimensionless radius so that for section A-A, the zero value of displacement at the outer surface satisfies the boundary condition at this point.

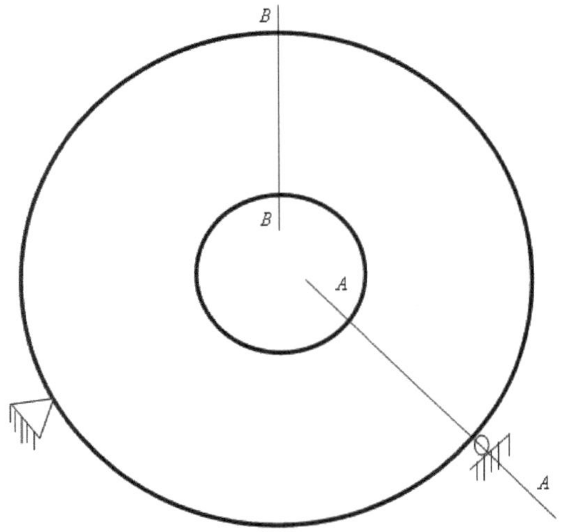

Figure 12. A schematic of asymmetric thick-walled sphere with simply-simply supported.

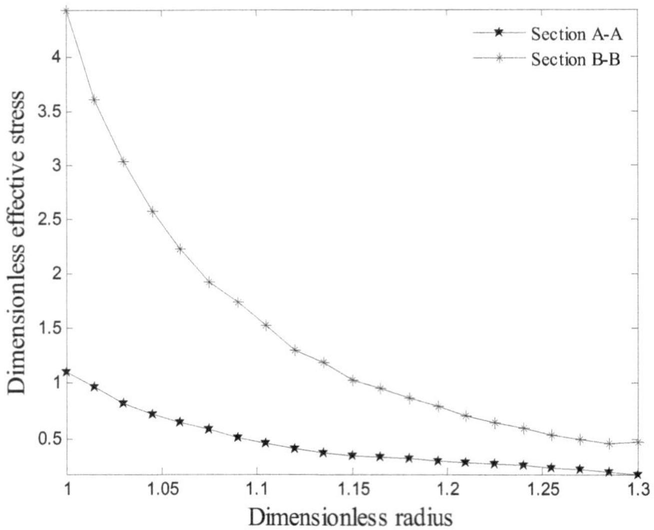

Figure 13. Effective stress distribution along the radius of asymmetric closed sphere.

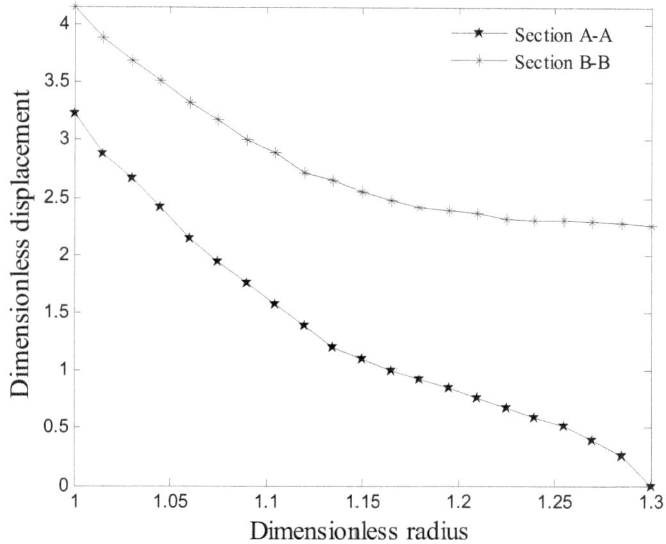

Figure 14. Total displacement distribution along the radius of asymmetric closed sphere.

5.3.2. Three- dimensional open sphere

The geometry and loading condition as well as its boundary conditions are shown in Fig. 15. Three different boundary conditions are considered in this case. These boundary conditions are clamped-clamped, clamped-simply and simply-simply supported respectively.

The solution obtained by the software clearly indicates the most critical region of the sphere. In the most critical region normalized effective stress distribution and the total displacements are plotted in Figs. 16 and 17 along normalized radius at all node points for the above mentioned three boundary conditions. Fig 16 shows that in general the effective stresses are decreasing along radius to an absolute minimum and then increasing to their maximum values located at the outer surface of the vessel. For simply-simply supported boundary condition this absolute minimum is located near the outer surface of the vessel, however for the clamped-clamped condition it is nearly at the middle surface of the vessel. For the clamped-simply supported condition this minimum is somewhere between the previous two cases.

It has been found that the magnitude of effective stresses at all node points are higher for the clamped-clamped condition and are lower for the simply-simply supported condition. It can be observed from Fig. 17 that the maximum displacements for the three boundary conditions are located at the inner surface and they are decreasing to zero value at the outer surface of the sphere. It is also found that the displacement curve for simply-simply support condition is higher than other boundary conditions.

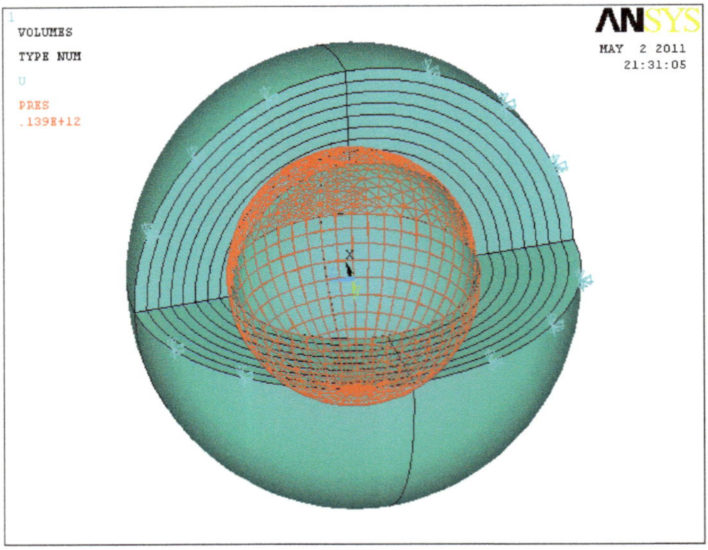

Figure 15. The three-dimensional finite element model for open sphere subjected to internal pressure with clamped-clamped boundary conditions.

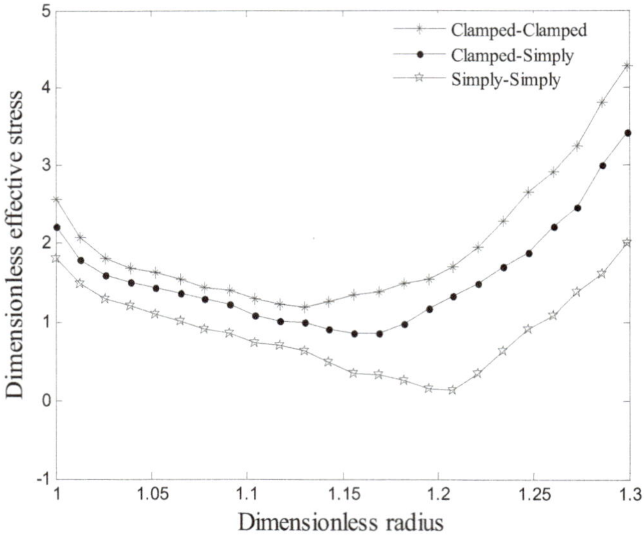

Figure 16. Effective stresses distribution along the radius of the open sphere with different boundary conditions.

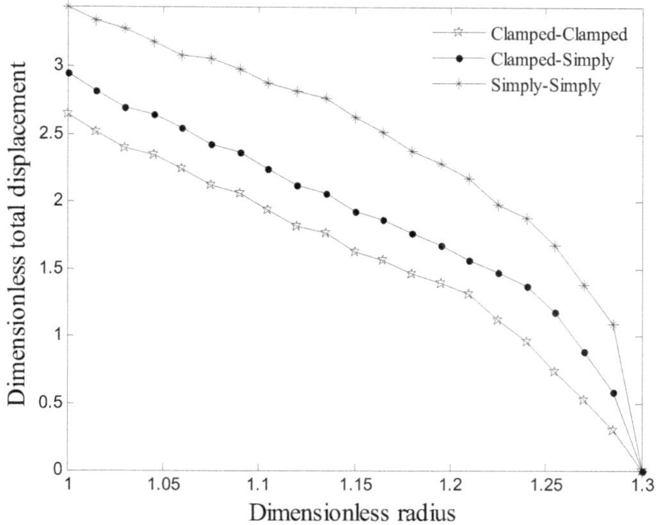

Figure 17. Total displacement distribution along the radius of the open sphere with different boundary conditions.

6. Conclusions

In this research, the electro-thermo-mechanical behavior of radially polarized FGPM hollow sphere was investigated. An analytic solution technique was developed for the electro-thermo-mechanical problem, where stresses were produced under combined thermomechanical and electrical loading conditions. Variation of normalized stresses, electric potential and displacement of four sets of boundary conditions for different material in-homogeneity parameters γ were plotted against dimensionless radius. In general, radial stresses and electric potentials satisfy the mechanical and electrical boundary conditions at the inner and outer surfaces of the FGPM sphere. It was concluded that higher absolute values of compressive radial stresses are associated with the higher induced electric potentials throughout the thickness in all cases. It was found that the induced radial and circumferential stresses of an imposed electric potential is similar to the residual stresses locked in the sphere during the autofrettage process of these vessels. Therefore, one might concluded that by easily imposing an electric potential there is no need to autofrettage these vessels. It was interesting to see that the compressive circumferential stresses due to an external pressure were very similar to the induced circumferential stresses resulted from imposing an electric potential. Moreover a three-dimensional finite element analysis of an asymmetric sphere subjected to an internal pressure and a uniform temperature field has been carried out using ANSYS software. In this study closed and opened spheres with different boundary conditions were considered. The finite element analysis indicated that the values of effective stress and total displacement at all node points along the thickness of the open sphere were the highest and lowest for the clamped-clamped condition, respectively.

Appendix A

$$D_1 = \frac{\gamma\left(C_1 + E_1^2\right) + 2C_1 + 2E_1^2}{C_1 + E_1^2}, D_2 = \frac{\gamma\left(2C_2 + 2E_1E_2\right) + 2C_2 + 2E_1E_2 - 2 - 2C_3 - 4E_2^2}{C_1 + E_1^2}, D_3 = \frac{2E_2}{C_1 + E_1^2}$$

$$D_4 = \frac{(\gamma+1)\left(2C_2\alpha_\theta + C_1\alpha_r + E_1\left(2E_2\alpha_\theta + E_1\alpha_r\right)\right) + 2\left(2C_2\alpha_\theta + C_1\alpha_r\right) + 2\left(E_1 - E_2\right)\left(2E_2\alpha_\theta + E_1\alpha_r\right)\left(\left(1 + C_3\right)\alpha_\theta + C_2\alpha_r\right)}{\left(C_1 + E_1^2\right)(\gamma+1)}$$

$$D_5 = \frac{\gamma\left(4C_2\alpha_\theta - 2C_1\alpha_r - 2E_1\left(2E_2\alpha_\theta + E_1\alpha_r\right)\right) - 2\left(2C_2\alpha_\theta + C_1\alpha_r\right) - \left(2E_1 - E_2\right)\left(2E_2\alpha_\theta + E_1\alpha_r\right) + \left(2\left(1 + C_3\right)\alpha_\theta + 2C_2\alpha_r\right)}{\left(C_1 + E_1^2\right)}$$

Appendix B

m_{ij} $(i, j = 1,\ldots 4)$ are the coefficient of K_1, K_2, A_1 and A_2 which defined as

$m_{11} = 2C_2 + C_1q_1 + 2E_1E_2 + E_1^2 q$, $m_{12} = 2C_2 + C_1q_2 + 2E_1E_2 + E_1^2 q_2$,

$$m_{13} = \frac{\left(2C_2 + 2E_1E_2 + (1-\gamma)(E_1^2 + C_1)\right)D_3}{(q_2 + \gamma - 1)(q_1 + \gamma - 1)} - 2(E_1)$$

$m_{21} = \left(2C_2 + C_1q_1 + 2E_1E_2 + E_1^2 q_1\right)\eta^{q_1 - 1 + \gamma}$, $m_{22} = \left(2C_2 + C_1q_2 + 2E_1E_2 + E_1^2 q_2\right)\eta^{q_2 - 1 + \gamma}$,

$$m_{23} = \frac{\left(2C_2 + 2E_1E_2 + (1-\gamma)(E_1^2 + C_1)\right)D_3}{(q_2 + \gamma - 1)(q_1 + \gamma - 1)} - 2(E_1)\eta^{-2}, \quad m_{31} = \left(\frac{2E_2}{q_1} + E_1\right),$$

$$m_{33} = \frac{(\gamma E_1 - E_1 - 2E_2)D_3}{(q_2 + \gamma - 1)(q_1 + \gamma - 1)(1-\gamma)} + \frac{1}{(1+\gamma)}, \quad m_{41} = \left(\frac{2E_2}{q_1} + E_1\right)\eta^{q_1}, \quad m_{42} = \left(\frac{2E_2}{q_2} + E_1\right)\eta^{q_2}$$

$$m_{43} = \frac{(\gamma E_1 - E_1 - 2E_2)D_3\,\eta^{1-\gamma}}{(q_2 + \gamma - 1)(q_1 + \gamma - 1)(1-\gamma)} + \frac{\eta^{-\gamma-1}}{(1+\gamma)}, \quad m_{41} = 0, \quad m_{24} = 0, \quad m_{34} = 1, \quad m_{44} = 1$$

b_i $(i = 1,2)$ are correspond to cases 1 and 24 which denoted as follows: $b_1 = \sigma_r(1)$, $b_1 = \sigma_r(\xi)$

Author details

A. Ghorbanpour Arani, R. Kolahchi, A. A. Mosalaei Barzoki and A. Loghman
Department of Mechanical Engineering, Faculty of Engineering, University of Kashan, Kashan, Iran

F. Ebrahimi
Department of Mechanical Engineering, Faculty of Engineering, International University of Imam Khomeini, Qazvin, Iran

7. References

[1] P. Destuynder, A few remarks on the controllability of an aeroacoustic model using piezo-devices, Int. J. Holnicki-Szulc. (1999) 53–62.

[2] H.W. Jiang, F. Schmid, W. Brand, G.R. Tomlinson, Controlling pantograph dynamics using smart technology, Int. J. Holnicki-Szulc. (1999) 125–132.

[3] W.Q. Chen, Problems of radially polarized piezoelastic bodies, Int. J. solids struct. 36 (1998) 4317-4332.

[4] D.K. Sinha, Note on the radial deformation of a piezoelectric, polarized spherical shell with a symmetrical distribution, J. Acoust. Soc. 34 (1962) 1073–1075.

[5] A. Ghorbanpour, S. Golabi, M. Saadatfar, Stress and electric potential fields in piezoelectric smart spheres, J. Mech. Sci. Tech. 20 (2006) 1920–1933.

[6] M. Saadatfar, A. Rastgoo, Stress in piezoelectric hollow sphere under thermal environment, J. Mech. Sci. Tech, 22 (2008) 1460–1467.

[7] H.L. Dai, X. Wang, Thermo-electro-elastic transient response, Int. J. solids struct. 42 (2005) 1151–1171.

[8] H.L. Dai, Y.M. Fu, Electromagnetotransient stress and perturbation of magnetic field vector in transversely isotropic piezoelectric solid sphere, Mater. Sci. Eng. B 129 (2006) 86–92.

[9] L.H. You, J.J. Zhang, X.Y. You, Elastic analysis of internally pressurized thick-walled spherical pressure vessels of functionally graded materials, Int. J. Press. Vess. Pip. 82 (2005) 347–354.

[10] H.J. Ding, H.M. Wang, W.Q. Chen, Analytical solution for a non-homogeneous isotropic piezoelectric hollow sphere, Arch. Appl. Mech. 73 (2003) 49–62.

[11] A. Ghorbanpour Arani, R. Kolahchi, A.A. Mosallaie Barzoki, Effect of material in-homogeneity on electro-thermo-mechanical behaviors of functionally graded piezoelectric rotating cylinder, J. Appl. Math. Model. 35 (2011) 2771–2789.

[12] H.M. Wang, Z.X. Xu, Effect of material inhomogeneity on electromechanical behaviors of functionally graded piezoelectric spherical structures, Comput. Mater. Sci. 48 (2010) 440–445.

[13] A. Ghorbanpour, M. Salari, H. Khademizadeh, A. Arefmanesh, Magnetothermoelastic problems of FGM spheres, Arch. Appl. Mech. (2010) 189-200.

[14] V. Sladek, J. Sladek, Ch. Zhang, Transient heat conduction analysis in functionally graded materials by the meshless local boundary integral equation method, Comput. Mater. Sci. 28 (2003) 494-504.

[15] J. Sladek, V. Sladek, P. Solek, A. Saez, Dynamic 3D axisymmetric problems in continuously non-homogeneous piezoelectric solids, Int. J. solids struct. 45 (2008) 4523–4542.

[16] Institute of Electrical and Electronics Engineers. Standard on Piezoelectricity, Std (1978) 176-1978 IEEE, New York.

[17] Y.C. Fungn, Foundations of Solid Mechanics, Prentice-Hall, New York, 1965.

[18] H.F. Tiersten, Linear Piezoelectric Plate Vibrations, Plenum Press, New York, 1969

[19] A. Manonukul, F.P.E. Dunne, D. Knowles, S. Williams, Multiaxial creep and cyclic plasticity in nickel-base superalloy C263, Int. J. Plasticity, 21 (2005) 1–20.
[20] H. Martin, ELASTICITY Theory, Applications and Numerics, Elsevier Inc, London, 2005.
[21] H.J. Ding, W.Q. Chen, Three Dimensional Problems of Piezoelasticity, Nova Science, New York, 2001.
[22] M. Sadeghian, H. Ekhteraei Toussi, Axisymmetric yielding of functionally graded spherical vessel under thermo-mechanical loading, Comput. Mater. Sci. 50 (2011) 975–981.
[23] A. Salehi-Khojin, N. Jalili, A comprehensive model for load transfer in nanotube reinforced piezoelectric polymeric composites subjected to electro-thermo-mechanical loadings. Compos: Part B. 39 (2008) 986–998.
[24] Zh. Li, Ch. Wang, Ch. Chen, Effective electromechanical properties of transversely isotropic piezoelectric ceramics with microvoids, Comput. Mater. Sci. 27 (2003) 381–392.
[25] N. Jalili, Piezoelectric-Based Vibration Control from Macro to Micro/Nano Scale Systems, Boston, 2010.
[26] M.H. Babaei, Z.T. Chen, Analytical solution for the electromechanical behavior of a rotating functionally graded piezoelectric hollow shaft, Arch. Appl. Mech. 78 (2008) 489–500.
[27] H.L., Dai, L. Hong, Y. M., Fu, X. Xiao, Analytical solution for electro-magneto-thermo-elastic behaviors of a functionally graded piezoelectric hollow cylinder, J. Appl. Math. Model. 34 (2010) 343–357.
[28] M. Saadatfar, A.S. Razavi, Piezoelectric hollow cylinder with thermal gradient, J. Mech. Sci. Tech. 23 (2009) 45–53.

Chapter 4

Finite Element Analysis of Bias Extension Test of Dry Woven

Samia Dridi

Additional information is available at the end of the chapter

1. Introduction

In the composite industry, the shearing behaviour of dry woven plays a crucial role in fabric formability when doubly curved surfaces must be covered [1-9]. The ability of fabric to shear within a plain enables it to fit three-dimensional surfaces without folds [10-12].

It has been proved that shear rigidity can be calculated from the tensile properties along a 45° bias direction. Bias Extension tests are simple to perform and provide reasonably repeatable results [13-14]. Extensive investigations have been carried out on the textile fabric in Bias Extension test [15]

The tests were conducted simply using two pairs of plates, clamping a rectangular piece of woven material such that the two groups of yarns are orientated ±45° to the direction of external tensile force. The ratio between the initial length and width of the specimen is defined as aspect ratio:

$$\lambda = l_0/w_0 \text{ (see Figure 1a).}$$

In the case of λ =2, the deformed configuration of the material can be represented by Figure1b, which includes seven regions. Triangular regions C adjacent to the fixture remain undeformed, while the central square region A and other four triangular regions B undergo shear deformation [16-17].

The present chapter focuses on numerical analysis of Bias Extension test using an orthotropic hyperelastic continuum model of woven fabric.

In the first, analytical responses of the Bias Extension test and the traction test on 45° are developed using the proposed model. Strain and stress states in specimen during these tests are detailed.

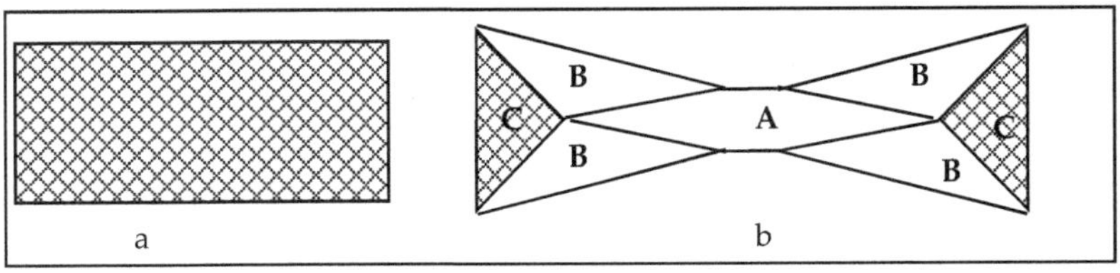

Figure 1. Kinematic of Bias Extension test, a: Initial state, b: Deformed state

In the second, the proposed model is implanted into Abaqus/Explicit to simulate the Bias Extension test of three aspect ratios.

Exploiting numerical results, we studied the effect of the ratio between shearing and traction rigidities on homogeneities of stress and strain in the central zone of three Finite Element Models (FEM).

2. The proposed hyperelastic model

One of significant characteristics of the woven structure is the existence of two privileged material directions: warp and weft. We considered that the fabric is a continuous structure having two privileged material directions defined by the two unit tensors $\boldsymbol{M_1}$ and $\boldsymbol{M_2}$ as follows:

$$\boldsymbol{M_1} = \vec{M}_1 \otimes \vec{M}_1 \ ; \ \boldsymbol{M_2} = \vec{M}_2 \otimes \vec{M}_2 \tag{1}$$

Where $\vec{M}_1$ and $\vec{M}_2$ are two unit vectors carried by two yarns directions. The sign ⊗ indicate the tensor product. In the reference configuration, these privileged material directions are supposed to be orthogonal and they are defined by g_1 and g_2 presented by Equation 2.

$$g_1 = \vec{g}_1 \otimes \vec{g}_1 \ , \ g_2 = \vec{g}_2 \otimes \vec{g}_2 \tag{2}$$

In Lagrangian formulation, the hyperelastic behavior is defined by the strain energy function $W(\boldsymbol{E})$ depending of Green-Lagrange tensor components [18-21].

The second Piola Kirchhoff stress tensor $\boldsymbol{S}$ derives is presented in Equation 3:

$$S = \frac{\partial W}{\partial E} \tag{3}$$

The physical behaviour is completely defined by the choice of $W(\boldsymbol{E})$. The woven structures is very thin, we are interested more particularly in plane solicitations (plane stress or strain) in the plan $(\vec{g}_1, \vec{g}_2)$. We supposed that $W(\boldsymbol{E})$ is an isotropic function of variables $(\boldsymbol{E}, \mathbf{g_1}, \mathbf{g_2})$. Using the representation theorems of isotropic functions, strain energy function $W(\boldsymbol{E})$ depends of invariants:

$$g_i{:}E \ , \ g_i{:}E^2 \ , \ tr(E^3) \ (i = 1..2) \tag{4}$$

We choose following invariants to present the strain energy function:

$$W(\boldsymbol{E}) = W(I_1, I_2, I_{12}) \tag{5}$$

Where

$$I_i = \boldsymbol{g_i}:\boldsymbol{E} \ \left(i = 1..2\right) \ ; \ I_{12} = \frac{1}{2}(\boldsymbol{g_1}\boldsymbol{E}:\boldsymbol{E}\boldsymbol{g_2})^{1/2} \tag{6}$$

I_i measured elongations along directions $\vec{g}_i$. I_{12} measured the sliding in the plane ($\vec{g}_1$, $\vec{g}_2$) witch is the angle variation between warp and weft direction. Components $E_{g_{ij}}$ of **E** in the reference system ($\vec{g}_1$, $\vec{g}_2$), are defined as follows:

$$I_i = E_{gij} = \frac{1}{2}(\delta_1^2 - 1) \ ,i = 1..2, I_{12} = \left|E_{g12}\right| = \frac{1}{2}\delta_1\delta_2\left|\cos(\theta)\right| \tag{7}$$

δ_1 and δ_2 are yarns extensions (ratio between deformed and initial lengths) along directions of $\vec{g}_1$ and $\vec{g}_2$. θ is the angle between $\vec{M}_1$ and $\vec{M}_2$.

The second Piola Kirchhoff stress tensor **S** can be written as:

$$\boldsymbol{S} = \frac{\partial W}{\partial I_1}\boldsymbol{g_1} + \frac{\partial W}{\partial I_2}\boldsymbol{g_2} + \frac{\partial W}{\partial I_{12}}\frac{1}{2I_{12}}(\boldsymbol{g_1}\boldsymbol{E}\boldsymbol{g_2} + \boldsymbol{g_2}\boldsymbol{E}\boldsymbol{g_1}) \tag{8}$$

A simplified hyperplastic model is proposed. It is based on following assumptions:

- The coupling between I_{12} and I_i is neglect,
- The strain energy function $W(\boldsymbol{E})$ is expressed by Equation 9:

$$W = \frac{1}{2}k_1 I_1^2 + \frac{1}{2}k_2 I_2^2 + k_{12}I_1 I_2 + k_3 I_{12}^2 \tag{9}$$

This leads to the constitutive equation:

$$S = (k_1 I_1 + k_{12}I_2)\boldsymbol{g_1} + (k_2 I_2 + k_{12}I_1)\boldsymbol{g_2} + k_3(\boldsymbol{g_1}\boldsymbol{E}\boldsymbol{g_2} + \boldsymbol{g_2}\boldsymbol{E}\boldsymbol{g_1}) \tag{10}$$

So k_1 and k_2 presented tensile rigidities in yarns directions. k_{12} described the interaction between two groups of yarns. k_3 presented the shearing rigidity of woven.

The relation between components S_{gij} of second Piola Kirchhoff stress tensor **S** and E_{gij} of Green Lagrange strain tensor **E** in the base $\vec{g}_i$ can be presented by one of flowing expressions

$$\begin{bmatrix} S_{g11} \\ S_{g22} \\ S_{g12} \end{bmatrix} = \begin{bmatrix} k_1 & k_{12} & 0 \\ k_{12} & k_2 & 0 \\ 0 & 0 & k_3 \end{bmatrix} \begin{bmatrix} E_{g11} \\ E_{g22} \\ E_{g12} \end{bmatrix} \tag{11}$$

$$\begin{bmatrix} E_{g11} \\ E_{g22} \\ E_{g12} \end{bmatrix} = \begin{bmatrix} c_1 & c_{12} & 0 \\ c_{12} & c_2 & 0 \\ 0 & 0 & c_3 \end{bmatrix} \begin{bmatrix} S_{g11} \\ S_{g22} \\ S_{g12} \end{bmatrix} \tag{12}$$

Where:

$$c_1 = \frac{k_2}{k_1 k_2 - k_{12}^2},\ c_2 = \frac{k_1}{k_1 k_2 - k_{12}^2},\ c_3 = \frac{1}{k_3},\ c_{12} = \frac{-k_{12}}{k_1 k_2 - k_{12}^2} \tag{13}$$

2.1. Out-axes tensile test: Tensile test on 45°

In tis parts the proposed hyperelastic model is used to study the mechanical behaviour during the out-axes tensile test of the dry woven.

Out-axes tensile test is a tensile test exerted on a fabric but according to a direction which is not necessarily warp or weft directions [22]. In the case of anisotropic behavior stress and strains tensors have not, in general, the same principal directions. During this test, the simple is subjected to a shearing. Particular precautions must be taken to ensure a relative homogeneity of the test [23].

We considered a tensile test along a direction $\vec{E}_1$ forming an angle ψ_0 with orthotropic direction $\vec{g}_i$ (Figure.2).

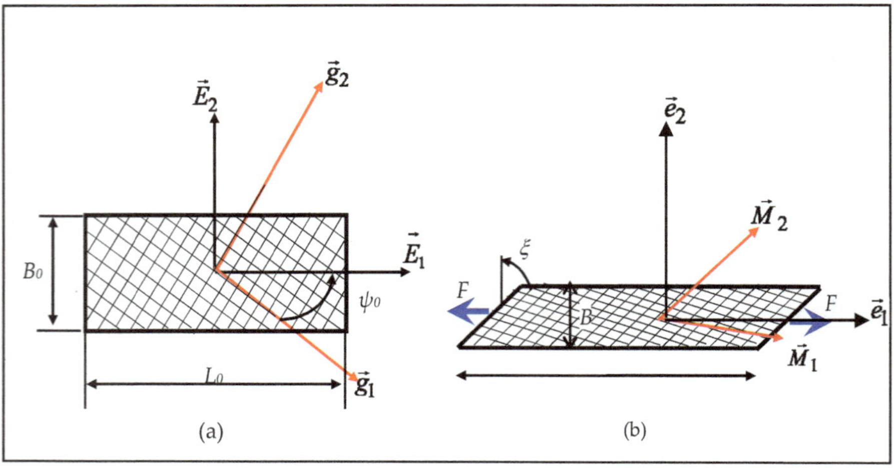

Figure 2. Kinematics of Out-axes tensile test, a: Reference configuration, b: Deformed configuration.

In the base $\vec{e}_i$, components of the second Piola Kirchhoff tensor S and the Gradient of transformation tensor F are as follows [23]

$$S_{/\vec{e}_i} = \begin{bmatrix} S & 0 \\ 0 & 0 \end{bmatrix}, F_{/\vec{e}_i} = \begin{bmatrix} f_1 & f_1\gamma \\ 0 & f_2 \end{bmatrix} \tag{14}$$

Where:

$$f_1 = \frac{L}{L_0}; f_2 = \frac{B}{B_0}; \gamma = \frac{f_2}{f_1} tg(\xi) \tag{15}$$

Let $P = \frac{F}{S_0}$ where F is the tensile force and S_o is the initial cross section of the specimen. P is related to S by:

$$P = \frac{F}{S_0} = f_1 S \tag{16}$$

The components of the Green–Lagrange strain tensor **E**, in the base $\vec{e}_i$, are as follows:

$$E_{/\vec{e}_i} = \begin{bmatrix} E_{11} & E_{12} \\ E_{12} & E_{22} \end{bmatrix} \tag{17}$$

Where

$$2E_{11} = f_1^2 - 1;\ 2E_{22} = f_2^2 - 1 + f_1^2\gamma^2;\ 2E_{12} = f_1^2\gamma \tag{18}$$

The response of the model presented by Equation 8 for this solicitation can be summarised as follows:

$$P = \frac{f_1 E_{11}}{C(\psi_0)};\ E_{22} = -\nu(\psi_0)E_{11};\ E_{12} = g(\psi_0)E_{11} \tag{19}$$

Where:

$$\begin{aligned} C(\psi_0) &= c_1\cos^4(\psi_0) + c_2\sin^4(\psi_0) + \frac{1}{2}(c_3 + c_{12})\sin^2(2\psi_0) \\ \nu(\psi_0 &= \frac{\left[(2c_3 + 2c_{12} - c_1 - c_2)\sin^2(2\psi_0) - c_{12}\right]}{4C(\psi_0)} \\ g(\psi_0) &= \frac{\sin(2\psi_0)\left[c_1\cos^2(\psi_0) - c_2\sin^2(\psi_0) - (c_3 + c_{12})\cos(2\psi_0)\right]}{2C(\psi_0)} \end{aligned} \tag{20}$$

The tensile test on 45° is a particular case of out-axes tensile tests where ψ_0=45). To replacing ψ_0 by 45°, Equation 20 became like the following:

$$C_{45} = \frac{1}{2k_3} + \frac{k_1 + k_2 + 2k_{12}}{4(k_1k_2 - k_{12}^2)},\ \nu_{45} = \frac{\left[(2c_3 + 2c_{12} - c_1 - c_2) - c_{12}\right]}{4C_{45}},\ g_{45} = -\frac{c_2}{2C_{45}} \tag{21}$$

S_1 and S_2 are respectively the maximum and the minimum Eigen values of Piola Kirchhoff tensor $\boldsymbol{S}$ .In Tensile test on 45°, Equation 14 shows that:

$$\frac{S_2}{S_1} = 0 \tag{22}$$

The expression of the applied force F is deducted from Equation 16:

$$F = PS_0 = \frac{2k_3 S_0 f_1 (f_1^2 - 1)(k_1 k_2 - k_{12}^2)}{2k_1 k_2 - 2k_{12}^2 + k_3 (k_1 + k_2 - 2k_{12})} \tag{23}$$

For a balanced woven ($k_1=k_2=k$) where the interaction between yarns is neglected ($k_{12}=0$), the expression of F became:

$$F = \frac{k_3 k S_0 f_1 (f_1^2 - 1)}{k + k_3} \tag{24}$$

The ratio between the minimum and the maximum Eigen values of Green Lagrange tensor $\boldsymbol{E}$., in the tensile test with 45°, is given by Equation 25:

$$\frac{E_2}{E_1} = -\frac{k - k_3}{k_3 + k} \tag{25}$$

2.2. Bias extension test

To explainer the pure shearing test of woven fabric, it has been noted that woven cloths in general deform as a pin-jointed-net (PJN) [24-28]. Yarns are considered to be inextensible and fixed at each cross-over point, rotating about these points like it is shown in Figure 3.

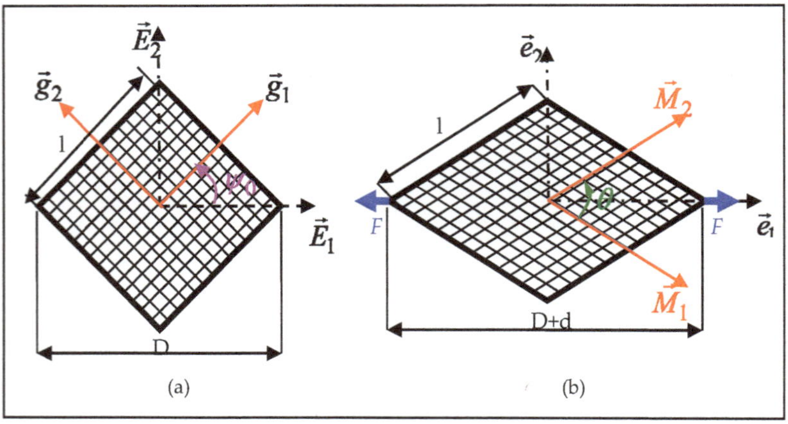

Figure 3. Kinematics Pure shear a: Reference configuration, b: Deformed configuration

During the Bias Extension test, the pure shearing occurred in the central zone A and the shear angle φ is defined by Equation 26:

$$\varphi = \frac{\pi}{2} - \theta = \frac{\pi}{2} - 2a\cos(\frac{D+d}{\sqrt{2}D}) \tag{26}$$

The Gradient of Transformation tensor $\boldsymbol{F}$ is presented by Equation 27:

$$\boldsymbol{F}_{/Ei,ei} = \begin{bmatrix} f_1 & 0 \\ 0 & f_2 \end{bmatrix} = \begin{bmatrix} \cos(\frac{\varphi}{2}) + \sin(\frac{\varphi}{2}) & 0 \\ 0 & \cos(\frac{\varphi}{2}) - \sin(\frac{\varphi}{2}) \end{bmatrix} \tag{27}$$

Using the proposed model, components S_{ij} ,E_{ij} of the second Piola Kirchhoff stress and Green Lagrange strain tensors are given, in the base $\vec{e}_i$, as follows:

$$\boldsymbol{S}_{/\vec{e}_i} = S(\varphi)\begin{bmatrix} 1 & 0 \\ 0 & -1 \end{bmatrix} \text{ where } S(\varphi) = \frac{1}{2}k_3 \sin(\varphi) \tag{28}$$

$$\boldsymbol{E}_{/e_i} = E(\varphi)\begin{bmatrix} 1 & 0 \\ 0 & -1 \end{bmatrix} \text{ where } E(\varphi) = \frac{1}{2}\sin(\varphi) \tag{29}$$

Thus

$$\frac{S_2}{S_1} = -1 \tag{30}$$

And

$$\frac{E_2}{E_1} = -1 \tag{31}$$

Where S_1 and S_2 are respectively the maximum and the minimum Eigen values of the second Piola Kirchhoff tensor $\boldsymbol{S}$ and E_1 and E_2 are respectively the maximum and the minimum Eigen values of Green Lagrange tensor $\boldsymbol{E}$.

The internal power per unit of volume in zone A is defined by Equation 32:

$$\omega a = \boldsymbol{S}_A : \dot{\boldsymbol{E}}_A = 2S(\varphi)\dot{E}(\varphi) = \frac{1}{4}k_3 \sin(2\varphi)\dot{\varphi} \tag{32}$$

To calculate to internal power per unit of volume in zone B we replace φ by $\frac{\varphi}{2}$ in Equation 32:

$$\omega b = \boldsymbol{S}_B : \dot{\boldsymbol{E}}_B = 2S(\frac{\varphi}{2})\dot{E}(\frac{\varphi}{2}) = \frac{1}{8}k_3 \sin(\varphi)\dot{\varphi} \tag{33}$$

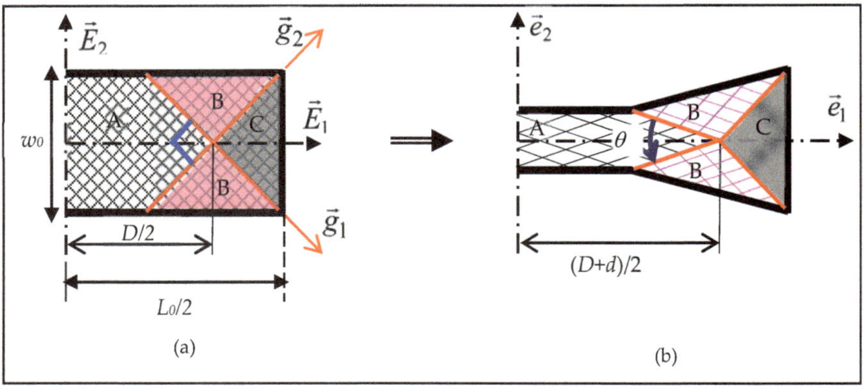

Figure 4. Kinematic of Bias Extension Test, a: initial configuration, b: deformed configuration

The total internal power in the specimen is given by Equation 34:

$$P\text{int} = Va.\omega a + Vb.\omega b \tag{34}$$

Where Va and Vb are respectively the initial volume in zones A and B defined as follows

$$\begin{aligned} Vb &= e_0 w_0^2 \\ Va &= e_0(Dw_0 - \frac{w_0^2}{2}) = e_0 Dw_0 - \frac{1}{2}Vb \end{aligned} \tag{35}$$

The External power is defined as:

$$Pext = F.\dot{d} = \frac{1}{2} FDf_2\dot{\varphi} \tag{36}$$

The equality between internal and external powers allows to determinate the expression of applied force F given by Equation 37:

$$F = e_0 w_0 f_1 k_3 \sin(\varphi)[1 + \frac{1}{4(\lambda - 1)\cos(\varphi)}(1 - 2\cos(\varphi))] \tag{37}$$

Where $\lambda = \dfrac{L_0}{w_0}$ is the aspect ratio.

3. Numerical simulation of Bias Extension test

In this section, we simulated the Bias Extension test (BE) using the hyperelastic proposed model implanted into Abaqus/Explicit thought user material subroutine (VUMAT). Out put of the VUMAT are stress components of Cauchy tensor projected in the Green-Nagdi basis, component of the second Piola Kirchhoff tensor *S*, and the Green Lagrange tensor *E* projected in $(\vec{g}_1, \vec{g}_2)$. We can also drew curves of Fore versus displacement.

The fabric is modelling by rectangular part meshed by continuum element (M3D4R).The boundary condition of model is presented in Figure 5a.

[29-30] compared the numerical results for the biased mesh and the aligned mesh and they proved that by using the biased mesh (Figure 5b), where the fibres are run diagonally across the rectangular element, neither the deformation profile nor the reaction forces are predicted correctly, for this we used the aligned mesh (Figure 5c).

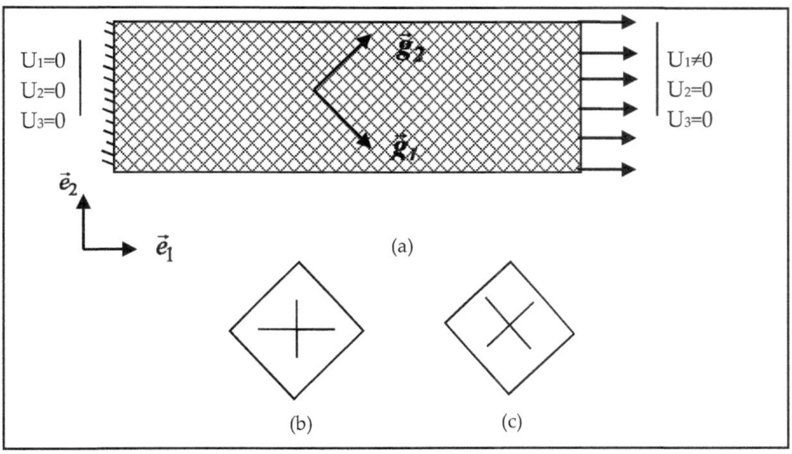

Figure 5. FEM mesh for the Bias Extension simulation, a: boundary condition of FEM, b: biased mesh, c: aligned mesh.

In order to simplify the problem, we used a balanced woven (k_1=k_2=k=700 N/mm^2) and we ignored the interaction between extension in yarns direction (k_{12}=0). The analysis is done for three different FEM with the same thickness of 0.2mm. Dimensions of FEM are presented in table 1.

MEF	Length(mm)	Width(mm)	Aspect ratio: λ
1	100	50	2
2	150	50	3
3	200	50	4

Table 1. Dimensions of samples

This analysis is realised on four values of the ratio between shearing and tensile rigidities $(\frac{k_3}{k} = 0.007, 0.02, 0.1, 0.3, 1)$ along three paths in FEM (see Figure 6).

The first path is longitudinal line in the middle of FEM. It joined zones A and C, the second path is along the yarn direction and the third path is transversal middle line Flowing results are illustrated for a displacement of 10% of initial length.

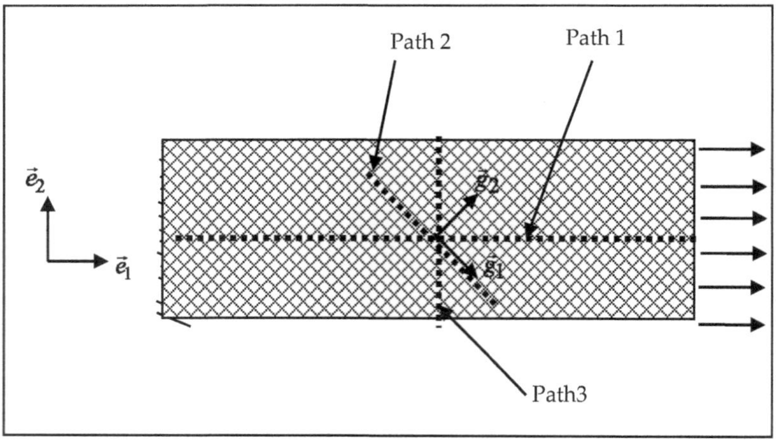

Figure 6. Different paths used in analysis

The deformed mesh with the contour of the Green Lagrange shear strain is shown in Figure7. We noticed that appearance of three discernible deformation zones of the Bias Extension test in three FEM. No significant deformation occurred in zone C. The main mode of deformation in zone A is the shearing. The most deformation of the fabric occurs in this zone.

In to order to study homogeneities of stress and strain states, we compared the analytical and the numerical results of strain and stress along three paths of Figure (6).

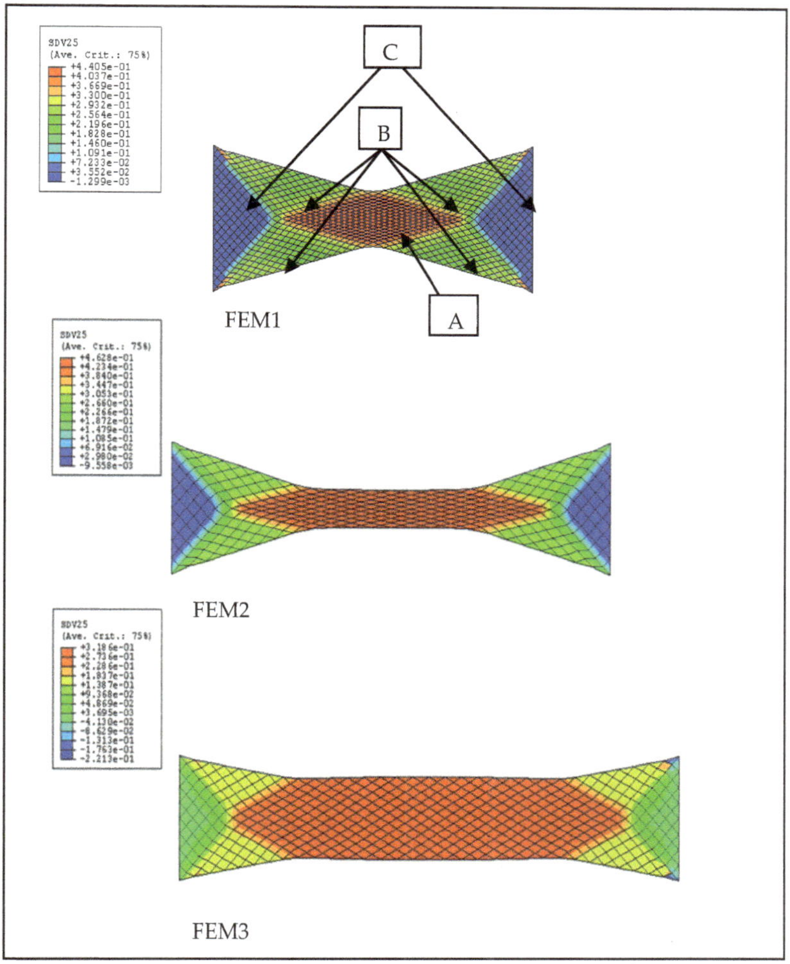

Figure 7. Deformed mesh with contour of Green Lagrange shear strain E_{12} for $\frac{k_3}{k}$ =0.007 and U_1=40mm.

3.1. Strain state

Figure 8 shows the variation of the maximum principal E_1 of Green Lagrange along the first path. We noticed that E_1 is symmetric with regard to the centre of the FEM. For the higher value of ratio of rigidities ($\frac{k_3}{k}$ =1), E_1 is homogenous and it conformed to the predicted value in the case of isotropic elastic material. To decreasing the ratio of rigidities ($\frac{k_3}{k}$), the central zone characterised by the higher value of E_1. En addition, we observed the appearance of two zones where the strain is not more important. In the first hand, to comparing with the analytical value of E_1 in the central zone, the numerical values of E_1 is closely to that predicted in the Bias Extension test for the few shearing rigidity. Zones C coincided with ends of the path where the deformation was not more significant. In another hand, we remarked that in the central zone of the path, the deformation is not homogenous especially in FEM1and FEM2. For more analyse the strains state in FEM, Figure 9 presented the evolution of $\frac{E_2}{E_1}$, along the first path. It is clear that to decreasing $\frac{k_3}{k}$, the value of $\frac{E_2}{E_1}$ tends to (-1) in three FEM. This proved that, in spite of the low displacement, the deformation in Bias Extension test is influenced by the ratio between shearing and tensile rigidities of the woven.

3.2. Stress state

Comparing the numerical and the analytical values of $\frac{S_2}{S_1}$, we determinate the stress state in different FEM for an displacement of 10% along the first path .

Figure10 show that to decreasing $\frac{k_3}{k}$, the value of $\frac{S_2}{S_1}$ decrease but never achieved (-1).

Indeed, if this simulation is interpreted like a Bias Extension test, $\frac{S_2}{S_1}$ should be verifying Equation 30 in the central zone. However the ratio of principal strain is approximately equal to 0. So it is conforming to Equation 22, and the stress state is the traction state.

In addition to varying the value of $\frac{k_3}{k}$, we evaluated the ratio of strain versus the ratio of stress in the central element of FEM. In Figure12, it can be noticed that in FEM1, to reducing the value of $\frac{k_3}{k}$, $\frac{E_2}{E_1}$ tend to (-1) and it conformed to the predicted value by Equation 31 for a few values of $\frac{k_3}{k}$. But $\frac{S_2}{S_1}$ have a negative value and it remain different to (-1). In FEM2, it was visibly that $\frac{S_2}{S_1}$ stayed proximity null for different value of $\frac{k_3}{k}$ thus it verified Equation 22 but $\frac{E_2}{E_1}$ tend to (-1) for few values of $\frac{k_3}{k}$. In FEM3, it was clear that for few value of $\frac{k_3}{k}$, $\frac{E_2}{E_1}$ tend to (-1), but the $\frac{S_2}{S_1}$ had positive values. Consequently, the shearing deformation in Bias Extension test depends of the ratio of rigidities between shearing and tensile, but the stress state is always the tensile stress.

FEM1

FEM2

FEM3

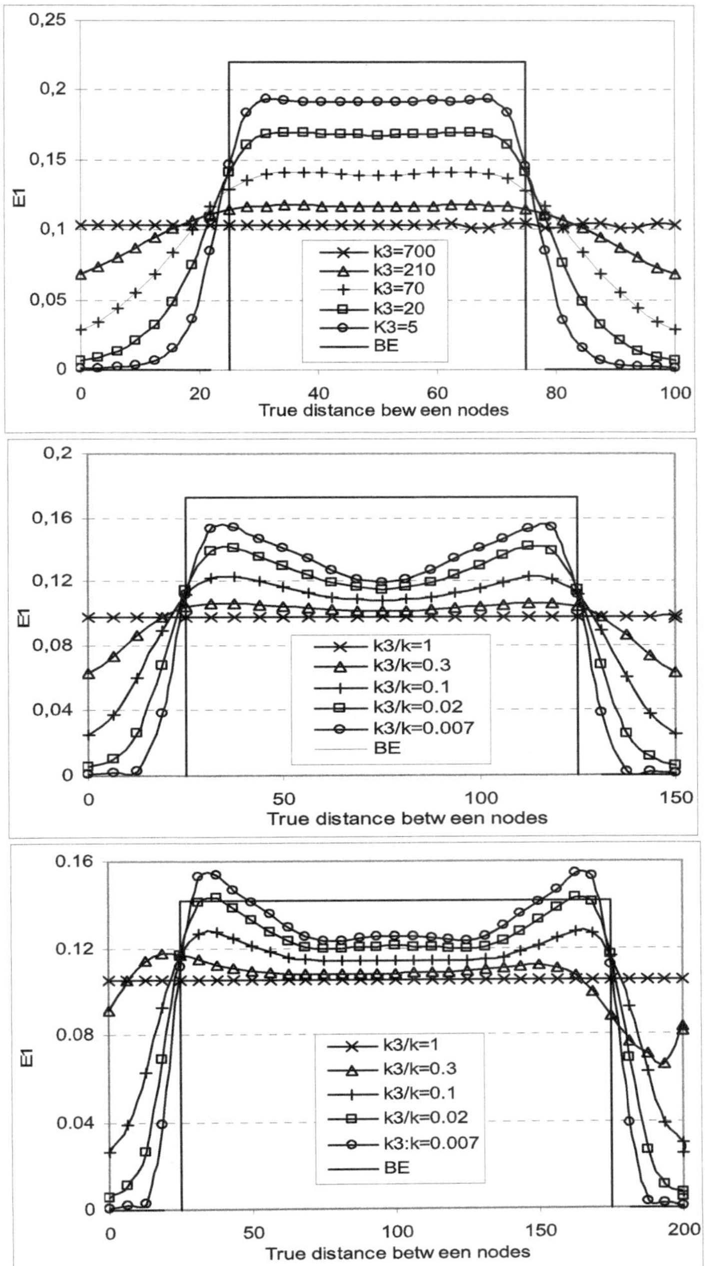

Figure 8. Variation of Maximum principal of Green Lagrange strain E_1 along the path1

FEM1

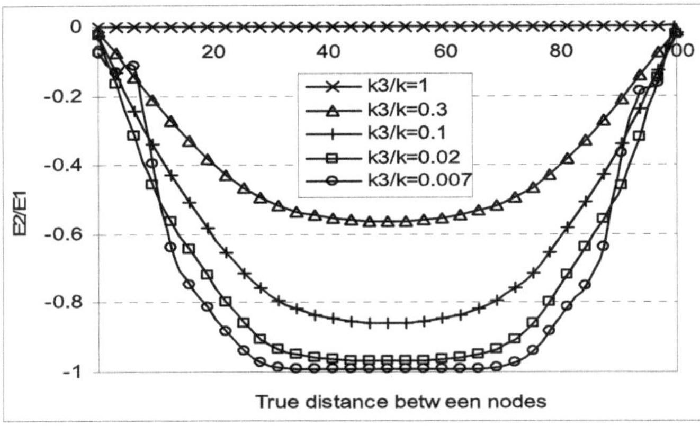

FEM2

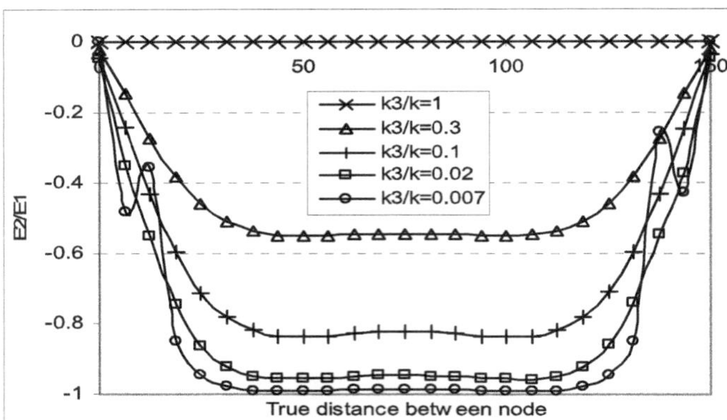

FEM3

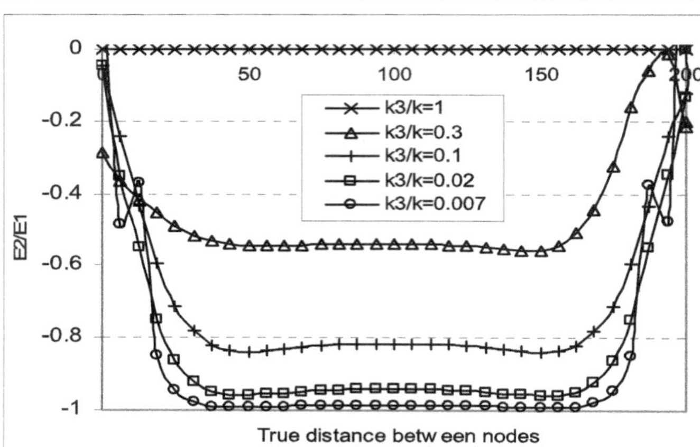

Figure 9. Variation of $\frac{E_2}{E_1}$ along the path 1

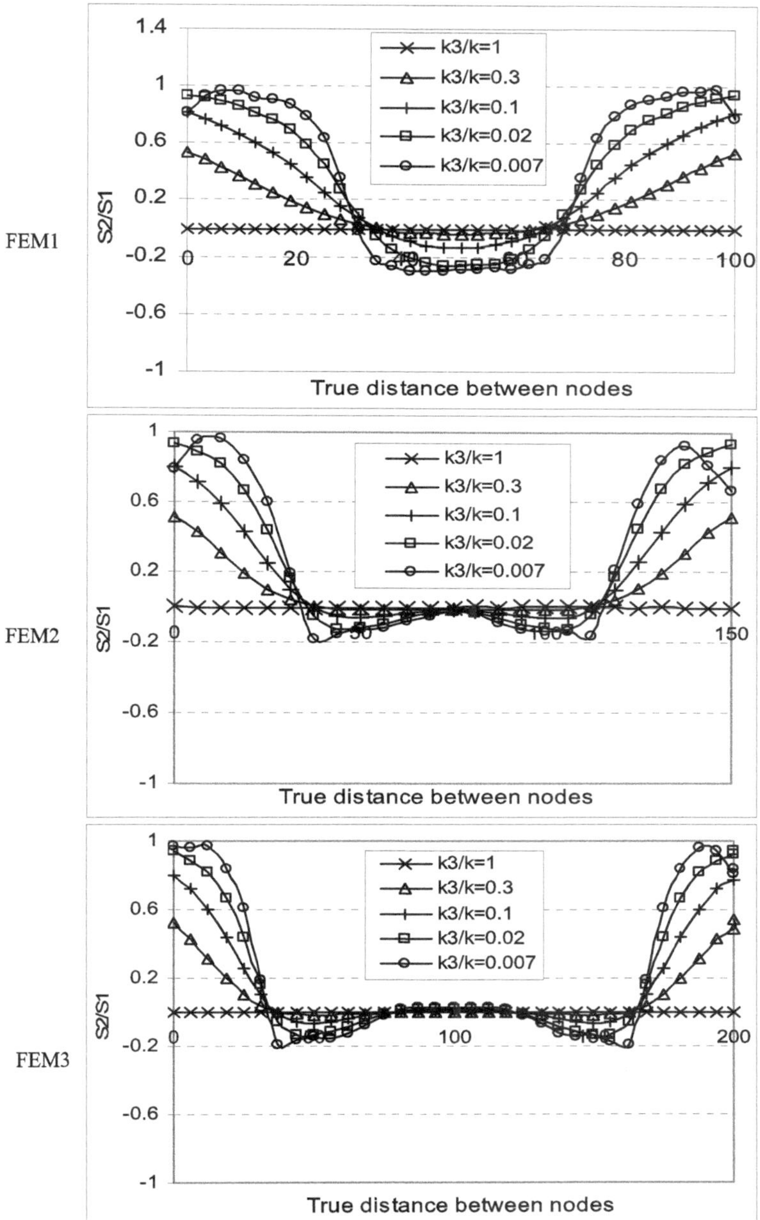

Figure 10. Variation of $\frac{S_2}{S_1}$ along path 1.

FEM1

FEM2

FEM3

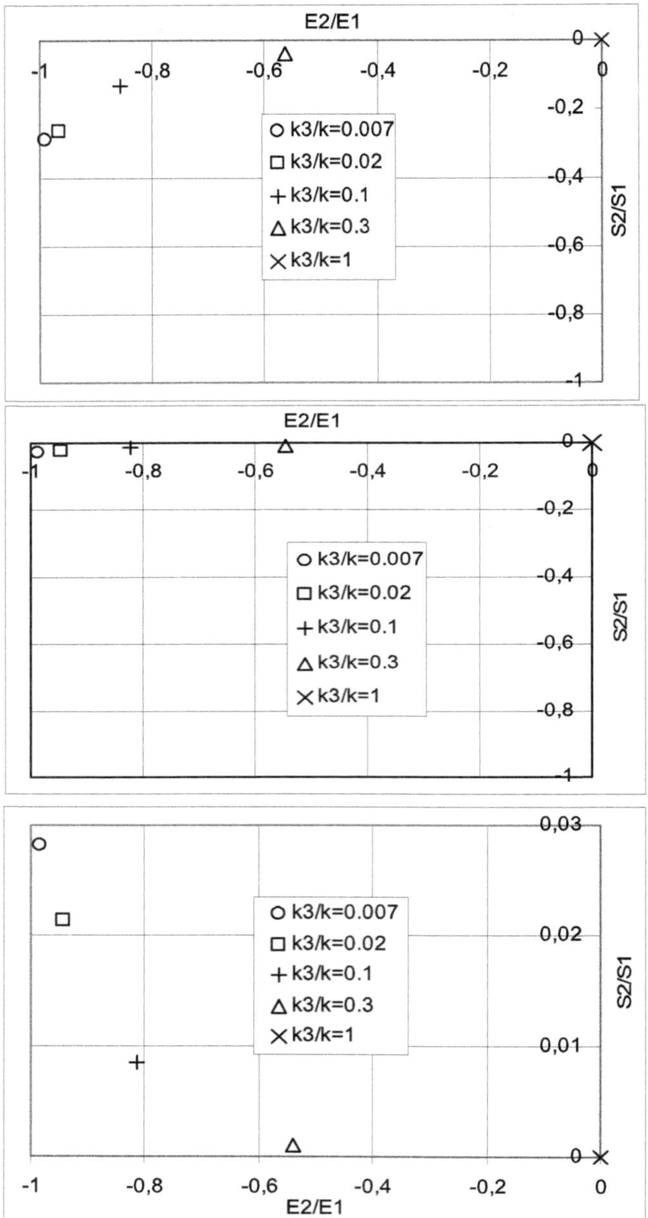

Figure 11. Variation of $\frac{S_2}{S_1}$ versus $\frac{E_2}{E_1}$ along path 1.

3.3. Angle between yarns

In this section, we compared between the numerical and the predicted values of the angle between yarns, along the first path.

Using the proposed model, the numerical angle between yarns is given by the following expression:

$$\theta_N = \arcos(\sqrt{\frac{(2E_{g12})^2}{(2E_{g22}+1)(2E_{g11}+1)}}) \tag{38}$$

In the case of the Bias Extension test, the predict angle between yarns in the central zone A is given by Equation 39:

$$\theta_B = 2\arcos(\frac{D+d}{\sqrt{2}D}) \tag{39}$$

The predict angle between yarns in the Tensile test in 45° is given by Equation 41:

$$\theta_T = \arcos(\frac{E_{11}(1-\nu_{45})}{\sqrt{(E_{11}+E_{22}+1)^2}}) \tag{40}$$

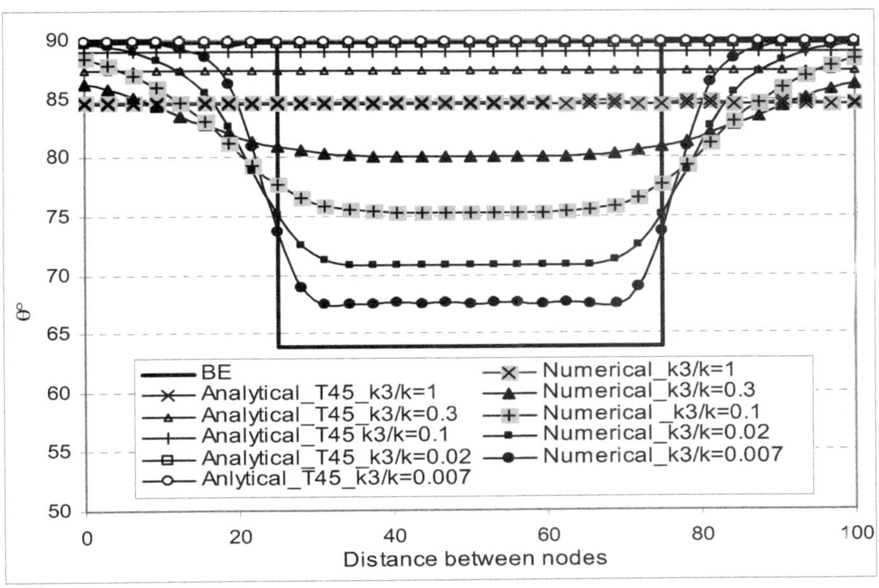

Figure 12. Comparison between Numerical and Predicted angles between yarns along the path 1 in FEM1.

Figure 12 demonstrated that the value of the angle between yarns was not uniform in the central zone of the FEM and it was not null in ends of the path1. For three FEM, the numerical angle between yarns tend to verify the predict angle (solid line) in the Bias Extension test for the lower value of $\frac{k_3}{k}$ This is another reason to justify the influence of the ration of rigidities on the shearing deformation of woven.

3.4. Elongation of yarns

Under the pin-joint assumption for trellising deformation mode, the edge length of the membrane element should remain unchanged during the deformation; thus the Green Lagrange stretch E_{g11} and E_{g22} should be null in Bias Extension test:

$$E_{g11} = E_{g22} = 0 \tag{41}$$

In Tensile test on 45°, warp and weft yarns are submitted respectively to Green Lagrange deformations $E_{g_{11}}$ and $E_{g_{22}}$ as follows:

$$E_{g_{11}} = \frac{1}{2}(\left|F.\vec{g}_1\right|^2 - 1) = \frac{k_1 - k_{12}}{2(k_1 k_2 - k_{12}^2)} \frac{E_{11}}{C_{45}} \tag{42}$$

$$E_{g_{22}} = \frac{1}{2}(\left|F.\vec{g}_2\right|^2 - 1) = \frac{k_2 - k_{12}}{2(k_1 k_2 - k_{12}^2)} \frac{E_{11}}{C_{45}} \tag{43}$$

In the case of balanced fabric without coupling between elongations in yarns directions, the warp and weft yarns are submitted to the same elongation:

So

$$E_{g11} = E_{g22} = E \tag{44}$$

Where

$$E = \frac{k_3}{k + k_3} E_{11} \tag{45}$$

In Figure 13, we compared numerical stretch deformation along the second path and the predicted elongation in yarn direction.

In the first hand, we noticed that the numerical elongation was not null. It became more important by increasing the value of $\frac{k_3}{k}$ in all FEM. In another hand, numerical value of elongation is closely conforming to the expected value in the tensile test in 45° for different values of $\frac{k_3}{k}$ in all FEM. This analysis provided that during Bias Extension test, yarns are subjected a few elongation. These stretches depend of the value of the ratio between shearing and tensile rigidities of woven. Same previous analyses are taken also along the third vertical path (Path3) and same results are verified.

Figure13 represented the evolution of $\frac{S_2}{S_1}$ versus $\frac{E_2}{E_1}$ along the third path. Like the first path, for few values of $\frac{k_3}{k}$, the shearing is the utmost deformation. But in all cases, the Bias Extension test is characterized by the tensile state.

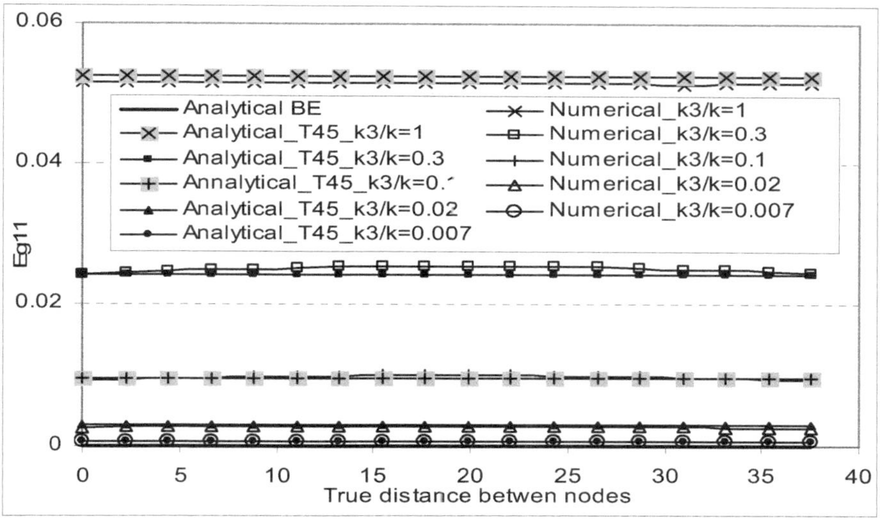

Figure 13. Variation of E_{g11} along the path 2in FEM1.

4. Conclusion

In this work, an orthotropic hyperelastic model test of woven fabric is developed and implanted into Abaqus/explicit to simulate Bias-Extension at low displacement. The analysis of numerical answers along longitudinal and transversal middle paths, proved, in the first hand, that to decreasing the ratio between shearing and tensile rigidities, the state deformation became to be conform to that predicted by the proposed model in the Bias Extension test for all FEM. In another hand, the angle between yarns tends to verify the predicted angle during the Bias Extension test. Although the stress state, is conform to the expected analysis of Traction test on 45°. The analysis of Green Lagrange stretching strain in the yarns direction, demonstrated that there was an elongation of yarns during test for different shearing rigidity. This elongation was exactly conforming to the predicted analytical elongation in the Traction test in 45°. Curves of Force versus displacement of the

Traction test in 45° applied to of the central zone A is closely to the numerical answers. We are able to adjust both curves by coefficients of adjustment.

This study allowed to verify analytical hypothesis adopted to interpret the Bias Extension test. The comparison between in Bias Extension test, the shearing deformation depends of the ratio between shearing and tensile rigidities of fabric. In Spite of the low displacement, this test presented always a stress state.

Author details

Samia Dridi
Department of Mechanics of Structures and Materials of the ISAE,
Institute Superieur of the Aeronautique and the Space, Edouard Belin, Toulouse, France

5. References

[1] Naujokaitytė L, Strazdienė E, Domskienė J (2008) Investigation of Fabric Behaviour in Bias Extension at Low Loads. Fibres & Textiles 16:59-63

[2] Zheng J, Komatsu T, Yazaki Y, Takater M, Inui S, Shimizu Y. (2006) Evaluating Shear Rigidity of Woven Fabrics. Textile Research Journal 76:145-151.

[3] Alamdar-Yazdi A (2004) A New Method to Evaluate Low-stress Shearing Behavior of Woven Fabrics. Fibres& Textiles 29:333-338.

[4] Dobb B, Russell S (1995). A System for the Quantitative Comparison of Wrinkling in Plain Fabrics. The Journal of the Textile Institute 86:495-497.

[5] Domskienė J, Strazdienė E (2005) Investigation of Fabric Shear Behavior. Fibres & Textiles 13:26-30.

[6] Mahar T.J, Ajiki I, Dhingra R.C, Postle R(1989) Fabric Mechanical and Physical Properties Relevant to Clothing Manufacture. International journal of clothing science and technologies 1:6-13.

[7] Lebrun G, Bureau M.N, Denault J (2003) Evaluation of bias-extension and picture-frame test methods for the measurement of intraply shear properties of PP/glass commingled fabrics. Composite Structures 61:341–352.

[8] Sharma S.B, Sutcliffe M.P.F, Chang S.H (2003) Characterisation of Material Properties for Draping of Dry Woven Composite Material. Composites Part A 34:1167–1175.

[9] Harrison P, Clifford M.J, Long A.C (2004).Shear Characterisation of Viscous Woven Textile Composites: a comparison between picture frame and bias extension experiments. Composites science and technology 64:1453–65.

[10] Potluri P, Perez Ciurezu D.A, Ramgulam R.B (2006) Measurement of Meso-scale Shear Deformations for Modelling Textile Composites. Composite Part A 37:303–314.

[11] Lomov S.V, Verpoest I (2006) Model of Shear of Woven Fabric and Parametric Description of Shear Resistance of Glass Woven Reinforcements. Composites Science and Technology 66: 919–933.

[12] Launay J, Hivet G, Duong A.V, Boisse P.(2008) Experimental Analysis of the Influence of Tensions on in Plane Shear Behaviour of Woven Composite Reinforcements. Composites Science and Technology 68:506–515.
[13] Kilby W.F (1963) Planar Stress-Strain Relationships in Woven Fabrics. The Journal of the Textile institute 54:9-27.
[14] Yick K.L, Chen K.P.S, Dhingra R.C, How Y.L(1966) Comparison of Mechanical Properties of Shirting Materials Measured on the KES-F and FAST Instruments. Textile Research Journal 66:622-633.
[15] Yu WR, Pourboghrat F, Chung K, Zampaloni M, Kang TJ (2002) Non Orthogonal Constitutive Equation for Woven Fabric Reinforced Thermoplastic Composites. Composite Part A 33:1095–1105.
[16] Yu X, Cartwright B, McGuckin D, Ye L, Mai Y.W (2006) Intra-ply Shear Locking In Finite Element Analyses of Woven Fabric Forming Process. Composites Part A 37:790-803.
[17] Sidhu R.M, Averill R.C, Riaz M, Pourboghrat F (2001) Finite Element Analysis of Textile Composite Perform Stamping. Composite Structures 52:483-497.
[18] Aimène Y, Vidal-Sallé E, Hagège B, Sidoroff F, Boisse P (2010) A Hyperelastic Approach for Composite Reinforcement Large Deformation Analysis. Journal of Composite Materials 44:5-26.
[19] Guo Z.Y, Peng X.Q, Moran B (2006) A Composite-based Hyperelastic Constitutive Model for Soft Tissue with Application to the Human Annulus Fibrosus. Journal of the Mechanics and Physics of Solids 54:1952-1971.
[20] Diani J, Brieu M, Vacherand J, Rezgui A (2004) Directional Model for Isotropic and Anisotropic Hyperelastic Rubber-like Materials. Mechanic of Materials 36:313-321.
[21] Bonet J, Wood R.D (1997) Non Linear Continuum Mechanics for Finite Element Analysis. Cambridge University Press.
[22] Zouari R, Amar S.B Dogui,A (2008) Experimental and Numerical Analyses of Fabric off-axes Tensile Test. The Journal of the Textile Institute 35:1-11.
[23] Dogui A (1988) Cinématique Bidimensionnelle en Grandes Déformations : Application à la traction hors axes et à la torsion. Journal de Mécanique Théorique et Appliqué 7:43-64.
[24] Mark C,Taylor H. M(1956) The fitting of Woven Cloth to Surfaces. The Journal of the Textile Institute 47:47-48.
[25] Lindberg J, Behre B, Dahlberg B (1961) Shearing and Buckling of Various Commercial Fabrics. Textile Research Journal 36:99-122;
[26] Skelton J (1976) Fundamentals of fabric shear. Textile Research Journal 46:862-869.
[27] Potter K(1979) The influence of Accurate Stretch Data for Reinforcements on the Production of Complex Structural Mouldings. Composites 10:161-173.
[28] Potter K (2002) Bias Extension Measurement on Cross Plied Unidirectionnel Prepreg. Composites Part A 33:63-67.
[29] Xiaobo Y, Bruce C, Damian M, Lin Y, Yiu W (2006) Intra-ply Shear Locking in Finite Element Analyses of Woven Fabric Forming Processes, Composites Part A 37:790–803.

[30] Tenthijie R.H.W, Akkeman R (2008) Solution to Inter Ply Shear Locking in Finite Element Analyses of Fibre Reinforced Material. Composite Part A 39:1167-1176.

Chapter 5

Nonlinear Large Deflection Analysis of Stiffened Plates

Khosrow Ghavami and Mohammad Reza Khedmati

Additional information is available at the end of the chapter

1. Introduction

Stiffened plates are basic structural members in marine structures as shown in Figure 1, and include also aeronautic and space shuttles among other structures. Due to the simplicity in their fabrication and high strength-to-weight ratio, stiffened plates are also widely used for construction of land based structures such as box girder and plate girder bridges. The stiffened plate has a number of one-sided stiffeners in either one or both directions, the latter configuration being also called a grillage (Figure 2). Ultimate limit state design of Stiffened plates' structures requires accurate knowledge about their behaviour when subjected to extreme loading conditions.

One of the most important loads applied on stiffened plates is the longitudinal in plane axial compression arising for instance from longitudinal bending of the ship hull girder as presented in Figure 3. The need to improve our knowledge of the buckling modes of such plates was emphasised after the collapse of several offshore structures and some ships in Brazil as well as the failure of several box girder bridges in the seventies of the twentieth century, Merrison Committee [1], Crisfield [2], Murray [3], Frieze, et.al. [4]. Stiffened plates are efficient structures, as a large increment of the strength is created by a small addition of weight in the form of stiffeners. However the collapse mechanisms of stiffened plates under predominantly compressive load present a complex engineering problem due to the large number of possible combinations of plate and stiffener geometry, materials, boundary conditions and loading. The design of such structure has to meet several requirements such as minimization of the weight and maximization of the buckling load. Thus, the designer of this structure is confronted with the problem of satisfying two conflicting objectives; such problems are called multi-objective or vector optimisation problems. In general, the objective-functions do not attain their optimum in a common point of the feasible points, Brosowski & Ghavami [5, 6].

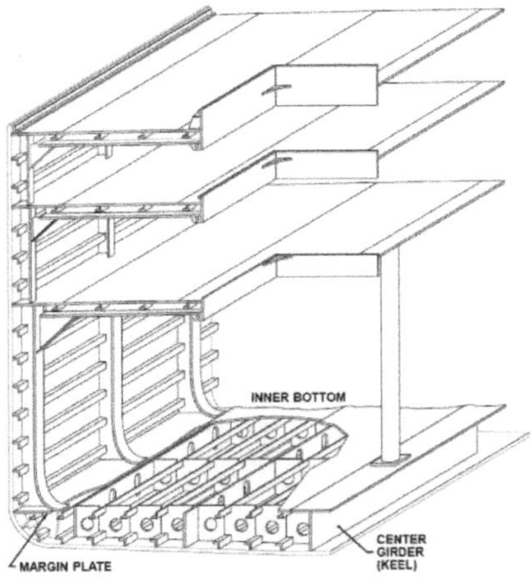

Figure 1. Some examples of thin-walled structures

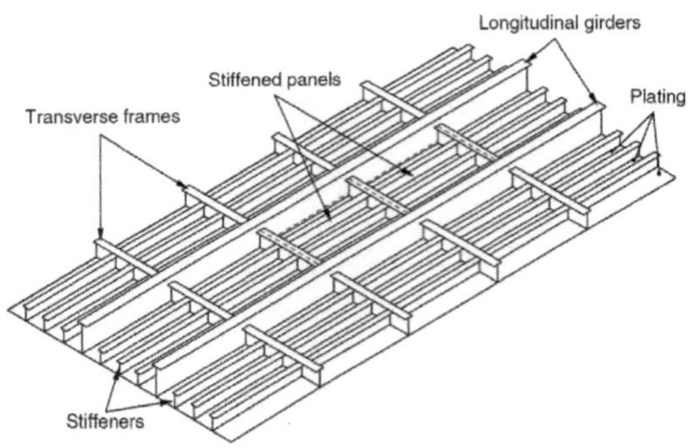

Figure 2. Structure of stiffened plates of the grillage type

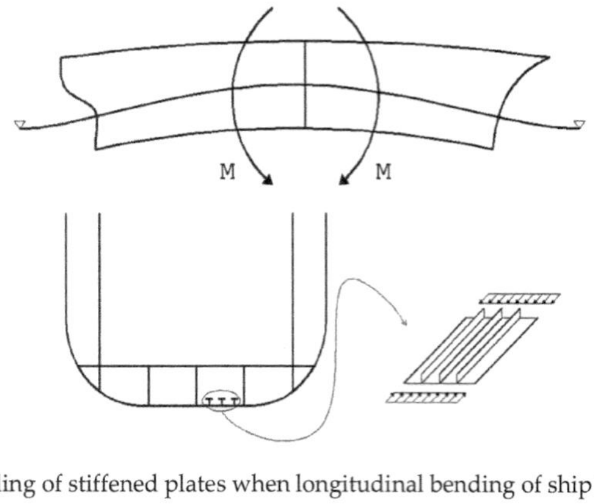

Figure 3. In-plane loading of stiffened plates when longitudinal bending of ship hull girder

For the analysis of such structural elements, the theory of orthotropic plate can be used to predict the global buckling stresses but not the local buckling and the interaction between the plate and the stiffeners, for the predominantly in-plane loading. In stiffened plates the initial imperfections due to the fabrication are inevitable. The buckling mechanism of stiffened plates depends, strongly, on the direction of initial bows, i.e. whether they are towards the plate or the stiffener. In the former case, the collapse is sudden due to buckling of the stiffener in contrast to the latter case, where a gradual failure occurs. Despite a substantial amount of theoretical research into the ultimate load behaviour of stiffened plates subjected to predominately in-plane loading, the accuracy and reliability of the predicted collapse load considering all the variables is not yet well confirmed. Specifically, in the available literature, no systematic theoretical and experimental investigation of the geometrical shape of the stiffeners cross-section on the ultimate buckling load behaviour of the stiffened plates, the interaction between the stiffeners and the plate, which was is the objectives of this chapter is being presented.

The buckling behaviour of stiffened plates under different loading conditions which has been the topic of the authors investigation, both experimentally and numerically, during last three decades has been reviewed concisely in this chapter. Chen et al. [7] carried out experimental investigations on 12 stiffened plates under in-plane longitudinal compression, purely or in combination with lateral load. The specimens were in different damage conditions: seven "as-built", two "dented" and three "corroded". Hu and Jiang [8] simulated some of the tests made by Chen et. al. [7], using the commercial program ADINA [9] and in-house program VAST [10], both based on the FEM. The former was used to analyse the "as-built" and "dented" stiffened plates, whereas the "corroded" specimens were analysed using VAST [4]. It was found, that in most cases the FEM produced similar responses to those of experimental results up to the loss of structural continuity. Grondin et al. [11] made a parametric study on the buckling behaviour of stiffened plates using the FEM-based commercial program ABAQUS [12]. Sheikh et. al. [13] extended the studies in [11] to investigate the combined effect of in-plane compression and bending using the same program. In these studies, only tee-shape stiffeners, plate aspect ratios, plate-to-stiffener cross-sectional area ratio with different initial imperfections of the plates were investigated.

All the cited studies, either experimentally or numerically, investigated the strength behaviour of longitudinally stiffened plates with specific boundary conditions. The continuity of both plates and stiffeners in thin-walled structures, composed of stiffened plates, leads to an interaction among the adjacent panels. Among the several available experimental investigations, two series of well executed experimental data on longitudinally multi-stiffened steel plates, with and without transversal stiffeners subjected to uniform axial in-plane load carried out to study the buckling and post-buckling up to final failure have been chosen. The first series are those of Ghavami [14] where the influences of stiffener cross-section of the type rectangular (R), L and T, as shown in Figure 4, have been investigated. The spacing of the stiffeners and the presence of rigid transversal stiffeners on the buckling behaviour up to collapse have also been studied. The second series of Tanaka &

Endo [15], where the behaviour of stiffened plates have three and two flat bars for longitudinal and transversal stiffeners respectively, were analysed. Besides, owing to the recent progress in the field of finite element method and available powerful FEM programs, it has been possible to assess the structural behaviour of the considered plates and stiffeners subjected to any combination of loads.

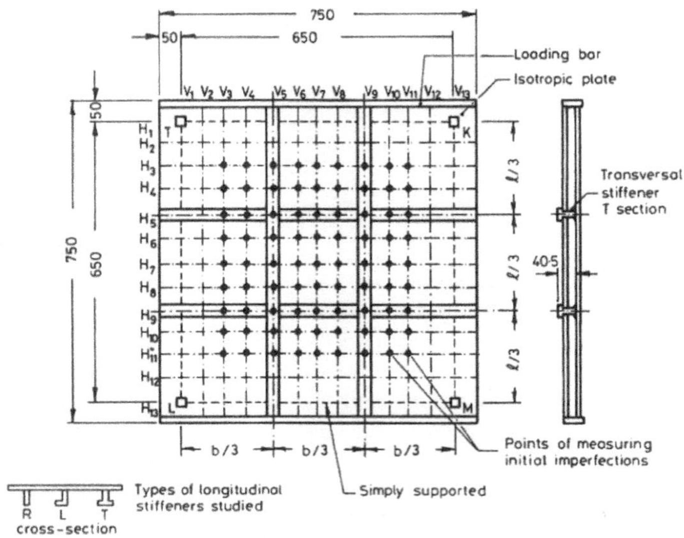

Figure 4. Ghavami's test models

Therefore one of the principal aims of this chapter is to present the applicability of the finite element method to simulate test results. The Finite Element Method (FEM) technique is employed to trace a full-range of elastic-plastic behaviour of the stiffened plates. It is seen that the FEM-based software is capable and accurate enough to simulate the test results. With the availability of high memory and high speed PCs', FEM programs become fast and cheap means to predict the buckling and post-buckling behaviour of stiffened plates with different configurations up to collapse. Successful simulations using FEM-based software means, that plate with different dimensions under various types of loading combinations and damages can be studied numerically. Besides, validated simulations using such programs enhance estimation of the ultimate strength analysis of box-like thin-walled structures composed of plates and stiffened plates.

2. Ghavami's experiments

Ghavami [14] tested a total number of 17 plate models of overall dimensions B=L=750 mm in a specially designed testing rig as shown in Figure 5. The models were divided into six series, with their definition and dimensions summarised in Tables 1 and 2. The average thickness of the plate was t=4.4mm for the longitudinally stiffened plates with one and two rectangular (R), L and T stiffeners, designated as P1R, P1L, P1T and P2R, P2L, P2T respectively. The thickness of the plates, stiffened longitudinally, as for the series II and III but with one or two transversal stiffeners of T sections, P1R1T, P1L1T, P1T1T, P2R1T, P2L1T, P2T1T and P2R2T, P2L2T, P2T2T respectively was equal to 4.8mm. The span between the simple supports for all models was 650mm in both directions. In each group one isotropic plate, P1, P2 was also tested as a reference model. However the supports for the longitudinally stiffened plates, series II and III were not continuously simply supported but were very closely discretized simply supported and those with transversal stiffeners had continuously simply supported boundary conditions. A summary of material properties and test results is given in Table 3.

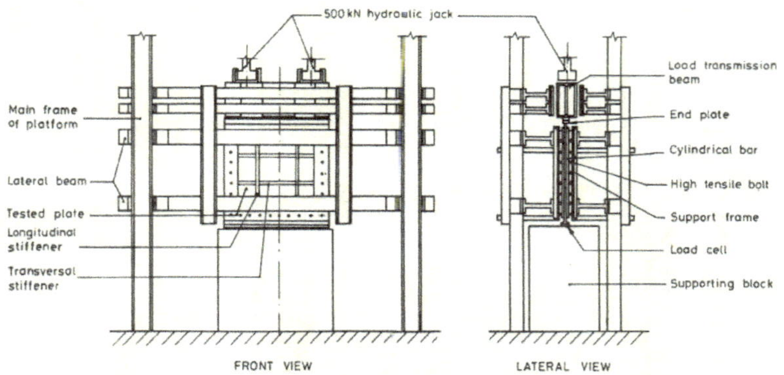

Figure 5. Ghavami's testing rig

Series No.	Definition	Test models
I	Unstiffened plate	P1, P2
II	Plate with one longitudinal stiffener of R, L and T cross-section	P1R, P1L, P1T
III	Plate with two longitudinal stiffeners of R, L and T cross-section	P2R, P2L, P2T
IV	Plate as series II but with addition of one transversal stiffener at the mid-span	P1R1T, P1L1T, P1T1T
V	Plate as series III but with addition of one transversal stiffener at the mid-span	P2R1T, P2L1T, P2T1T
VI	Plate as series II but with addition of two transversal stiffeners at 1/3 of span	P2R2T, P2L2T, P2T2T

Table 1. Definition of Ghavami test models

Test model	Plate			Longitudinal stiffener				Transverse stiffener			
	L	b	t	t_w	h_w	t_f	b_f	t_{wt}	h_{wt}	t_{ft}	b_{ft}
	mm	mm	mm	mm	mm	mm	mm	mm	mm	mm	mm
P1	650	650	4.4	----	----	----	----	----	----	----	----
P1R	650	325	4.4	7.0	30.0	----	----	----	----	----	----
P1L	650	325	4.4	6.4	30.0	3.9	16.4	----	----	----	----
P1T	650	325	4.4	6.4	30.0	4.8	26.4	----	----	----	----
P2R	650	217	4.4	7.0	30.0	----	----	----	----	----	----
P2L	650	216	4.4	6.4	30.0	19.5	16.4	----	----	----	----
P2T	650	217	4.4	6.4	30.0	20.0	26.4	----	----	----	----
P2	650	650	4.8	----	----	----	----	----	----	----	----
P1R1T	325	325	4.8	5.1	30.0	----	----	4.7	41.1	4.1	35.3
P1L1T	325	325	4.8	5.2	30.2	3.4	14.8	4.8	40.4	4.1	34.2
P1T1T	325	325	4.8	4.6	30.0	3.8	25.3	4.9	40.4	4.2	35.2
P2R1T	325	216	4.8	5.1	30.0	----	----	4.7	40.7	3.8	35.7
P2L1T	325	217	4.8	5.1	30.2	17.2	14.6	4.6	40.6	4.1	35.9
P2T1T	325	217	4.8	4.7	28.8	13.5	25.0	4.7	39.6	4.1	34.8
P2R2T	216	216	4.8	5.0	30.1	----	----	4.7	40.4	4.1	35.7
P2L2T	217	217	4.8	5.1	30.0	17.0	14.9	4.7	40.6	4.1	35.5
P2T2T	216	216	4.8	4.6	29.8	13.0	24.8	4.8	40.6	4.1	35.5

Table 2. Dimensions of plate and stiffeners in Ghavami test models

The testing rig was constructed within the Structural and Material Laboratory of PUC-Rio and is shown in Figs. 5 and 6. Out of plane deflections of plates and stiffeners were measured principally by mechanical dial gauges fixed at specific points mounted on the testing rig, as shown in Figure 6. In all models electrical linear strain gauges or rosettes measured the strains. More details on the test rig, test models and the process of the tests can be found in reference [14]. In each test the maximum ultimate collapse stress σ_{ult} was calculated by dividing the ultimate load P_u to the overall cross-section of the plate A_p and stiffeners A_s as given by eqn (1):

$$\sigma_{\text{ult}} = P_u / \left(A_p + A_s\right) \tag{1}$$

The squash load P_{sq} was calculated by multiplying the yield stress of the plate σ_{Yp} and the stiffener σ_{Ys} with their appropriate cross-section areas as eqn (2):

$$P_{sq} = \sigma_{Yp} A_p + \sigma_{Ys} A_s \tag{2}$$

The test results together with those of maximum initial W_0 , final W_{max} deflections and in-plane shortening U_{max} are given in Table 3.

Test model	Material properties			Measured deflection	Maximum deflection	Maximum shortening	Collapse stress
	E	σ_{Yp}	σ_{Ys}	W_o / t	W_{max} / t	U_{max} / t	$\sigma_{ult} / \sigma_{Yp}$
	$MPa \times 10^5$	MPa	MPa	%	%	%	%
P1	1.81	218	----	61	278	0.38	42.2
P1R	1.81	218	390	69	188	0.31	70.2
P1L	1.99	227	270	36	123	0.34	66.5
P1T	1.99	227	170	9	33	0.27	60.0
P2R	1.95	224	390	25	117	0.41	66.0
P2L	2.21	223	270	19	142	0.30	74.0
P2T	2.21	223	270	3	128	0.37	74.0
P2	1.78	220	----	20	121	0.40	48.2
P1R1T	1.85	219	326	21	123	0.33	74.0
P1L1T	1.91	225	326	27	27	0.48	71.1
P1T1T	1.75	219	273	33	121	0.33	72.1
P2R1T	1.75	219	326	70	52	0.64	88.6
P2L1T	1.89	227	326	40	33	0.60	84.6
P2T1T	1.78	220	273	23	30	0.51	89.1
P2R2T	1.91	225	326	32	18	0.63	86.2
P2L2T	1.89	227	326	28	53	0.56	97.4
P2T2T	2.09	218	273	32	16	0.51	103.2

Table 3. Summary of material properties and results for Ghavami test models

Figure 6. Stiffened plate model positioned in the Ghavami's testing rig

3. Tanaka & Endos' experiments

Tanaka & Endo [15] carried out a series of experimental and numerical investigations on the ultimate compressive strength of plates having three and two flat-bars stiffeners welded longitudinally and transversally respectively. A total of 12 tests were performed. The test specimen was designed so that the longitudinally stiffened plates located in the middle of whole test specimens could fail. The test specimens were intended to fail by local plate buckling or tripping of longitudinal stiffeners. A typical test rig from the Tanaka & Endo study is shown in Figure 7. A stiffened plate model positioned in their testing rig is presented schematically in Figure 8. To account for the effect of adjacent panels on the collapse behaviour of central panel, three-span models with two adjacent (dummy) stiffened panels and supported by two transverse frames were employed. The thickness of plate and stiffeners in two adjacent panels was 1.2-1.3 times that of plate and stiffeners in the central panel. Table 4, where a=1080mm is the span length of the plate with average plate thickness between t=4.38mm to t=6.15mm, represents geometric and material properties for the Tanaka & En'os' test structures. The boundaries of stiffened plates were continuously simply supported and the in-plane axial compression load was applied longitudinally. The maximum measured initial deflections in the plate were ranging between 0.1-0.4 mm. The ultimate collapse strength and squash load were calculated in the same manner using equations 1 and 2 as those considered by Ghavami [14].

Structure No.	a (mm)	b (mm)	t (mm)	h_w (mm)	t_w (mm)	$\frac{h_w}{t_w}$	A_{03} (mm)	σ_{Yp} (MPa)	σ_{Ys} (MPa)	E (GPa)
D0		1440	6.15	110.0	9.77	11.26	0.101	234.2	287.1	205.8
D0A		1440	5.65	110.0	10.15	10.84	0.250	249.9	196.0	205.8
D1		1200	5.95	110.0	10.19	10.79	0.143	253.8	250.9	205.8
D2		1560	5.95	110.0	10.19	10.79	0.288	253.8	250.9	205.8
D3	1080	1440	5.95	103.5	11.84	8.74	0.312	253.8	326.3	205.8
D4		1440	5.95	118.5	7.98	14.85	0.119	253.8	284.2	205.8
D4A		1440	5.65	118.5	8.08	14.67	0.379	249.9	274.4	205.8
D10		1200	4.38	65.0	4.38	14.84	0.515	442.0	442.0	205.8
D11		1200	4.38	90.0	4.38	20.55	0.503	442.0	442.0	205.8
D12		1440	4.38	65.0	4.38	14.84	0.523	442.0	442.0	205.8

Table 4. Geometric and material properties of Tanaka & Endo tests

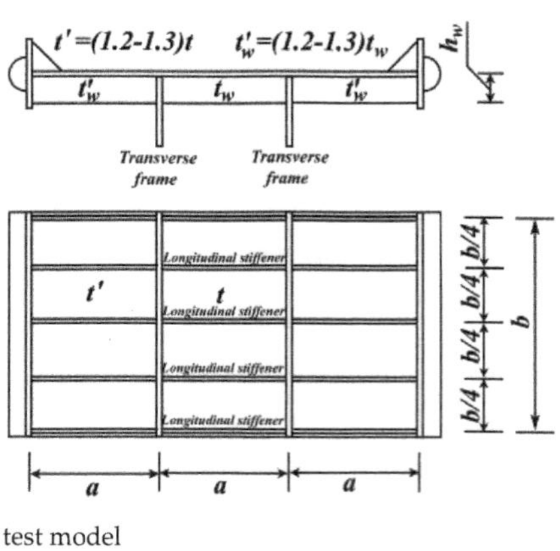

Figure 7. Tanaka & Endo's test model

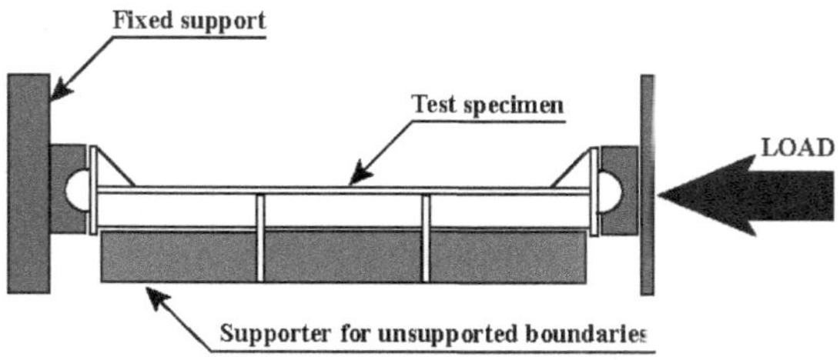

Figure 8. Stiffened plate model positioned in the Tanaka & Endo's testing rig

4. Finite element simulations

Since the test specimens in all above-reported experiments, had large deflections and plastic deformations, finite element analyses had to be performed using the software offering combined geometrical and material non-linear capabilities. In this study, the commercially available finite element code, ANSYS [16] was adopted. In the control menu of ANSYS solver, the options of "large deflection" and "arc-length method" are activated. The arc-length method is used to trace the non-linear large deflection response of the models.

4.1. Shell element formulation

Both plate and stiffeners are modelled using SHELL43 elements selected from ANSYS library of elements. The SHELL43 element in Figure 9 is a so-called plastic large strain element and categorised in the family of four-node quadrilateral elements. Each node has three translational degrees of freedom in the nodal x, y and z directions as well as three rotational degrees of freedom about the nodal x, y and z-axes. The chosen element allows for elastic, perfectly plastic, with strain hardening or strain softening, large strain and large deflection response [16].

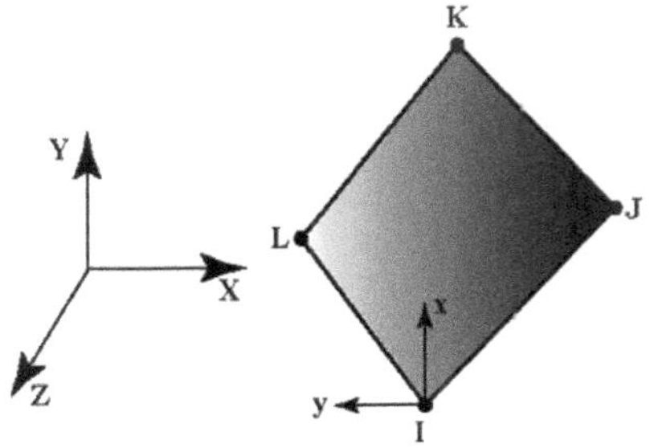

Figure 9. Shell43 element of the ANSYS FEM program

4.2. Finite element mesh and boundary conditions

A convergence study indicated that in the finite element mesh of isotropic and stiffened plates respectively, assuming $10\times(a/b)$ mesh divisions along local plate panels and 10 mesh divisions across them is sufficient to capture accurately the buckling and plastic collapse behaviour. Respectively *a* and *b* represent the length and breadth of local plate panels. In order to model the stiffener's web and flange, respectively 6 to 7 and 5 to 6 elements are sufficient. However, the purpose of this study was to simulate the testing results and finer meshes were therefore used. In the case of Tanaka & Endo tests, to reduce the number of mesh divisions and speed up the time of analysis, a rational assumption was made. The transverse stiffeners or frames for the case of Tanaka & Endo tests were not modelled for simplicity; instead the nodes on the line of attachment of the transverse stiffeners were constrained from translational movement out of plate plane. Furthermore, the translational movements of these nodes along the axis perpendicular to the line of attachment of transverse stiffeners were coupled with each other. Transverse frames were modelled in the case of Ghavami's tests. In both Ghavami and Tanaka & Endo's tests, the stiffened plates were loaded in axial compression along the stiffeners. Also in their tests the simply supported boundary conditions were assumed in the models. Figs. 10 and 11 show typical finite element models with the simulated boundary conditions, used for the analysis of Tanaka & Endo test specimens and Ghavami P2L2T test specimen (as an example).

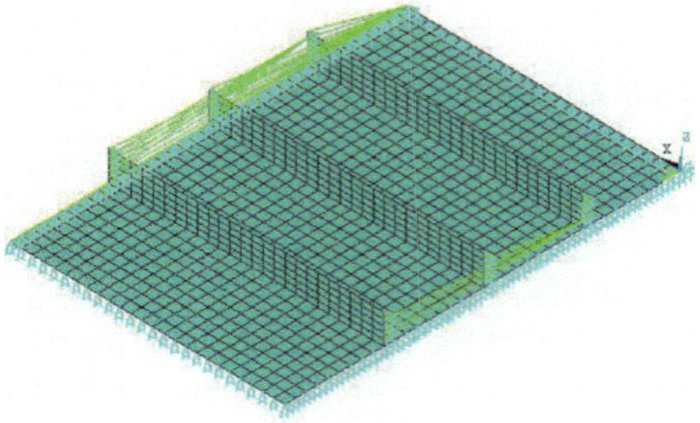

Figure 10. Finite element model of Tanaka & Endo's test specimens

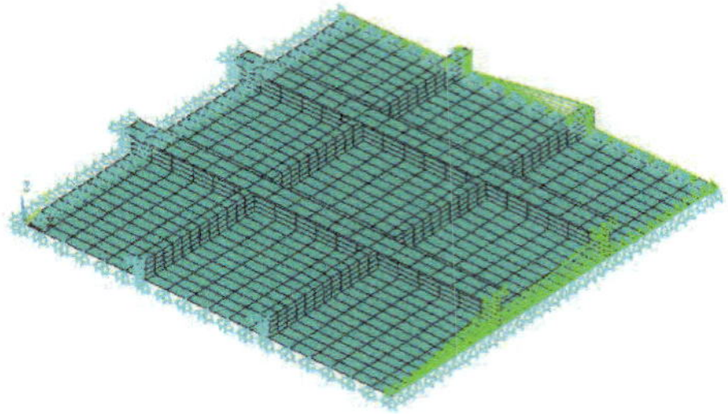

Figure 11. Finite element model of Ghavami's P2L2T test specimen

4.3 Imperfections

Welding residual stresses were not modelled specifically in this study. However, in order to simulate the complex pattern of residual welding stresses and initial deflections stated in references [14] and [15], a special procedure was employed. Uniform lateral pressure was applied first on the stiffened plate model and a linear elastic finite element analysis was carried out. This analysis was repeated in a trial and error sequence of calculations so that the magnitude of maximum deflection of plate reached that, measured by Ghavami. It is assumed that this procedure would simulate both the residual welding stresses and initial geometrical imperfection. After satisfying this condition, the information concerning the coordinates of nodal points, element coordinates and boundary conditions was transferred to a new finite element mesh for the geometrical and material non-linear response analysis under the action of longitudinal in-plane compression.

It should be emphasised that the pattern of initial deflections induced in the Ghavami's specimens [14] were nearly matching the pattern produced by this procedure. For the case of Tanaka & Endo's tests, first an eigenvalue buckling analysis was made using ANSYS, in order to capture the three-wave buckling mode deflection of the specimens [15]. Then the deflection pattern in this mode was scaled to the same pattern with the maximum magnitude of initial deflection, A_{03}, (Table 4) before testing, which has been reported by Tanaka & Endo [15]. Nonlinear response analysis under the action of longitudinal in-plane compression was performed on this model.

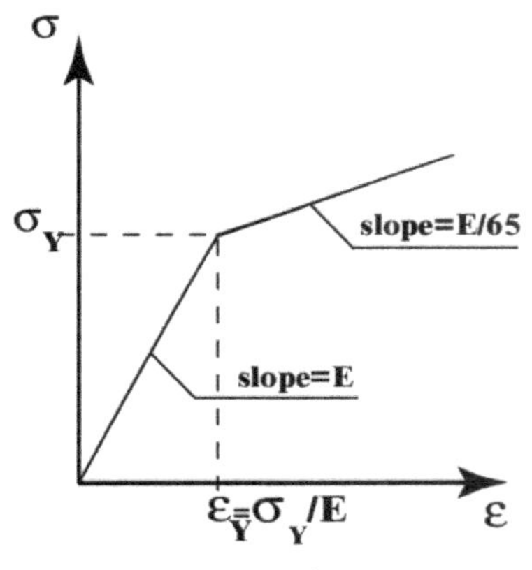

Figure 12. Assumed bi-linear behaviour for the material

4.4. Material properties

It is evident that strain-hardening effect has an important influence on the non-linear behaviour of isotropic and stiffened plates respectively. The degree of such an influence is a function of several factors including plate and stiffener slenderness. In this chapter, experimental material behaviour for both plate and stiffener are modelled as a bi- linear elastic-plastic with strain-hardening rate of $E/65$, as seen in Figure 12. E is modulus of elasticity of material. This value was obtained through an extensive study of elastic-plastic large deflection analyses made by Khedmati [17] and presents an average value for the strain-hardening rate. The application of $E/65$ predicts the collapse load with sufficient accuracy. Poisson's ratio, ν, in all experimental investigation and FEM analysis was considered to be equal to 0.3.

5. Large deflection behaviour of the tested plates

A summary of the results obtained through the finite element simulation of the experimental research carried out by Ghavami is given in Tables 5 and 6 and that of Tanaka & Endo is presented in Table 7. In these tables, the collapse modes from FEM analyses are, also, presented. A comparison of the experimental and those obtained results from FEM results present a very good agreement. The maximum differences varied between 16 percents and 22 percents for the series II and III (Table 5) of Ghavami's experimental result. These two extreme differences are related to the plates with L shape stiffener, which does not have a symmetrical geometrical shape. The simple assumption considered in the FEM simulation of complex pattern of initial imperfections (including both initial deflections and welding residual stresses) inherent in the experimental investigation, in addition to not having perfect simply supported boundary conditions in these two series must have led to those higher discrepancies. It should be emphasized that it was possible to trace the curve of average stress-average strain relationship for any combination of plate and stiffener. Finite element simulation results for Ghavami's test models without transverse frame show that the collapse has occurred following the buckling instability of local plate panels (Table 5). This was well predicted by FEM for test specimen P2R with only 5 percent difference. Detailed information concerning the behaviour of each of the Ghavami's test specimens are well documented in the References [18-22].

In the analysis of Tanaka & Endo's tests, the longitudinally stiffened plate located in the middle of the test specimens were simulated assuming all edges straight and having simply supported conditions. The same boundary conditions were considered in FEM analysis. In such cases, finite element simulation results, described also well the interactive buckling of plates and stiffeners in most of the cases, (Table 7). The smallest value of stiffener web height-to-web thickness ratio belongs to model D3, while the biggest value of this ratio corresponds to models D4 and D4A. Model D3 has failed due to local deformation in the plate, while in the case of models D4 and D4A the collapse has been produced by large plastic deformations both in the plate and stiffeners. Interactive buckling in both plate and stiffeners can be observed in other models, where the level of plastic deformations, in the plate varies among them. The ultimate strength predicted by FEM are well consistent as compared with those obtained by Tanaka & Endo [15]. This could be related to the initial deflection of the test specimens which was presented in FEM with a good accuracy.

A summary of results for three tests from each series of VI, V and VI that had perfect simply supported boundary is presented in Table 6. It can be noted that the difference between FEM and those of experimental results had only a difference of up to 5 percent. In the following, the results of FEM for P1R1T, P2R1T and P2L2T of the Ghavami's models with transverse frame are discussed in details.

Test model	$\frac{(\sigma_{ult})_{FEM}}{(\sigma_{ult})_{EXPERIMENT}}$	Collapse mode
P1R	1.11	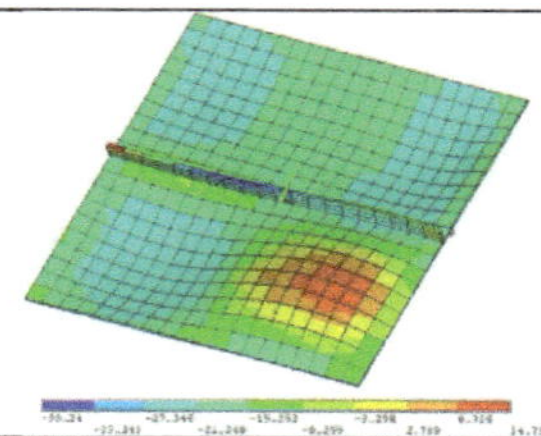
P1L	0.84	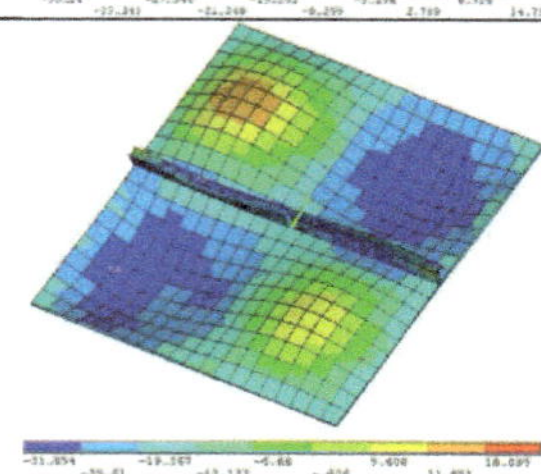
P1T	0.87	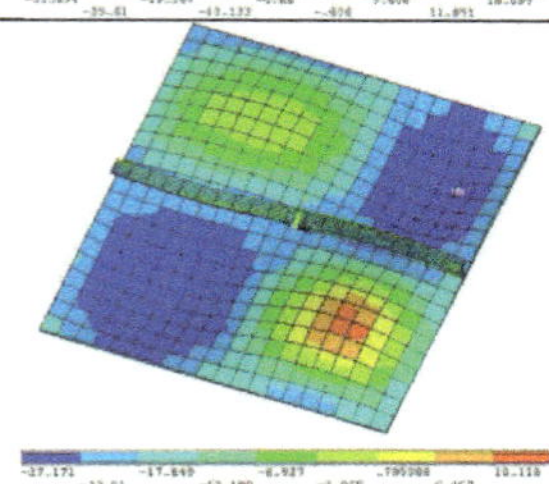
P2R	1.05	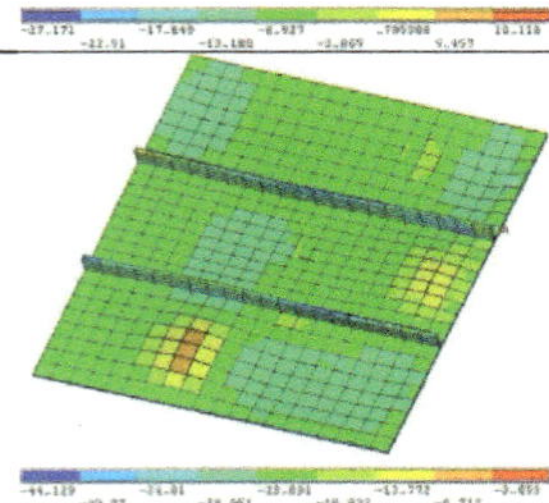
P2L	1.22	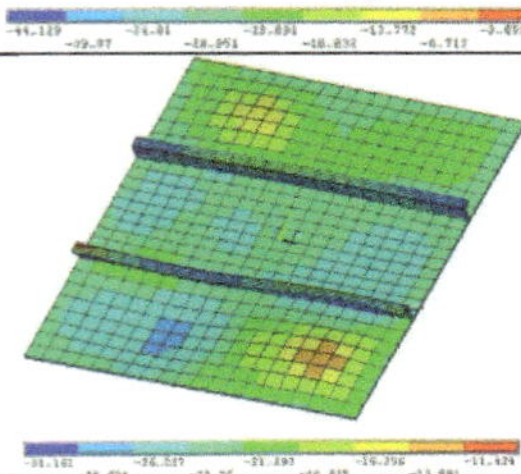

Table 5. Summary of finite element simulation results for some of Ghavami test models without transverse frame

Test model	$\frac{(\sigma_{ult})_{FEM}}{(\sigma_{ult})_{EXPERIMENT}}$	Collapse mode
P1R1T	1.05	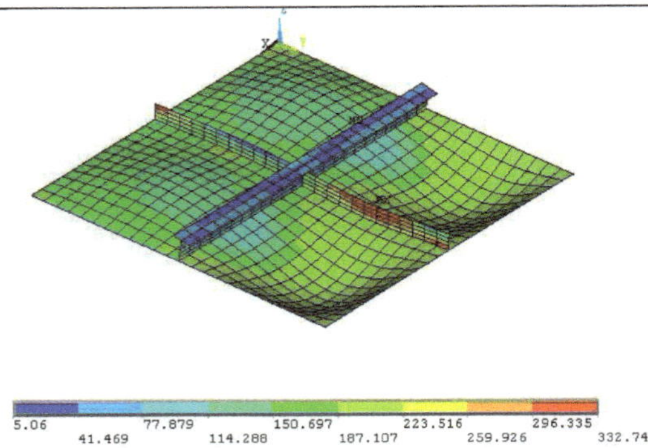
P2R1T	1.02	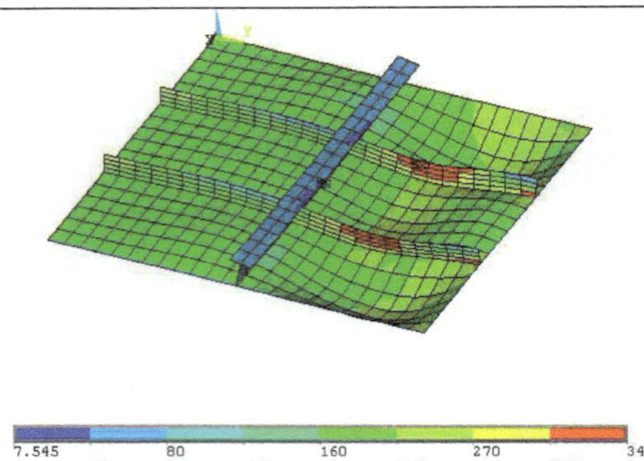
P2L2T	1.02	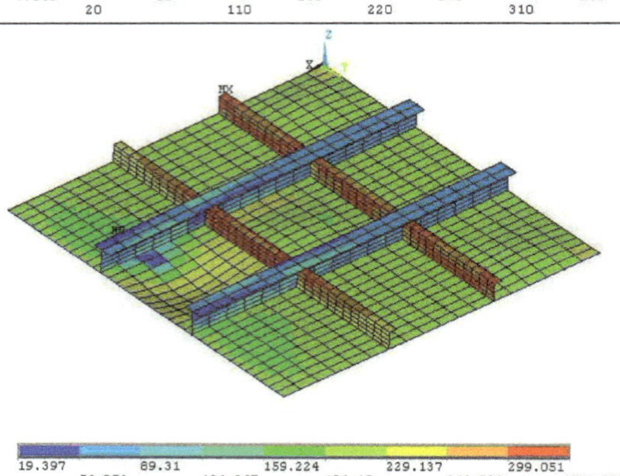

Table 6. Summary of finite element simulation results for some of Ghavami test models with transverse frame

Structure No.	Tanaka & Endo	Present	
	$\frac{(\sigma_{ult})_{FEM}}{(\sigma_{ult})_{EXPERIMENT}}$	$\frac{(\sigma_{ult})_{FEM}}{(\sigma_{ult})_{EXPERIMENT}}$	Collapse mode
D0	0.977	1.014	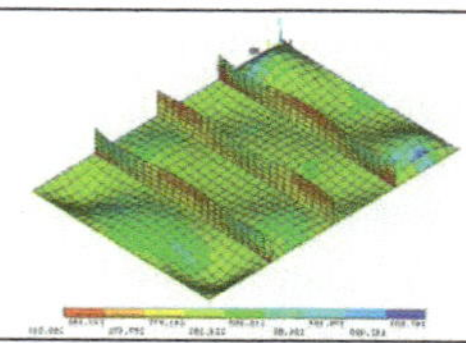
D0A	1.028	1.065	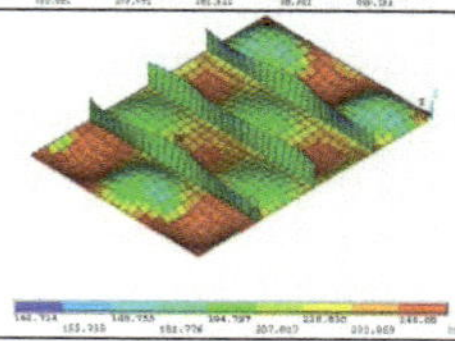
D1	0.869	0.911	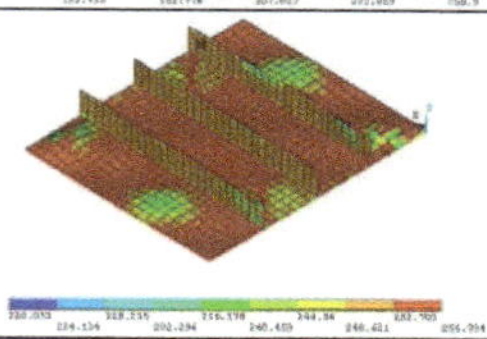
D2	0.936	0.944	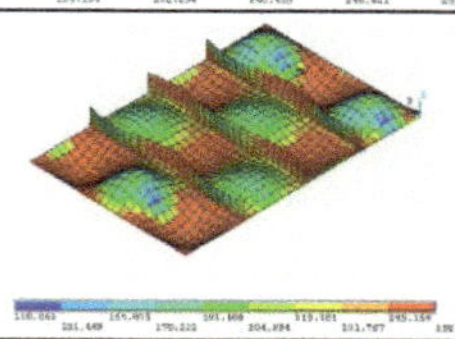
D3	0.860	0.853	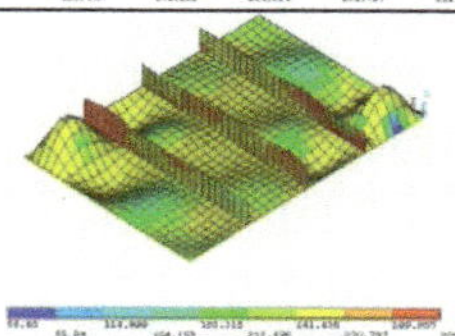
D4	0.792	0.866	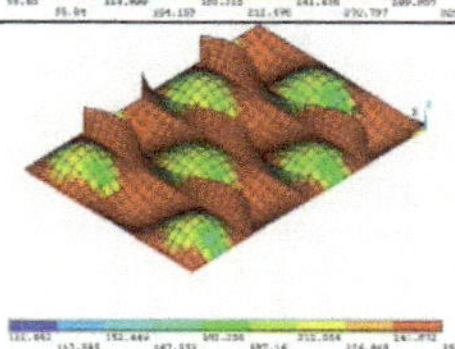
D4A	0.866	0.960	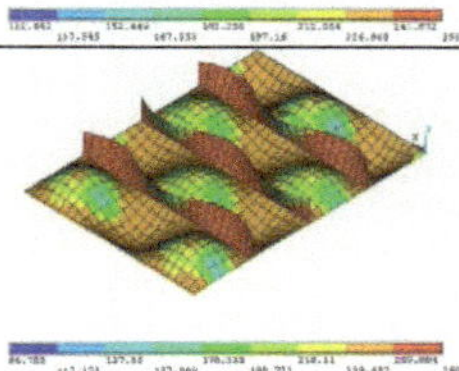

Table 7. Summary of results for some of Tanaka & Endo tests

5.1. P1R1T Ghavami model

The relative undimensional average stress-average strain relationship obtained by FEM analysis for P1R1T model is shown in Figure 13. The P1R1T model failed because of torsional buckling and plastic failure mechanism of the longitudinal stiffener (R). The torsional failure of the stiffener is induced in the FEM model shortly before the collapse of the model due to work softening as can be seen in Figure 13. A comparison between the collapse modes of the experimental model, Figure 14 (left) and that of FEM analysis, Figure 14 (right) is presented. It can be observed that the simulation of plate deformations by FEM analysis is almost identical to the failure mode occurred in the test specimen. The work hardening of the model started at about $\varepsilon/\varepsilon_y$ =0.8 and reached the ultimate buckling stress at $\varepsilon/\varepsilon_y$ =1.0 (ε, ε_y is the average strain and the yield strain respectively). The ultimate buckling strength of this model is about 80 percent of the plate yield strength, as can be seen in Figure 13 in turn it is close to the experimental results presented in Table 3. The FEM result overestimated the experimental one by only 5 percent. This mainly could be related to the discrepancy in the consideration of initial welding and initial deflection in REM analysis.

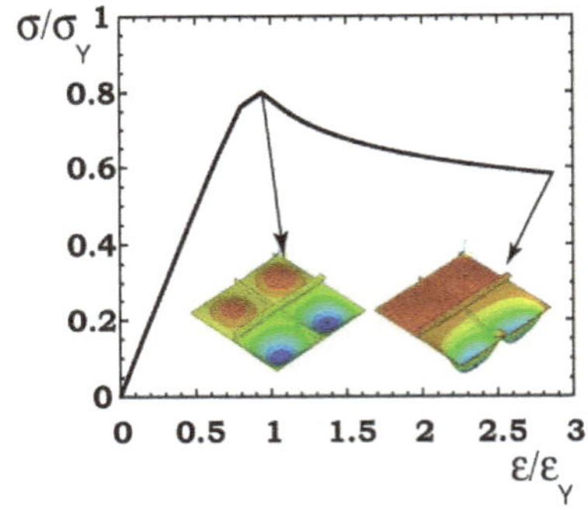

Figure 13. Average stress-average strain relationship and spread of yielding at collapse and final step of calculation for Ghavami P1R1T model

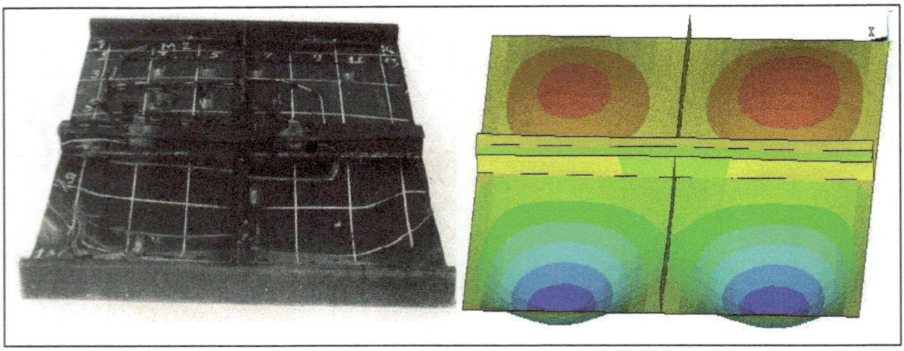

Figure 14. Deflected mode at collapse for Ghavami P1R1T model obtained by experiment (left) and FEA (right)

5.2. P2R1T Ghavami model

As it can be seen from the relative average stress-average strain relationship of P2R1T model (Figure 15), the work hardening of the test model started at about $\varepsilon/\varepsilon_Y$ =0.8 and reached the ultimate buckling stress at $\varepsilon/\varepsilon_Y$ =0,93 percent in relation to the plate material yield strength. Then the work softening or unloading started at $\varepsilon/\varepsilon_Y$ =1.0 together with the local plastic deformations in the post-ultimate buckling region. The P2R1T model failed under axial compression load due to the buckling in both plate and stiffeners. Such a failure was predominant in upper part of the transverse T frame, as can be observed in Figure 16 (left). In the lower part of the transverse T stiffener, the plastic deformation in the plate and stiffeners was not very large. The comparison of FEM results with that of the experimental one, presented in Figure 16 present a relatively perfect prediction of the ultimate buckling modes. The FEM result overestimated the experimental one by only 2 percent. This can be also related mainly to consideration of the initial welding and initial deflection in the FEM analysis.

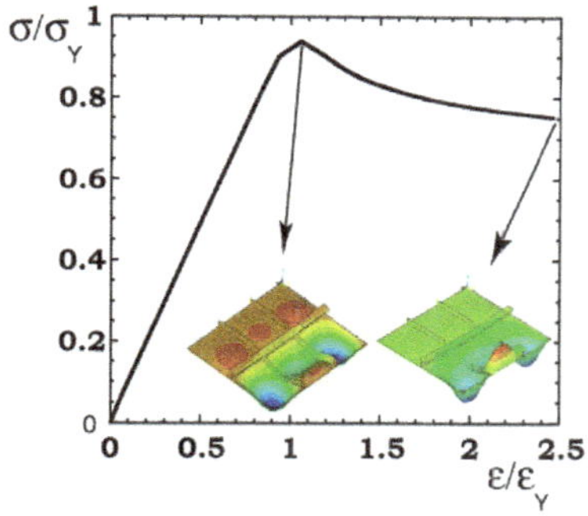

Figure 15. Average stress-average strain relationship and spread of yielding at collapse and final step of calculation for Ghavami P2R1T model

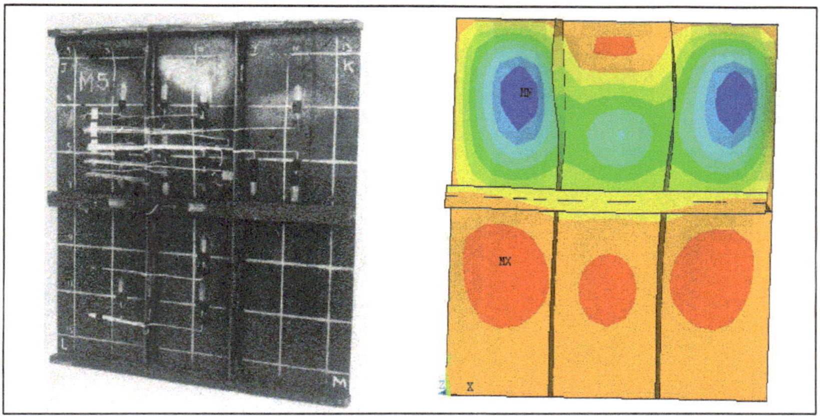

Figure 16. Deflected mode at collapse for Ghavami P2R1T model obtained by experiment (left) and FEA (right)

5.3. P2L2T Ghavami model

As it can be seen in Figure 17 which presents the relative average stress-average strain, relationship of P2R2T model, a small work hardening started at about $\varepsilon/\varepsilon_Y$=0.88 of the plate yield stress and reached the ultimate buckling stress of 100 percent. Then a plastic deformation started at the $\varepsilon/\varepsilon_Y$ =1.0 up to $\varepsilon/\varepsilon_Y$ =1.7 generating several local plate. After this stage the work softening or unloading started with the expansion of local plastic deformations in the post-ultimate buckling region. The P2R2T model finally failed due to the buckling induced in both plate and longitudinal L stiffeners in the centre of the stiffened plate as can be noted well in Figure 18 (left). The P2L2T model showed a high strength under in-plane compression load. The FEM deflected form in Figure 18 (right) simulated well the experimental results. The FEM result overestimated the experimental one by only 2 percent as can be seen in Table 6. This could be related principally to the initial welding and initial deflection.

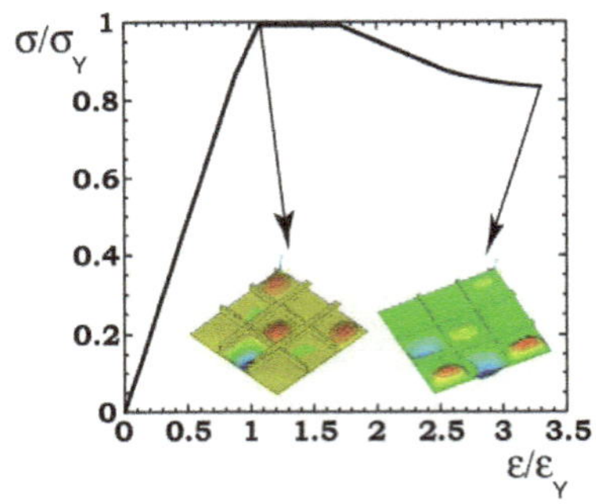

Figure 17. Average stress-average strain relationship and spread of yielding at collapse and final step of calculation for Ghavami P2L2T model

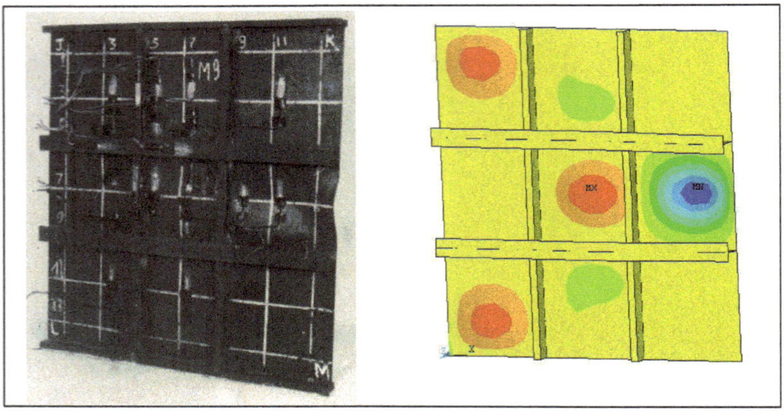

Figure 18. Deflected mode at collapse for Ghavami P2L2T model obtained by experiment (left) and FEA (right)

6. Large deflection behaviour of Stiffened plates subjected to combined in-plane compression and lateral pressure

For the stiffened plates in the bottom structure of ships, the basic load case for buckling design consists of the following loads applied simultaneously (Figure 19):

- longitudinal compression arising from the overall hull girder bending,
- transverse compression arising from the bending of double bottom under lateral pressure, and
- local bending arising from the direct action of lateral pressure.

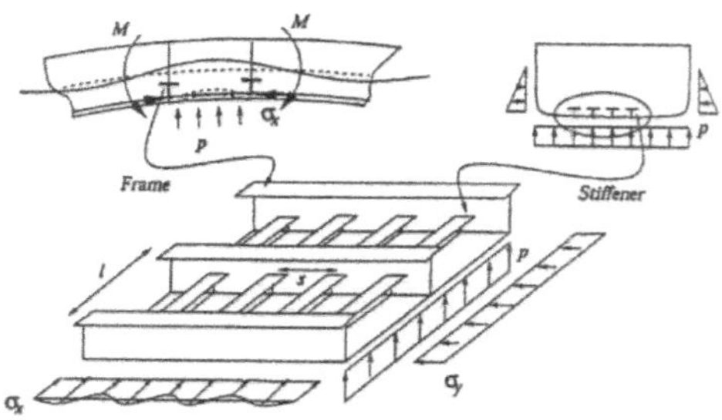

Figure 19. Basic loads applied on ship stiffened plates

The continuous plate was assumed to be simply-supported along the stiffener lines with no out-of-plane deflection. In reality, however, the stiffener is also subjected to lateral pressure, and it may collapse prior to the failure of the panels. The focus of the present chapter is concentrated on the buckling and plastic collapse behaviour of continuously stiffened plates subjected to combined biaxial compression and lateral pressure with the main objective of identification of the collapse modes of the plates subjected to mentioned combination of loading condition.

A series of elasto-plastic large deflection FEM analyses is performed on continuous stiffened plates with flat-bar, tee-bar, and angle-bar stiffeners of the same flexural rigidity. The buckling/plastic collapse behaviour and ultimate strength of stiffened plates are hereby assessed so that both the material and geometrical nonlinearities are taken into account.

Local plate panels with length, a, of 2400 mm and breadth, b, of 800 mm are considered, and their thickness, t, changes from 13mm, 15mm, and 20 mm. Yield stress of the material, σ_Y, is taken as 313.6 MPa, and bilinear stress-strain relationship is assumed with the kinematical strain-hardening rate of $E/65$, where E is Young's modulus of the material. E is considered as 205.8GPa. The cross-sectional geometries of stiffeners are given in Table 8. In each group, the stiffeners have the same moment of inertia. A triple span-double bay model is applied for the analysis of buckling/plastic collapse behaviour of continuous stiffened plate with symmetrical stiffeners (ABDC in Figure 20). When a stiffener has an unsymmetrical geometry, a triple span-triple bay model is used (ABFE in Figure 20) [23].

Type	Model	shape	h_w	t_w	b_f	t_f	h_w/t_w
1	F1	flat	150	17	-	-	8.82
1	T1	tee	150	9	90	12	16.67
1	A1	angle	150	9	90	12	16.67
2	F2	flat	250	19	-	-	13.16
2	T2	tee	250	10	90	15	25.00
2	A2	angle	250	10	90	15	25.00
3	F3	flat	350	35	-	-	10.00
3	T3	tee	400	12	100	17	33.33
3	A3	angle	400	12	100	17	33.33

h_w: Height of stiffener web (mm)
t_w: Thickness of stiffener web (mm)
b_f: Breadth of stiffener flange (mm)
t_f: Thickness of stiffener flange (mm)

Table 8. Cross-sectional geometries of stiffeners

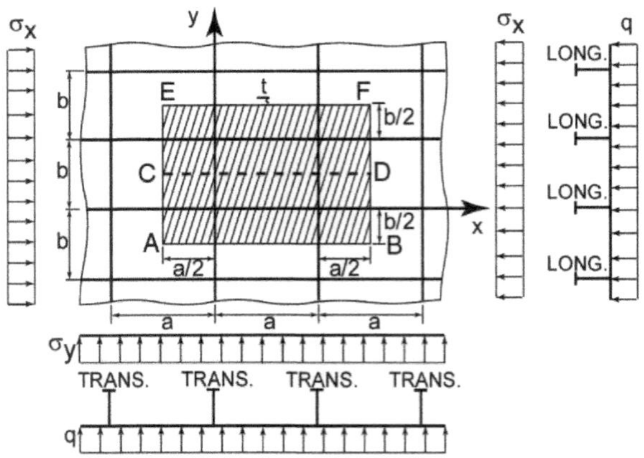

Figure 20. Stiffened plate model for FEM analysis

The considered boundary conditions are as follows:

- Periodically continuous conditions are imposed at the same y-coordinate along the transverse edges (i.e. along AC and BD in double bay model and along AE and BF in triple bay model).
- Symmetry conditions are imposed along the longitudinal edges of double bay model (i.e. along AB and CD). But periodically continuous conditions are defined at the same x-coordinate along the longitudinal edges of triple bay model (i.e. along AB and EF).
- Although transverse frames are not modelled, the out-of-plane deformations of plate and stiffener are restrained along the junction lines of them and the transverse frame.
- To consider the plate continuity, in-plane movement of the plate edges in their perpendicular directions is assumed to be uniform.

The lateral pressure ranging from 0 to 60 metres water head initially is applied up to a specified value always perpendicularly to the plate surface. Then biaxial compression is exerted proportionally by uniform forced displacements.

Three types of initial imperfections as described in the following are accounted for:

- initial deflection in the plate with the maximum magnitude of $t/100$ (Figure 21(a)):

$$W_{p0} = \frac{t}{100}\sin\frac{m\pi x}{a}\sin\frac{\pi y}{b} \tag{3}$$

where m is the number of buckling half-waves in the plate,

- initial deflection in the stiffener with the maximum magnitude of $a/1000$ (Figure 21(b)):

$$W_{s0} = \frac{a}{1000}\sin\frac{\pi x}{a} \tag{4}$$

- and angular distortion of the stiffener which is taken as (Figure 21(c)):

$$\phi_0 h_w = \frac{a}{1000}\sin\frac{\pi x}{a} \tag{5}$$

The welding residual stresses are not considered.

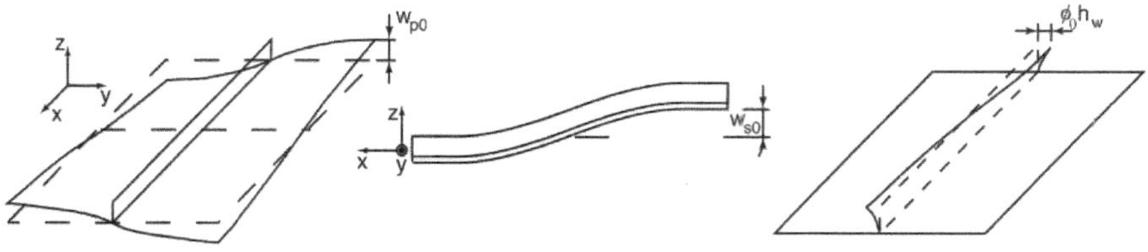

(a) Initial deflection in the plate (b) Initial deflection in the stiffener (c) Angular distortion of the stiffener

Figure 21. Initial imperfections in the stiffened plate models

6.1. Plates with flat-bar stiffener subjected to combined longitudinal compression and lateral pressure

Average stress-average strain relationships for continuous stiffened plates with flat-bar stiffeners subjected to combined longitudinal compression and variable levels of lateral pressure, are shown in Figure 22 for the plate thickness of t =13 mm. The deflection mode and spread of yielding at ultimate strength are presented in Figure 23.

The characteristics of the collapse behaviour can be summarised as follows:

- When there is no lateral pressure (water head, h =0 m), the stiffened plate under longitudinal compression collapses in Eulerian buckling mode, preceded by the local buckling of plate with three buckling half waves.

- With increase in lateral pressure, the deflection mode at the ultimate strength changes from the Eulerian buckling mode to a both-ends clamped mode, and the tripping deformation of stiffener gets decreased.
- Under very high lateral pressure, the stiffener web is fully yielded at both ends of each span, and subsequently it is deflected entirely to opposite sideward directions in neighbouring spans. Therefore, a kind of simply-supported flexural-torsional deformation is produced in the stiffener web.
- With an increase in the flexural rigidity of the stiffener, ultimate strength of the stiffened plate is increased with a decrease in the post-ultimate strength.

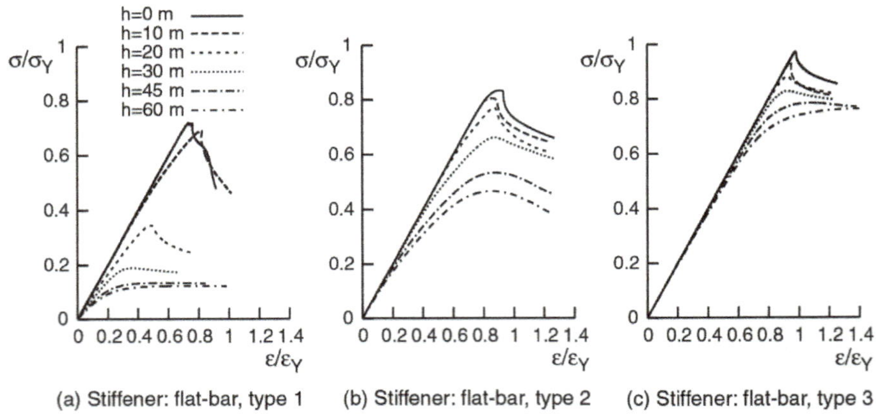

Figure 22. Comparison of average stress-average strain relationships for a continuous stiffened plate under combined longitudinal thrust and lateral pressure (plate: 2400x800x13 mm)

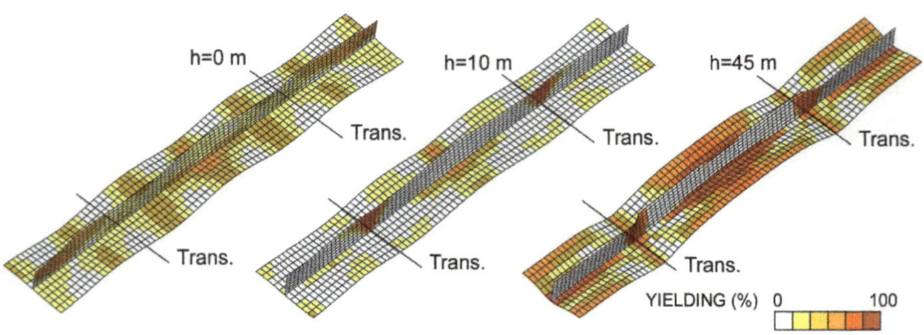

Figure 23. Change in the deflection mode at ultimate strength for a continuous stiffened plate under combined longitudinal thrust and lateral pressure (plate: 2400x800x13 mm, stiffener: flat-bar of type 2)

For plates with flat-bar stiffeners of type 1 having smaller flexural rigidity, as the plate thickness is increased, the ultimate strength is increased with the increase of lateral pressure up to a certain value. This is because the collapse mode changes from Eulerian buckling mode to a clamped mode in which the plate itself exhibits a higher resistance to longitudinal compression. With a further increase in the applied lateral pressure, however, the deteriorating effect of lateral pressure, i.e. enhancing yielding at stiffener becomes more predominant and the ultimate strength starts to decrease considerably.

6.2. Plates with tee-bar stiffener subjected to combined longitudinal compression and lateral pressure

Average stress-average strain relationships for continuous stiffened plates with tee-bar stiffeners of type 2 subjected to combined longitudinal compression and variable levels of lateral pressure, are shown in Figure 24(a) for the plate thickness of t =13 mm. Fundamental collapse behaviours and ultimate strength of stiffened plates with tee-bar stiffeners are almost the same as those for the flat-bar stiffener, but strength reduction in the post-ultimate range is smaller comparing with Figure 22(b). This is because the horizontal bending rigidity of tee-bar is much greater than that of flat-bar.

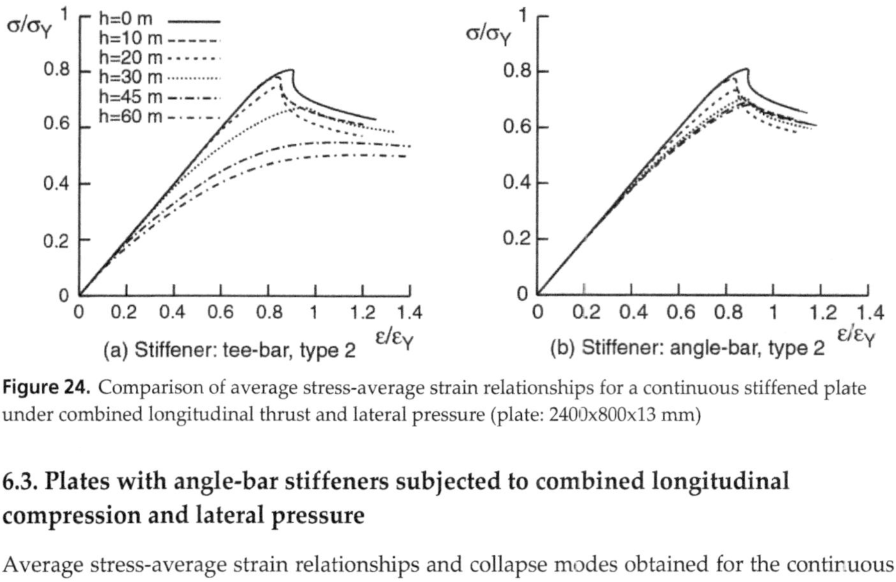

Figure 24. Comparison of average stress-average strain relationships for a continuous stiffened plate under combined longitudinal thrust and lateral pressure (plate: 2400x800x13 mm)

6.3. Plates with angle-bar stiffeners subjected to combined longitudinal compression and lateral pressure

Average stress-average strain relationships and collapse modes obtained for the continuous stiffened plates with angle-bar stiffeners are shown in Figure 24(b) and Figure 25, respectively, for the plate thickness of t =13 mm.

Unlike the flat-bar or tee-bar stiffeners having symmetrical cross-sectional shape, the angle-bar stiffener deflects to the same horizontal and vertical directions in all adjacent spans (Figure 25). This flexural-torsional deflection of stiffener clamped at both ends constrains the panel deformation, resulting in larger ultimate strength and smaller strength reduction in the post-ultimate range than those for flat-bar or tee-bar stiffeners.

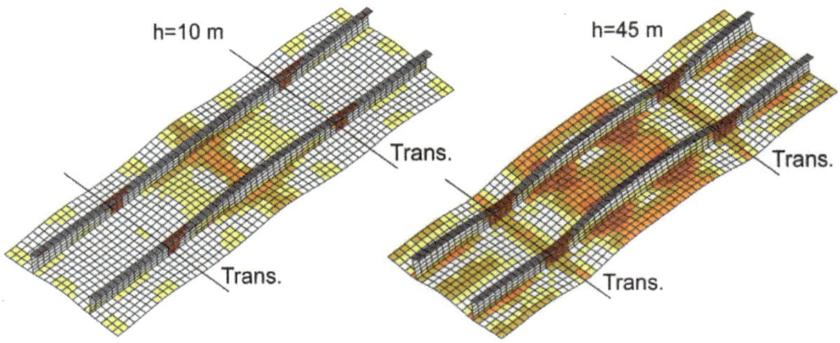

Figure 25. Change in the deflection mode at ultimate strength for a continuous stiffened plate under combined longitudinal thrust and lateral pressure (plate: 2400x800x13 mm, stiffener: angle-bar of type 2)

It is to be noted here that although an angle-bar stiffener is quite effective from the viewpoint of buckling/plastic collapse strength, it should be carefully used from the view point of fatigue strength [24].

6.4. Stiffened plates subjected to combined transverse compression and lateral pressure

The results for the continuous stiffened plates with flat-bar stiffeners of type 2 subjected to combined transverse compression are shown in Figs. 26 and 27.

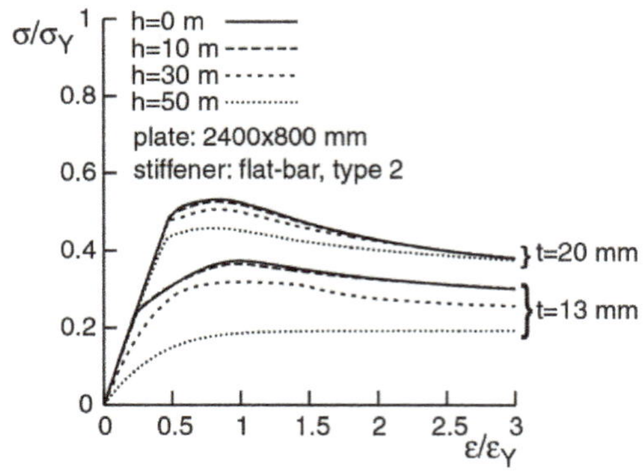

Figure 26. Comparison of average stress-average strain relationships for a continuous stiffened plate under combined transverse thrust and lateral pressure

When lateral pressure is small, the local rectangular panels collapse as if they were simply-supported along the edges, accompanied by some rotation of stiffeners. With an increase in lateral pressure, the collapse mode changes from the simply-supported mode to the all-edges clamped mode. These behaviours are basically the same as those observed for continuous plate simply-supported along stiffener lines. Since the stiffener is not subjected to compression, its deflection is small compared to the panel deflection.

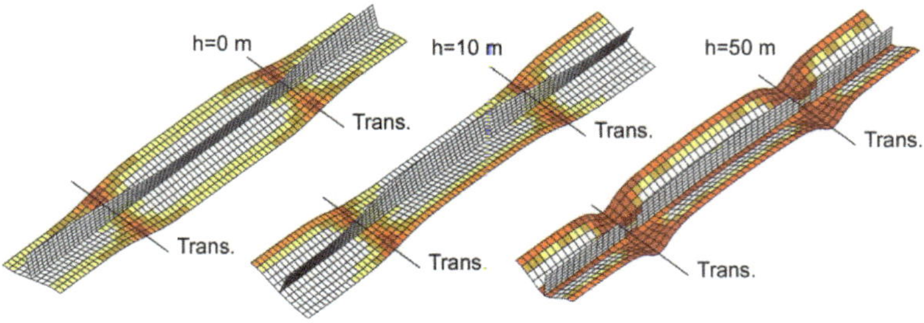

Figure 27. Change in the deflection mode at ultimate strength for a continuous stiffened plate under combined transverse thrust and lateral pressure (plate: 2400x800x13 mm, stiffener: flat-bar of type 2)

6.5. Stiffened plates subjected to combined biaxial compression and lateral pressure

A series of FEM analyses is performed on a continuous stiffened plate with flat-bar stiffeners subjected to combined biaxial compression and lateral pressure. The results are shown in Figure 28. The dotted lines are loading paths for different ratios of applied biaxial displacements. The solid line is the obtained envelope of all loading paths representing the ultimate strength interaction curve.

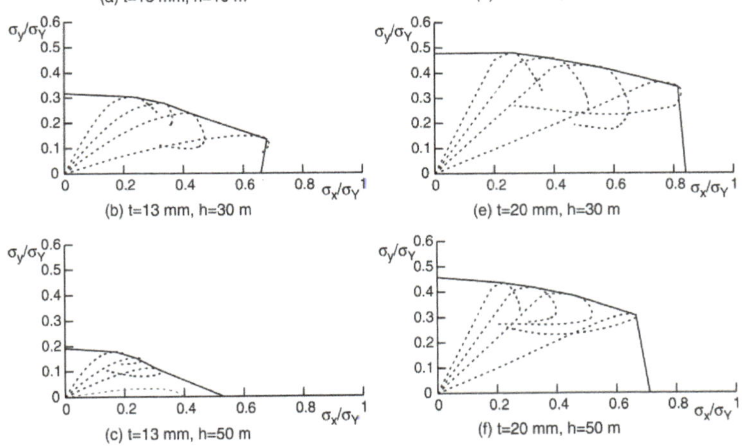

Figure 28. Interaction curves for a continuous stiffened plate subjected to combined biaxial thrust and lateral pressure (plate: 2400x800 mm, stiffener: flat-bar of type 2)

It is seen that each interaction curve basically consists of two parts; a semi-horizontal region in which the stiffened plate behaves as if it were under combined transverse compression and lateral pressure, and a semi-vertical region where the behaviour as in the case of combined longitudinal compression and lateral pressure is dominant.

7. Conclusion

Basing the results of this chapter on the analysis of 29 experimental investigation, on stiffened steel plates subjected to uniform axial compression load up to final failure, by the Finite Element program ANSYS, the following conclusions may be drawn. The selected element SHELL43, could trace full-range, elastic-plastic behaviour of the stiffened plates. The capability of the non-linear FEM to perform the analysis of stiffened plates has been demonstrated through the accurate simulation of the Ghavami and Tanaka & Endo tests. Although some simplifying assumptions for the simulation of initial imperfections and residual welding stresses were made for reducing the calculation volume and speeding up the analysis, the accuracy of the collapse load obtained through FEM simulations is relatively in good consistency with the test results. The differences were higher in cases of not having perfect simply supported boundary conditions as in series II and III of Ghavami's test. It was shown also, that obtaining deflection mode is possible at any step of loading. This allows predicting the local buckling of stiffened plates with relatively good precision.

For small value of lateral pressure, the local panel and stiffener tend to collapse in a simply-supported mode. With an increase in the applied pressure, they are likely to fail in a clamped mode. Angle-bar stiffener has larger stiffening effects than those of flat-bar and tee-bar stiffeners having the same flexural rigidity, from the view point of ultimate strength.

Author details

Khosrow Ghavami
Department of Civil Engineering, Pontificia Universidade Católica (PUC-Rio), Rio de Janeiro, Brazil

Mohammad Reza Khedmati*
Faculty of Marine Technology, Amirkabir University of Technology, Tehran, Iran

Acknowledgement

The authors would like to thank the sponsoring organizations for their financial supports and special thanks are also due to the students who executed the projects through the years.

8. References

[1] Merrison Committee (1973) Inquiry into Basis of Design and Methods of Erection of Steel Box Girder Bridges. Report of the Committee – Appendix 1: Interim Design and Workmanship Rules. Her Majesty's Stationary Office, London.

* Corresponding Author

[2] Crisfield MA (1975) Full-range Analysis of Steel Plates Stiffened Plating under Uniaxial Compression. Proc. Civ. Engrs. 59(2): 595-624.
[3] Frieze PA, Dowling PJ, Hobbs RE (1976) Ultimate Load Behaviour of Plates in Compression. Proc. International Conference on Steel Plated Structures, Imperial College, London.
[4] Brosowski B, Ghavami K. (1996) Multi-Criteria Optimal Design of Stiffened Plates. Part I, Choice of the Formula for Multi-Criteria for Buckling Load. Thin Walled Structures. 24(9): 353-369.
[5] Brosowski B, Ghavami K. (1997) Multi-Criteria Optimal Design of Stiffened Plates. Part II, Mathematical Modelling of the Optimal Design of Longitudinally Stiffened Plates. Thin Walled Structures. 28(2): 179-198.
[6] Murray NW. (1975) Analysis and Design of Stiffened Plates Collapse Load. The Structural Engineer. 53(3): 153-158.
[7] Chen Q, Zimmerman TJE, DeGeer D, Kennedy BW. (1997) Strength And Stability Testing of Stiffened Plate Components. Ship Structural Committee Report. SSC-399.
[8] Hu SZ, Jiang L. (1998) A Finite Element Simulation of the Test Procedure of Stiffened Plates. Journal of Marine Structures. 11: 75-99.
[9] ADINA-IN for ADINA Users Manual. (1990) Report ARD 90-4, ADINA R&D, Inc.
[10] Vibration and Strength Analysis Program (VAST) User's Manual, Version 7.2. (1996) Martec Limited, Halifax, Nova Scotia, Canada, 1996.
[11] Grondin GY, Elwi AE, Cheng JJR. (1999) Buckling of stiffened plates- a parametric study. Journal of Constructional Steel Research. 50: 151-175.
[12] Hibbitt et al. ABAQUS/Standard, Version 5.4. (1994) Hibbitt, Karlsson and Sorensen, Inc.: Pawtucket, RI.
[13] Sheikh IA, Elwi AE, Grondin GY. (2003) Stiffened Steel Plates under Combined Compression and Bending. Journal of Constructional Steel Research. 59: 911-930.
[14] Ghavami K. (1994) Experimental Study of Stiffened Plates in Compression up to Collapse. Journal of Constructional Steel Research. 28(2): 197-222 [Special Brazilian Issue, Guest Editor Khosrow Ghavami]
[15] Tanaka Y, Endo H. (1988) Ultimate Strength of Stiffened Plates with Their Stiffeners Locally Buckled in Compression. Journal of the Society of Naval Architects of Japan. 164 (in Japanese).
[16] ANSYS User's Manual (version 7.1). (2003) Swanson Analysis Systems Inc. Houston,.
[17] Khedmati MR. (2000) Ultimate Strength of Ship Structural Members and Systems Considering Local Pressure Loads. Dr. Eng. Thesis, Graduate School of Engineering, Hiroshima University.
[18] Ghavami K, Conci A, Rocha SAS. (1983) Resistencia Compressao De Placas De Aco Enrijecidas Longitudinalmente. In Proc. VII Congress Brasileiro de Engenharia Mecanica-COBEM 83, Uberlandia, Brasil: 339-349.
[19] Ghavami K, Conci A, Rocha SAS. (1983) Metodos De Calculo De Placas Enrijecidas Sob Carregamento De Compressao Axial. In Proc. VII Congress Brasileiro de Engenharia Mecanica-COBEM 83, Uberlandia, Brasil: 351-361.

[20] Conci A. (1983) Instabilidade Ate O Colapso De Placas De Aco Enrijecidas Em Dual Direcoes. MSc thesis, Pontificia Universidade Catolica, Rio de Janeiro.

[21] Rocha SAS. (1982) Comportamento Ultimo De Placas Enrijecidas. MSc thesis, Pontificia Universidade Catolica, Rio de Janeiro.

[22] Ghavami K. (1986) The Collapse of Continuously Welded Stiffened Plates Subjected to Uniaxil Compression Load. In Proc. Inelastic Behaviour of Plates and Shells, Simp. Rio de Janeiro, 1985, eds. L. Bevilacqua, R. Feijoo & R. Valid, Springer Berlin: 404-415.

[23] Yao T, Fujikubo M, Yanagihara D, Irisawa M. (1998) Consideration on FEM Modelling for Buckling/Plastic Collapse Analysis of Stiffened Plates. Trans. of the West-Japan Soc. Naval Arch. 85:121-128 (in Japanese).

[24] Kawano H, Kuramoto Y, Sakai D, Hashimoto K, Inoue S, Fushimi A, Hagiwara K. (1992) Some Considerations on Basic Behaviour of Asymmetric Sectional Frame Under Uniform Pressure. Trans. of the West-Japan Soc. Naval Arch. 83:161-166 (in Japanese).

Chapter 6

3D Nonlinear Finite Element Plastic Analysis of Cylindrical Vessels Under In-Plane Moment

B.H. Wu

Additional information is available at the end of the chapter

1. Introduction

Cylindrical vessels with nozzles are common structural components in many industries, such as power engineering, petrochemical, etc. Under the applied pressure and piping loads, a high stress concentration is caused by the geometric discontinuity. The connection region of the vessel and nozzle will become the weakest location of the entire structure. Therefore, it is necessary to have an accurate design method for the structure. The plastic limit design method is one possibility. In design, a gross plastic deformation is prevented by restricting the allowable load relative to the plastic limit load of the vessels. Therefore, the key step in limit design is to determine the plastic limit load of the vessels under different loads.

The plastic limit load estimate for practical engineering materials (with strain hardening and geometrical strengthening) can be determined by the twice-elastic-slope (TES) criterion [1]. Many approaches to determine the plastic limit load have been contributed by a number of authors employing analytical, experimental and finite element methods for components under internal pressure, and nozzle or branch pipe loading. Ellyin[2] [3] reported experimental results for the elastic-plastic behavior and plastic limit loads of five tee-shaped cylinder-cylinder intersections under internal pressure, and in-plane or out-of-the-plane moment. The results indicated that the out-of-plane loading case was the critical one. Schoreder[4] provided experimental limit branch moment loads on 4-in. ANSI B16.9 Tees using different limit load criteria. The results of the study showed that the branch moment capacity for the tee models was greater than the theoretical limit load of the equivalent nominal size straight pipe. Junker[5] performed inelastic finite element analyses to estimate the limit moment for a cylindrical vessel with a nozzle subjected to in-plane and out-of-plane moment loadings. The results indicated that the predicted limit moment levels agree to within 10% with the experimental results and that the finite element method gives a reasonably accurate determination of limit moments for cylindrical vessels with nozzles under in-plane and out-of-plane moments. Moffat[6] performed an experimental study of branch connections subjected to external moment loadings. Further, in 1991, Moffat et al.[7] provided extensive numerical results for the effective stress factor of branch junctions under internal pressure and external moment loads. An empirical formula was presented using polynomial equations. Rodabaugh[8] [9] contributed a valuable review of limit loads for pipe connections in pressure vessels and piping. In the review, a comprehensive overview of pipe connections was provided. Other studies [10]-[13] carried out important works on plastic limit analysis of cylindrical vessels under external nozzle loadings.

The objective of this paper is to determine the plastic limit moment by both experiment and finite element analysis for cylindrical vessels under in-plane moment loading on the nozzle. Based on these results, a parametric analysis is carried out and an empirical formula is proposed.

2. Experimental study

2.1. Model vessels

Three model vessels with different d/D ratios were designed and fabricated for the experimental study. Every model vessel consisted of a cylinder, nozzle, and flanges for fixing and loading bars. Figure 1 shows a typical configuration, and dimensions for the model vessels are listed in Table 1.

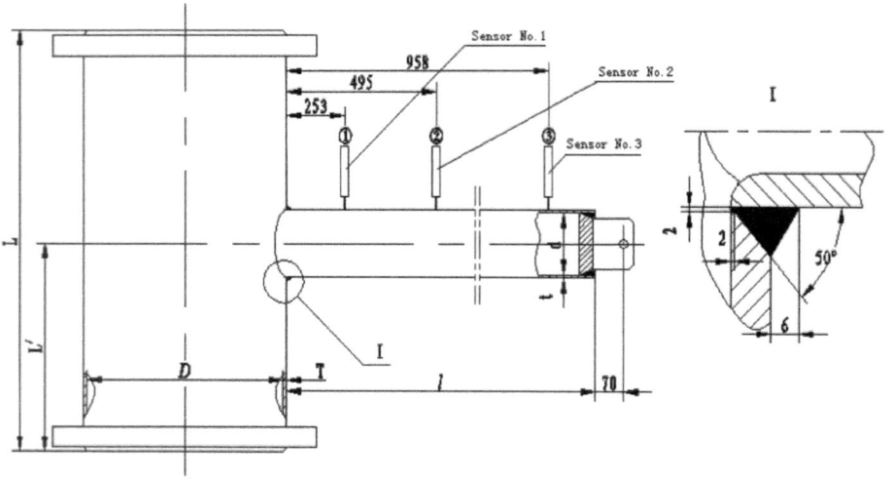

Figure 1. Arrangement of model vessels (mm)

	D(mm)	L(mm)	L' (mm)	T(mm)	d(mm)	l(mm)	t(mm)	d/ D	t /T	D / T
L_1	500	1000	500	8	86	1000	3	0.172	0.375	62.5
L_2	500	1000	500	8	123	1000	4	0.246	0.50	62.5
L_3	500	1000	500	8	214	1000	5	0.428	0.625	62.5

Table 1. Dimensions of model vessels

The materials of the cylinder and the nozzle are Q235-A (low carbon steel, similar to A36-77) and 20# (low carbon steel, similar to A106-80 GrA), respectively. Detailed chemical composition and mechanical properties of materials are given in Reference [14]. Figure 2 shows the engineering stress-strain curve of materials for Q235-A and 20# Steel.

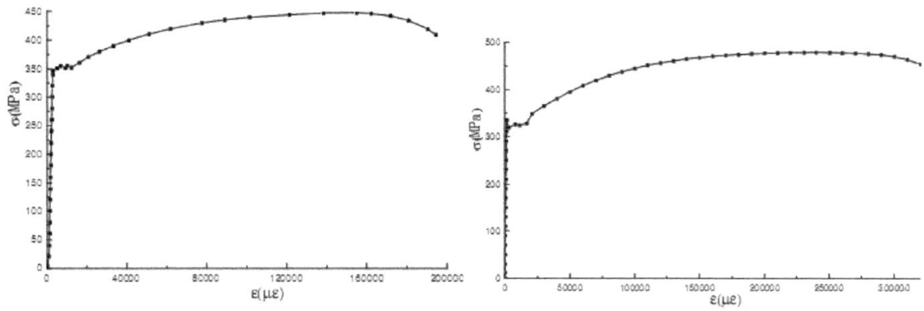

(a) Q235-A (b) $20^{\#}$

Figure 2. The curve of engineering stress-strain

2.2. Experimental setup

The local strains and nozzle displacements under in-plane moment on nozzle were measured to obtain the elastic stresses and deformation characteristics of the model vessels.The in-plane moment is applied as a force at the end of the nozzle, and the moment arm is the distance to the surface of the vessel.

Strain values at the typical measurement points of the cylinders and nozzles were measured using two element strain rosettes. Strain gages were mounted in the axial direction on the outside and inside surfaces for the cylinders, and on the outside surface only for the nozzles. Figure 3 illustrates the detailed locations of the strain gages for test vessel No. L_2.

Displacement values under in-plane moment (longitudinal loading direction) for the selected measurement points on the nozzle were measured using mechanical displacement sensors to obtain the load-displacement relationship in the elastic and plastic stages of the nozzles. A total of three displacement sensors were installed at locations along the nozzle length direction. The locations of the displacement sensors are indicated in Figure1.

Figure 4 is a photograph of test vessel No. L_1 during the test. The figure shows an obvious longitudinal deformation of the nozzle under the in-plane moment on the nozzle.

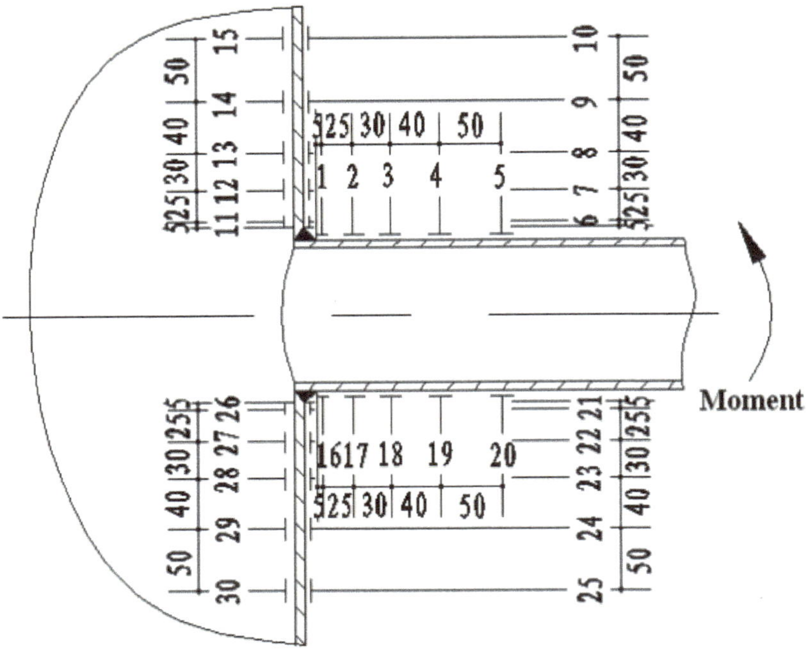

Figure 3. Locations of steain gages for model No. L2

Figure 4. Photo of model No L1 during the test

2.3. Experimental results

2.3.1. Plastic deformation behavior

Figure 5 indicates the elastic-plastic load deformation response at the three measurement points on the nozzle for model vessel No. L_2.

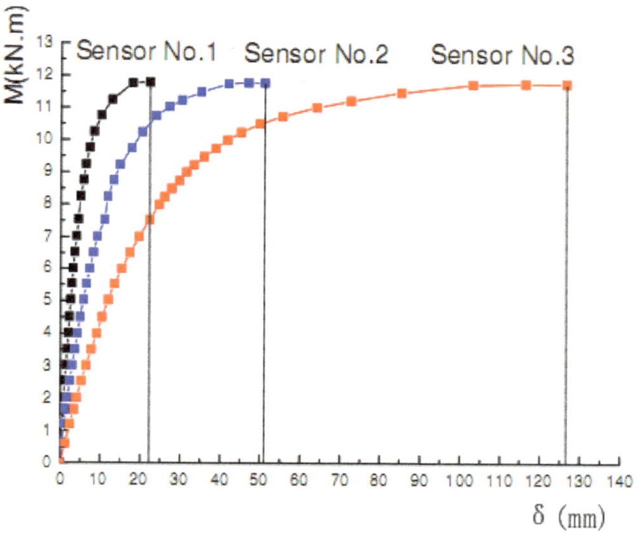

Figure 5. Load and elastic-plastic deformation response of nozzle

The intersection area of the nozzle and cylinder also produced an evident plastic deformation: the root of the nozzle moved into the vessel wall in the radial direction; as the cylinder yielded on the compression side of the cylinder and nozzle. Figure 6 shows an actual deformation state for test vessels No.$L_1$and No.L_2 under in-plane moment of 4.2kN.m and 7.5kN.m.

2.3.2. Plastic limit moment

Experimental plastic limit moments were obtained by the use of load versus displacement plots of the measurement points on the nozzle and load against strain curves of the key gauges located near the junction of the cylinder and nozzle. The plastic limit load is defined by applying the twice elastic slope criterion provided by the ASME Boiler and Pressure Vessel Code[1].

Figure 7 illustrates some typical load-displacement curves and corresponding limit moment for the sensor No.3.

Figure 8 shows some typical load-strain curves and relevant limit moment of the strain gage No.6 as an example for the experiment vessels.

(a) vessel No.L_1

(b) vessel No.L_2

Figure 6. Load-deformation of the test vessels

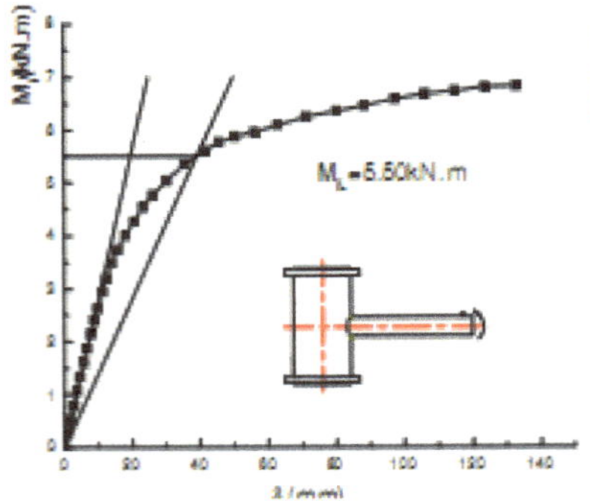

(a) Load-displacement curves for L_1 vessel

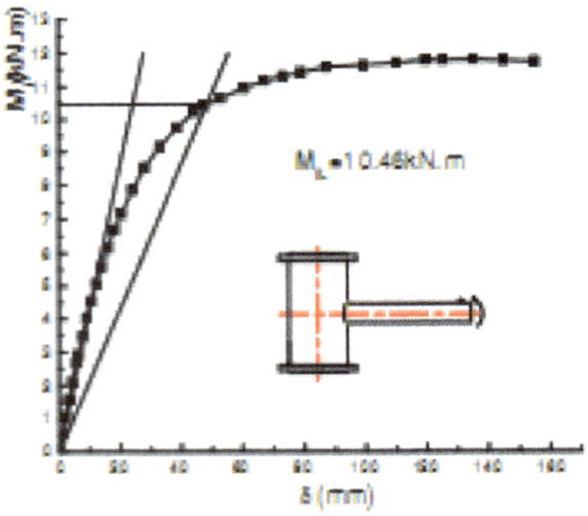

(b) Load-displacement curves for L_2 vessel

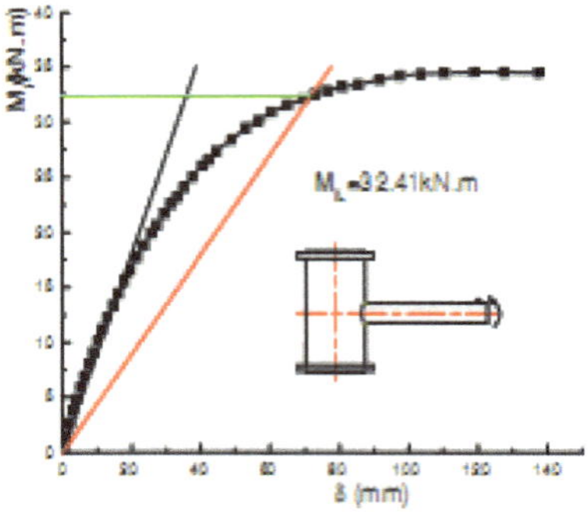

(c) Load-displacement curves for L_3 vessel

Figure 7. Load-displacement curves for test models (Sensor No.3)

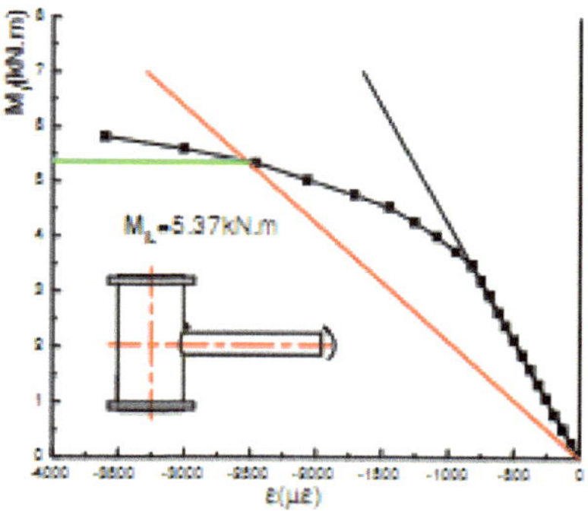

(a) Load-strain curves for L_1 vessel

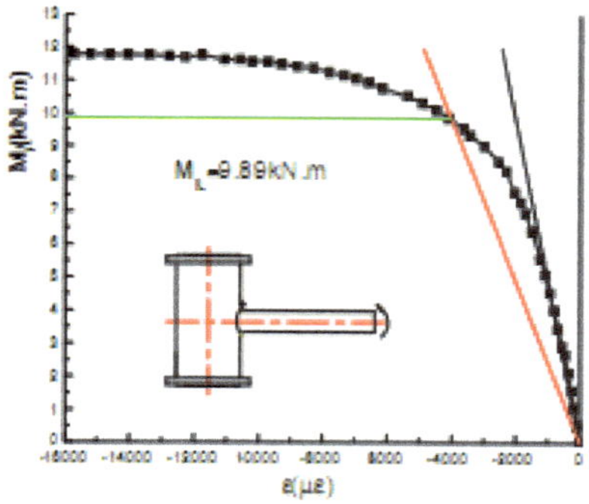

(b) Load-strain curves for L_2 vessel

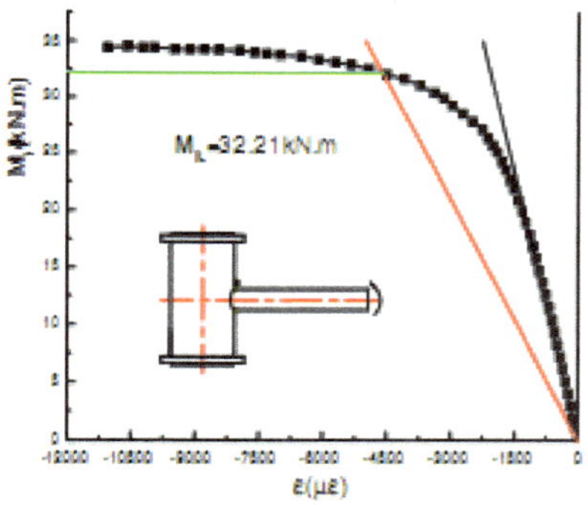

(c) Load-strain curves for L_3 vessel

Figure 8. Load-strains curves for test models (Strains gage No.6)

Table 2 is a summary and comparison of the experimental plastic limit moment for the three model vessels. From the table, it is seen that the limit moments obtained by two measurement methods (displacement measuring and strain measuring) are in agreement. The maximum difference between the limit moments obtained from the two methods is 9.76%. For the same model vessel, the limit moments obtained using different measuring points also are in good agreement regardless of whether the measuring method is displacement measuring or strain measuring.

Models		L_1	L_2	L_3
By load-displacement method	Sensor No.1	5.96	10.81	33.42
	Sensor No.2	5.77	10.53	33.22
	Sensor No.3	5.50	10.46	32.41
	Average value	5.74	10.60	33.02
By load-strain method	Strain gage No.1	5.05	9.73	33.22
	Strain gage No.6	5.37	9.89	32.21
	Strain gage No.11	5.15	10.33	31.09
	Strain gage No.16	5.16	9.77	32.86
	Average value	5.18	9.93	32.35
Difference of the average value (%)		9.76	6.32	2.03

Table 2. Experimental results of plastic limit moment (kN.m)

3. Finite element analysis

3.1. Models and mesh

A static nonlinear finite element analysis of the experimental model vessels was carried out using the ANSYS code [15].Three-dimensional isoperimetric solid elements defined by eight nodes were used to generate the FEA mesh of the cylinder, nozzle and weld seam between the cylinder and nozzle. Two elements were used across the thickness of the shell.Due to the symmetry of the structure and loading, only one-half of the test vessel was modeled. Figure 9 illustrates the finite element mesh of the model vessel No.L_2.

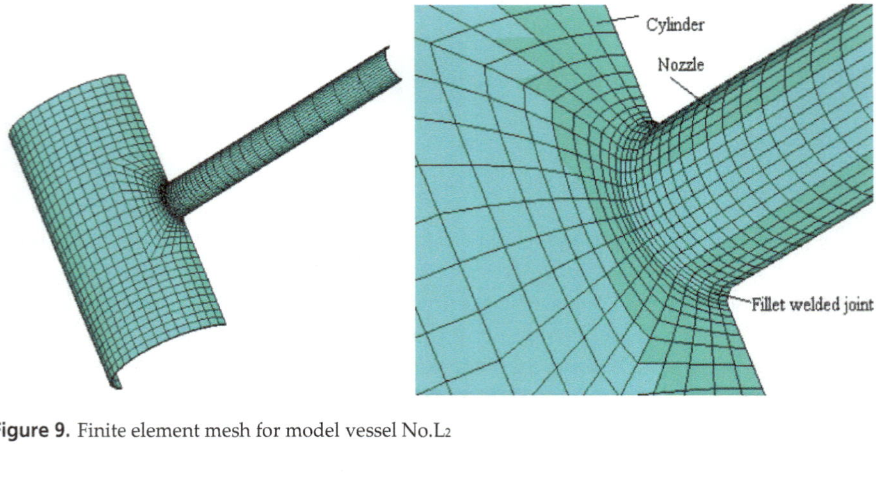

Figure 9. Finite element mesh for model vessel No.L_2

3.2. Boundary and loading conditions

The boundary conditions for the numerical simulation are the same as those used in the experimental portion of this study. One end of the cylinder is fixed, while the other end is free. All nodes on the symmetry section (longitudinal plane of the vessel) are constrained against displacement in the direction perpendicular to the symmetry plane. The FEA models were loaded a force at the end of the nozzle along the longitudinal direction of the vessels like in the experiment. For the purpose of comparing the results of the element analysis with those of the experiments, the same materials and loading increments as those of the test vessels were used. As a nonliner geometry analysis, the multilinear isotropic type of hardening behavior is used.A multi-linear elastic-plastic material model defined by nine points, as given in Table 3, was employed.

Material	Point	1	2	3	4	5	6	7	8	9
Q235-A	Stress (MPa)	355	357	378	403	432	463	485	510	519
	Strain (με)	1830	12232	20620	32698	49778	74576	96572	129445	158602
20#	Stress (MPa)	319	334	376	415	464	512	553	584	612
	Strain (με)	1477	16600	29567	48896	77066	113430	157103	198948	262455

Table 3. Material model

3.3. Analysis results

3.3.1. *Plastic deformation*

For the same loading condition as the experiment, the nozzle produced an obvious bending deformation and a depression of the cylinder on the compression side (in the longitudinal section) occurs. Figure 10 shows the local deformation characteristics which are consistent with those of the experiment (see Figure 4) for model vessel No. L_2.

3.3.2. *Plastic limit moment*

Figure 11 indicates the load-displacement plots of simulation vessels from the sensor No.3 and corresponding limit moments on the nozzles.

At the same location as that of the experiment, load strain curves were plated in Figure 12, and corresponding limit moments for simulation vessels of strain gage No.6 were also shown in Figure 12.

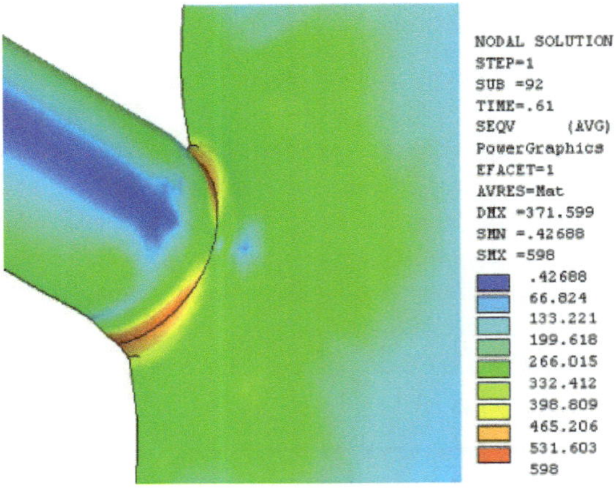

Figure 10. Local deformation for model NO. L_2 (M_i=13.05 kN.m)

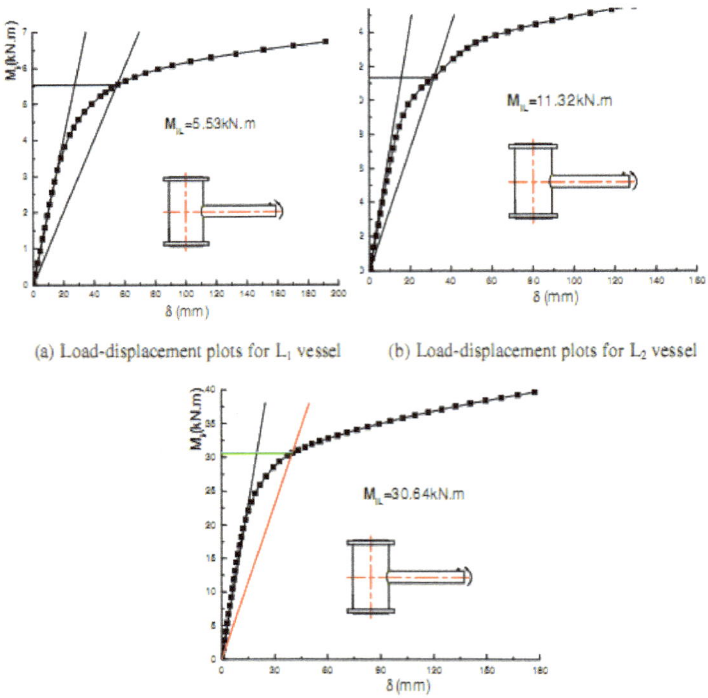

(c) Load-displacement plots for L_3 vessel

Figure 11. Load-displacement plots for simulation models (Sensor No. 3)

The plastic limit moments for the three model vessels by FEA are listed in Table 4.

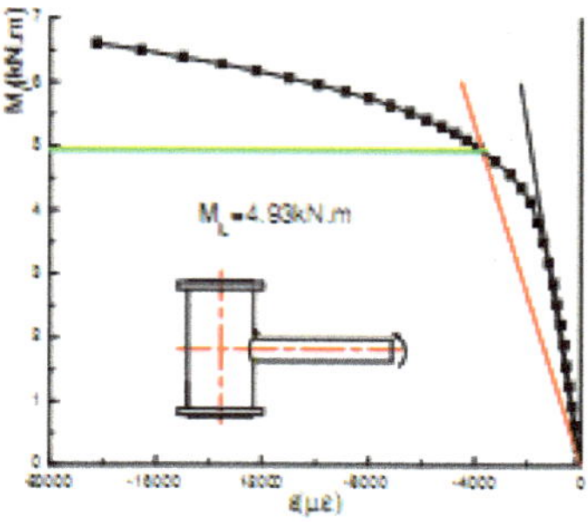

(a) Load-strain plots for L_1 vessel

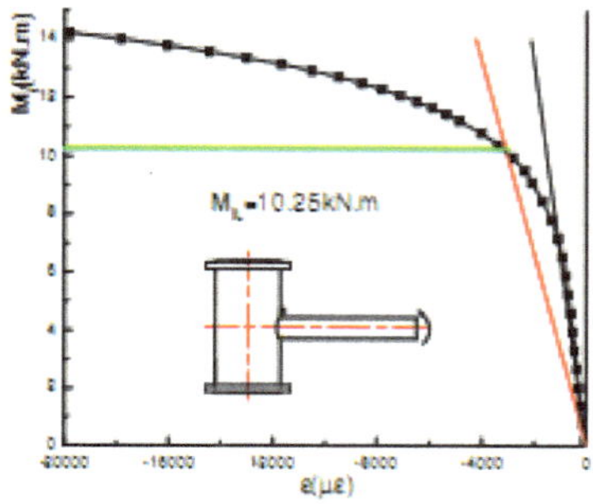

(b) Load-strain plots for L_2 vessel

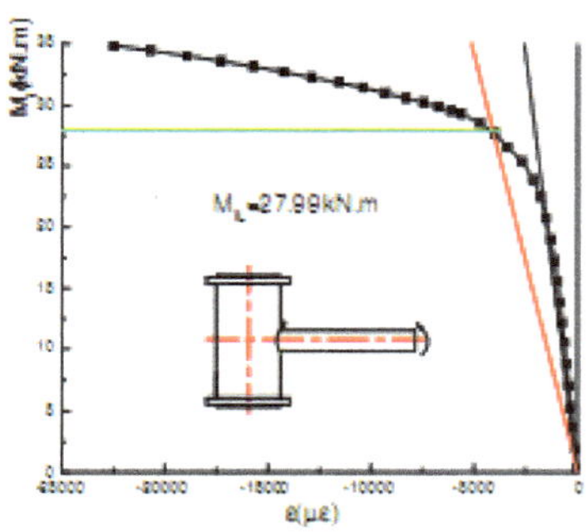

(c) Load-strain plots for L_3 vessel

Figure 12. Load-strain plots for simulation models (Strain gage No. 6)

Models		L_1	L_2	L_3
By load-displacement method	Sensor No.1	5.27	11.67	30.38
	Sensor No.2	5.39	11.49	30.52
	Sensor No.3	5.53	11.32	30.64
	Average value	5.40	11.49	30.51
By load-strain method	Strain gage No.1	4.95	10.37	28.62
	Strain gage No.6	4.93	10.25	27.99
	Strain gage No.11	4.81	10.46	28.11
	Strain gage No.16	5.02	10.36	28.94
	Average value	4.93	10.36	28.42
Difference of the average value (%)		8.70	9.83	6.85

Table 4. FEA Results for the Plastic Limit Moment (kN.m)

4. Summary and comparison

From Tables 2 and 4, it is seen that the plastic limit moments from the displacement measurements are consistent with those determined from the strain measurements, whether by experimental method or finite element analysis.

Using an average of the values from different measuring points and displacement sensors, a summary and comparison of the twice-elastic-slope plastic limit load by experiment and finite element analysis for the three model vessels are shown in Table 5.

Model No.	D /T	d/ D	t/T	Limit-moment by Experiment (kN.m) load-displacement method	load-strain method	average	Limit-moment by FEA (kN.m) load-displacement method	load-strain method	average	Difference of the average value) (%)
L_1	62.5	0.172	0.375	5.74	5.18	5.46	5.40	4.93	5.17	5.31
L_2	62.5	0.246	0.50	10.60	9.93	10.27	11.49	10.36	10.93	6.43
L_3	62.5	0.428	0.625	33.02	32.35	32.69	30.51	28.42	29.47	9.85

Table 5. Summary and comparison of plastic limit moment

From Table 5, it can be seen that the results for the plastic limit moment from experiment and finite element analysis including displacement and strain measuring techniques are in good agreement. The difference between experiment and finite element analysis is within 10%.

5. Parametric analysis and correlation equation

5.1. Parametric finite element modeling

On the basis of the previous studies, parametric modeling was performed. The sets of parameters of the analysis models used in this part of the study are given in Table 6. A total of 64 configurations were analyzed to investigate the relationship between the various geometric parameters and plastic limit moments of the vessel-nozzle structures. The data in Table 7 were obtained from the finite element analysis of 64 models.

5.2. Correlation equation

A correlation equation for plastic limit load of cylindrical vessels under in-plane moment on the nozzle can be obtained by using the data from Table 7 and the software package Statistica (non-linear regression). The resulting correlation equation for the plastic limit moment on the nozzle is as follows:

$$M_{iL} = [-0.0148 + 0.704\left(\frac{d}{D}\right) + 1.845\left(\frac{d}{D}\right)^2]\left(\frac{d}{D}\right)^{-1.761}\left(\frac{t}{T}\right)^{-0.555}\left(\frac{D}{T}\right)^{-0.476} M_b$$

Where, M_b is the limit load of a straight nozzle,

$$M_b = d^2 t \sigma_s{}^{[8]} \text{ (kN.m)}$$

This equation holds for the following range of parameters:

$$0.1 \le d/D \le 0.4 ， 0.25 \le t/T \le 1 ， 50 \le D/T \le 125 ， 500mm \le D \le 2000mm$$

Model	d/ D	t/T	D/T
1	0.1	0.25	50
2	0.1	0.5	75
3	0.1	0.75	100
4	0.1	1.0	125
5	0.2	0.25	75
6	0.2	0.5	50
7	0.2	0.75	125
8	0.2	1.0	100
9	0.3	0.25	100
10	0.3	0.5	125
11	0.3	0.75	50
12	0.3	1.0	75
13	0.4	0.25	125
14	0.4	0.5	100
15	0.4	0.75	75
16	0.4	1.0	50

Table 6. Primary parameters of the FE models

Models	Limit Moment			
	D=500mm	D=1000mm	D=1500mm	D=2000mm
1	2.25	16.43	64.37	156.52
2	3.11	22.56	88.98	216.47
3	3.52	25.58	97.96	231.89
4	3.12	24.27	79.05	185.34
5	5.63	43.73	161.53	398.26
6	18.31	142.05	539.24	1307.58
7	6.51	50.14	158.86	367.96
8	11.52	85.09	282.45	632.46
9	9.47	73.76	275.15	608.14
10	11.67	83.05	276.31	650.87
11	58.02	434.86	1536.42	3500.24
12	36.21	256.45	853.13	2068.57
13	13.37	109.21	363.01	548.12
14	27.52	204.18	680.76	1358.54
15	52.54	360.04	1221.85	2808.47
16	118.34	836.42	2804.57	6618.21

Table 7. Limit loads of the FE models (kN.m)

In order to verify the accuracy of the correlation equation, the plastic limit moments of three models were calculated and compared with those from the experimental and FEA studies as seen in Table 8.

Models	L1	L2	L3	Models	L1	L2	L3
Limit-moment by Regression Equation (M_{EQU})	5.55	11.53	33.63	Limit- moment by Regression Equation (M_{EQU})	5.55	11.53	33.63
Limit- moment by Experiment (M_{EXP})	5.46	10.27	32.69	Limit- moment by FEA (M_{FEA})	5.17	10.93	29.47
Difference $(\frac{\lvert M_{EQU}-M_{EXP}\rvert}{M_{EQU}})$	1.62%	10.93%	2.80%	Difference $(\frac{\lvert M_{EQU}-M_{FEA}\rvert}{M_{EQU}})$	6.85%	5.20%	12.57%

Table 8. Comparison of plastic limit moment (kN.m)

6. Conclusions

Experiments and comparative 3D-nonlinear finite element analyses of cylindrical vessel-nozzle connections subjected to an in-plane moment on the nozzle were carried out. In addition, an extensive geometric parameter study of such joints was performed with FEA.The following general conclusions can be drawn from these studies:

1. The experimental and 3D-nonlinear finite element analyses including load-strain and load-displacement of the nozzle responses provide effective and reliable determination methods for the pressure vessel-nozzle (or main-branch pipe) connections subjected to external loads on the nozzle.
2. The plastic limit moments determined from experimental and finite element analysis studies are in good agreement. The maximum error of the average experimental versus FE calculated limit moments is about 10%.
3. The limit moments from load-displacement and load-strain curves are in reasonable agreement for the analyzed geometry.
4. The plastic limit moments from experimental and finite element analysis studies are greatly increased with an increase of d/D, t/T ratios of the model vessels.
5. The correlation equation provided in this paper can be used to determine, within reasonable limits, the plastic limit moment of cylinder-nozzle connections subjected to an in-plane load on the nozzle.

Author details

B.H. Wu
Department of Mechanical and Power Engineering, Nanjing University of Technology, Nanjing, People's Republic of China

7. References

[1] ASME Boiler and Pressure Vessel Code, 2004 edition, section Ⅷ, Division 2, American Society of Mechanical Engineering, New York.
[2] Ellyin, F., 1976, "Experimental Investigation of Limit Loads of Nozzles in Cylindrical Vessels,"Welding Research Council Bulletin. No.219, WRC, New York.
[3] Ellyin, F., 1977, "An Experimental Study of Elastic-plastic Response of Branch Pipe Tee Connection Subjected to Internal Pressure, External Couples and Combined Loadings," Welding Research Council Bulletin, No.230, WRC, New York.
[4] Schroeder, J., 1985, "Experimental Limit Couples for Branch Moment Loads on 4-in. ANSI B16.9 Tees,"Welding Research Council Bulletin, No.304, WRC, New York.
[5] Junker, A.T., 1985, "Finite Element Determination of the Limit Load of a Nozzle in a Cylindrical Vessel Due to In-plane and Out-of-plane Moments," Swanson Engineering Associates Corporation (SEAC) Report No.SEAC-TR-312,Rev.2, to the Pressure Vessel Research Committee, WRC, New York.
[6] Moffat, D.G., 1985, "Experimental Stress Analysis of Four Fabricated Equal Diameter Branch Pipe Intersections Subjected to Moment Loadings and Implications on Branch Junction Design," Proc Instn. Mech. Engrs., Vol.199, No.4, PP. 261~284.
[7] Moffat, D.G., 1991, Mwenifumbo, J.A.M., Xu, S. H. and Mistry, T., "Effective Stress Factors for Piping Branch Junctions Due to Internal Pressure and External Moment Loads," Journal of Strain Analysis, Vol.26, No.2, PP.85~101.
[8] Rodabaugh, E.C., 1988, "A Review of Area Replacement Rules for Pipe Connections in Pressure Vessels and Piping,"Welding Research Council Bulletin, No.335, WRC, New York.
[9] Rodabaugh, E.C., Interpretive report on limit analysis and plastic behavior of piping products, Welding Research Council Bullition No.254,1976
[10] Kalnins, A., 2001, "Guidelines for Sizing of Vessels by Limit Analysis,"Welding Research Council Bulletin, No.464, WRC, New York.
[11] Chapuliot, S., Moulin, D., and Plancg, D., 2002, "Mechanical Behavior of a Branch Pipe Subjected to Out-of-plane Bending Load," ASME Journal of Pressure Vessel Technology, Vol.124, PP.1~14.
[12] Mourad Hashem, M., Maher, Y.A., 2002, "Limit-Load analysis of Pipe Bend under Out-of-plane Moment and Internal Pressure," ASME Journal of Pressure Vessel Technology, Vol.124, PP.7~32.
[13] Widera, G.E.O., Wei, Z., 1998, "Parametric Finite Element of Large Diameter Shell Intersection," Part 2, External Loading, PVRC Project No.96-20AS, WRC, New York.

[14] Sang, Z.F., Wang, Z.L., Xue, L.P., Widera, G..E.O., 2005, "Plastic Limit Loads of Nozzles in Cylindrical Vessels under Out-of-plane Moment Loading," International Journal of Pressure Vessels and Piping, 82, PP. 638-648.
[15] Swanson Analysis System Inc., "ANSYS Engineering Analysis Systems User's Manual".

Chapter 7

Flexural Behavior of Functionally Graded Sandwich Composite

Mrityunjay R. Doddamani and Satyabodh M Kulkarni

Additional information is available at the end of the chapter

1. Introduction

During World War II, the british made a bomber De Havilland *Mosquito* which served in Europe, Middle and Far East and on the Russian front. Designed as a bomber, it excelled not only in this field but also as a fighter aircraft, mine layer, path finder in military transport and photo reconnaissence. It was constructed during the *Battle of Britain* and the first prototype made its maiden flight in november 1940, less than a year after the design project is started. From an engineering viewpoint, it has one spectacular feature - the fuselage is made of a molded plywood-balsa sandwich material, which is strong and yet lightweight and equally important in times of war, its components are readily available unlike aluminium ones. The importance of the Mosquito in the war effort proved the value of the new sandwich materials [1]. Sandwich composites are popular due to high specific strength and stiffness. The concept of sandwiches came in as early as the year 1849 AD but their potential realized mainly during Second World War as mentioned earlier. Sandwiches are composed of two stiff, strong and thin faces (skins) bonded to a light, thick weaker core. Faces sustain in-plane and bending loads, while the core resist transverse shear forces and keep the facings in place. These provide increased flexural rigidity and strength by virtue of their geometry. The high specific strength and stiffness make them ideal in structural design [2-3]. Developments in aviation posed requirement of lightweight, high strength and highly damage tolerant materials. Sandwich composites, fulfilling these requirements became the first choice for many applications including ground transport and marine vessels [4].

Sandwich panels are used in a variety of engineering applications including aircraft, construction and transportation where strong, stiff and light structures are required [5]. The applicability of sandwiches could be improved if it contains a FG core which might help to distribute the stresses due to bending or in progressive absorption of energy under impact loading [6]. It is required to study the behavior of sandwich panels under these types of failures with a functionally graded material (FGM) as core to explore their new application in bullet proofing and crash worthiness. FGM's are new class of materials where property is function of geometry such as thickness, length etc [7]. These are the materials whose composition and microstructure are not uniform in space, but gradually vary following a predetermined law [8-11]. FGM's differ from composites in the sense that property is uniform in a particular direction throughout the composite. The concept of FGM's is proposed as early as 1984 by material scientists as a means of preparing thermal barrier materials [12]. Closest to FGM's is laminated composites with variation in laminate properties but they possess distinct interfaces across which properties change abruptly [13]. For example, a rocket motor casing can be made with a material system such that the inside is made of a refractory material, the outside is made of a strong metal, and the transition from the refractory material to the metal is gradual through the thickness [14]. FGM's possess a number of advantages that make them attractive in many applications, including a potential reduction of in-plane and transverse through-the-thickness stresses, an improved residual stress distribution, enhanced thermal properties, higher fracture toughness, and reduced stress intensity factors. It is worth mentioning that the distribution of the material in functionally graded structures may be designed to various spatial specifications (1). Currently, advanced processing methods to introduce compositional gradients into various material systems are being developed by materials scientists [15-17]. A typical particulate composite with prescribed variation in distribution of constituent phases could be a representative FGM. The FGM concept could be borrowed in making sandwiches with FG core which exhibit resistance (stiffness) proportional to the applied load can serve some applications better than regular sandwiches, like a spring with varying stiffness. Such a sandwich could be realized by using a particulate composite with varying volume fraction of constituents.

The flexural behavior of sandwich beams has been studied extensively by many investigators [18-23]. Studies on three point bend tests have been conducted in flexural [24-25] and short beam shear test configurations [26]. An experimental investigation of failure of piecewise FG of sandwiches subjected to three point bending is carried out by Avila [27]. In addition, fiber reinforced syntactic foams [28-30] and syntactic foam core sandwich composites have also been studied for bending properties [31]. Specific properties of sandwich with complaint FG core needs attention as it is yet to be reported.

2. Objectives and scope

From the foregoing literature survey, clear is the fact that the research reports on development of low cost materials for bullet proofing and energy absorption is hardly available. A low cost ash filled functionally graded polymer system is proposed for applications like ballistic energy absorption. The perusal of sandwich literature review prompted a thorough and systematic study on these sandwiches by performing experimental characterization of flexural properties. Therefore the work undertaken pursues the following objectives:

1. To prepare functionally graded rubber cores with varying fly ash reinforcement.
2. To plan the experiments using DOE for processing FG sandwiches with different factors (weight fraction of fly ash, core to total sandwich thickness - C/H ratio and jute skin orientation) as per L9 orthogonal array at three levels.
3. To study the effect of above parameters on mechanical properties of sandwich three point loading condition.
4. To identify the most influential factor governing the mechanical behavior of FG sandwiches.
5. To validate the gradation observed through finite element (FE) modeling using spring analogy for variations in property like uniform, linear and piecewise linear.
6. Comparison of Experimental and FE results for properties of sandwich under consideration.
7. Visual inspection of fractured FG sandwiches under different tests.

Developed FG cores are utilized in sandwiches to characterize FG sandwiches for their suitability in real world applications. Sandwiches are prepared as per design of experiments approach so that multiple factors (fly ash weight fraction, C/H ratio and jute skin orientation) at three different levels can be simultaneously analyzed. Further, these sandwiches are subjected to bending test. Another set of samples called confirmatory set is made with 25% and 35% filler by weight. Five samples are subjected to mechanical test and the response is averaged out for these five.

Furthermore, experimental values are compared with results of FE analysis. ANSYS 5.4 package is used to achieve this objective. Analysis are carried out with three gradation variations namely uniform, linear and piecewise linear. Young's modulus is computed for FG cores using FE approach and is compared with experimental result. Specific bending strength is the properties focused in simulating sandwich behavior. Finally, elaborate discussion on fractured samples is presented as the last segment of this work.

3. Processing details

This section presents properties of starting material used, procedures followed for preparing FG composites and their sandwiches. Details of reagents / chemicals used at different stages like for sample curing are also described. Characteristics of the reinforcements used are also enlisted. As outlined in the objectives and scope of the work in the preceding section, the objective of the present investigation is to study the properties of functionally graded sandwiches. This section lists materials and their properties and methods adopted for processing composites with varying content of the filler.

3.1. Plan of experiment

In this work experiments are designed based on Taguchi's DOE approach for FG sandwiches [32]. Factors and levels chosen for planning the experiments for FG sandwiches are presented in Table 1. Table 2 shows orthogonal array for sandwich. Table 3 presents coding of samples bearing varying content of filler, C/H ratio and jute orientation.

Details	Wt Fraction of Fly ash % (Factor 1)	Core to thickness ratio (Factor 2)	Orientation of Jute Fabric (Factor 3)
Level 1	20	0.4	$0^0/90^0$
Level 2	30	0.6	$30^0/60^0$
Level 3	40	0.8	$45^0/45^0$

Table 1. Factors and Levels selected for sandwich with FG core

Experiment No.	Parameters		
	Weight Fraction (%)	C/H Ratio	Orientation
1	20	0.4	$0^0/90^0$
2	20	0.6	$30^0/60^0$
3	20	0.8	$45^0/45^0$
4	30	0.4	$30^0/60^0$
5	30	0.6	$45^0/45^0$
6	30	0.8	$0^0/90^0$
7	40	0.4	$45^0/45^0$
8	40	0.6	$0^0/90^0$
9	40	0.8	$30^0/60^0$

Table 2. L9 Orthogonal array for FG Sandwich

Sample code	Description
$W_aR_bO_c$	Sandwich specification
W	Indicates factor 1 (Wt. fraction of fly ash)
a	Levels of factor 1 in % (20, 30, 40)
R	Indicates factor 2 (C/H ratio)
b	Levels of factor 2 (0.4, 0.6, 0.8)
O	Indicates factor 3 (Fiber Orientation in skin)
c	Levels of factor 3 ($0^0/90^0$, $30^0/60^0$, $45^0/45^0$)

Table 3. Description of sample codes used for sandwiches

Experimentation is done with due considerations to all the above parameters with both configurations of gradation namely rubber up and ash up. In each trial minimum of five replicates are tested. Average of the measured parameters for each set of replicates is subjected to statistical ANOVA to find the most influential factor governing the behavior using Minitab release 14 statistical analysis tool.

3.2. Materials

Details of materials used for main constituents of sandwiches (core and skin) are presented hereafter.

3.2.1. Core for FG sandwich

From the standpoint of cost, availability, and the scarce literature prompted for going in for an elastomeric material which is naturally occurring and known by the name 'natural rubber' as the matrix material. Further it is reinforced with fly ash and is used as core in sandwich.

As many of the polymeric systems for developing FGM's are generally with the tag of expensiveness associated, it is decided to examine the gradation in composition and its subsequent mechanical behavior when an abundantly available lower density possessing fly ashes are the filler materials for the core. Fly ashes are fine particulate waste products derived during generation of power in a thermal power plant. These have aspect ratios closer to unity and hence are expected to display near isotropic characteristics. These inexpensive and possessing good mechanical properties, when used with well established matrix systems help to reduce the cost of the system and at the same time either retain or improve specific and desirable mechanical properties. Fly ash has attracted interest [33-34] lately, because of the abundance in terms of the volume of the material generated and the environmental-linked problems in the subsequent disposal. Fly ash mainly consists of alumina and silica, which are expected to improve the composite properties. Fly ash also consists to some extent hollow spherical particles termed as cenosphere which aid in maintenance of lower density values for the composite, a feature of considerable significance in weight-specific applications [35-36]. Again, as the fillers do not come under irregular shape, the resin spread, is better and as the ashes are essentially a mixture of solid, hollow and composite particles displaying near isotropic properties, developing newer and utilitarian systems using them should be an interesting and challenging task [37]. Compositional details of a fly ash particle are tabulated in Table 4.

Constituent	Wt. %
SiO_2	63
Al_2O_3	26.55
CaO	0.42
Fe_2O_3	6.7
TiO_2	2.47

Table 4. Compositional details of fly ash particle

3.2.2. Skin used in sandwich

Further on, in this effort, for the skins too, it is decided to employ instead of the well explored man-made fibers like glass, carbon or aramid a fairly strong but naturally occurring one going by the name 'jute fiber' and known for its inexpensiveness. Jute is an attractive natural fiber for use as reinforcement in composite because of its low cost, renewable nature and much lower energy requirement for processing. In comparison to glass fibers jute has higher specific modulus and lower specific gravity as against that of glass fiber. Jute reinforced plastics offer attractive propositions for cost-effective applications [38]. These in the form of laminates have much better properties than their neat resin counterparts [39]. Better properties of woven jute fabric reinforced composites demonstrated their potential for use in a number of consumable goods in an earlier literature [40]. Substantial increases in flexural modulus and strength with small amounts of reinforcement of unidirectional jute have also been reported [41]. Keeping these things in mind a bi-directionally woven jute fabric is used in different orientations. Table 5 gives the brief overview of comparison between glass fibers and jute fibers.

Property	E-glass	Jute
Specific Gravity	2.5	1.3
Tensile Strength (MN/m^2)	3400	442
Young's Modulus (MN/m^2)	72	55.5
Specific Strength (MN/m^2)	1360	340
Specific Modulus (GN/m^2)	28.8	42.7

Table 5. Mechanical Properties of Glass and Jute Fibers

The major drawback of natural fiber reinforced composites is due to its affinity towards moisture. Many experimental studies have shown that compatible coupling agents are capable of either slowing down or preventing the de-bonding process and hence moisture absorption even under severe environmental conditions such as exposure to boiling water. Jute fibers/fabrics can be modified chemically through graft co-polymerization and through incorporation of different resin systems by different approaches.

3.2.3. Matrix for skin

For fabricating both the skins and core a matrix system is required. A thermosetting epoxy is chosen for this purpose as far as the skins are concerned. The adhesive used in present work consists of a medium viscosity epoxy resin (LAPOX L-12) and a room temperature curing polyamine hardener (K-6) supplied by ATUL India Ltd. Epoxy resin is selected as the material for the matrix system because of its wide application, good mechanical properties, excellent corrosion resistance and ease of processing. Some details including density of the constituents of the matrix system chosen are listed in Table 6.

Constituent	Trade name	Chemical name	Epoxide equivalent	Density (kg/m^3)	Supplier	Parts by weight
Resin	LAPOX L-12	Diglycidyl Ether of bisphenol A (DGEBA)	182 - 192	1162	ATUL India Ltd.	100
Hardener	K-6	Tri ethylene Tetra amine (TETA)	----	954	- do -	10-12*

*As suggested in the manufacturer's catalogue

Table 6. Details of the constituents of matrix used for skin in sandwich

With these materials in hand, FG sandwiches are prepared for mechanical testing.

4. Processing of FG sandwich

FG cores used in the present work are produced using the following procedure. The gradation in the core is expected due to differential settling of the particles with different densities at different depths in the rubber matrix. A measured quantity of natural latex is mixed with pre-weighed amounts of fly ash, sulphur (vulcanizer) and zinc oxide (catalyst) [42] by adopting gentle stirring for about 1 hour. The mold employed for preparation of core specimen is completely covered on all sides with teflon sheet. Subsequently, silicone releasing agent is applied to facilitate ease of removal of the cast sample at a later stage. The mixture is then slowly decanted into the mold cavity followed by curing at 90°C in an oven for about 5-6 hours. The cured rigid plate sample is withdrawn from the mold and the edges trimmed. Figure 1 presents one such FG sample which in turn will be used as core in sandwiches.

Figure 1. Functionally graded core sample drawn out of mold

As regards the sandwich skins, a bi-directional woven jute fabric procured from M/S Barde Agencies, Belgaum, Karnataka, India is used. This fabric is cut into layers of dimensions depending on the sandwich sample size in required orientation. Thickness of each fabric piece is 0.5 mm. All the layers of jute fabric are heated in an oven at 70^{0}C for 5-10 minutes to remove moisture present. The jute stack thickness to form the thin skin, on either side of FG core, is computed. This enables one to arrive at the required number of fabric layers to be used, as thickness of each layer is known. Based on required C/H ratio number of fabric layers to be used are determined (Table 7).

C/H Ratio	Core thickness - C (mm)	Number of jute layers below core	Number of jute layers above core	Sandwich thickness - H (mm)
0.4	4	6	6	10
0.6	6	4	4	10
0.8	8	2	2	10

Table 7. Jute layer arrangement for achieving C/H ratios in sandwich

With this background data on hand to begin with, the required fabric pieces are dipped in mixture of epoxy and K-6 hardener and placed on base plate forming the bottom stack of the sandwich. Now, the earlier mentioned procedure-wise made FG core dipped in resin mixture is placed on the bottom stack of skins. Finally, over such an arrangement, the remaining layers of jute fabrics having undergone the same procedure for fabrication are stacked to constitute the top skin. A procedure of this nature should help in ensuring a greater degree of spread of the resin on the fibrillar jute. Following this, the excess resin is made to come out by a squeezing operation that is aided by tightening of the mold top plate. The mold assembly is then cured at room temperature for about 24-26 hours. The sandwich sample is withdrawn from the mold and trimmed to the required size. Similarly numbers of samples are made with various core thickness and orientation in skin as schematically illustrated in Figure 2. Figure 2 (a) shows top view with different orientations and while the front view with varying core thickness to total sandwich thickness (C/H ratio) is presented in Figure 2 (b).

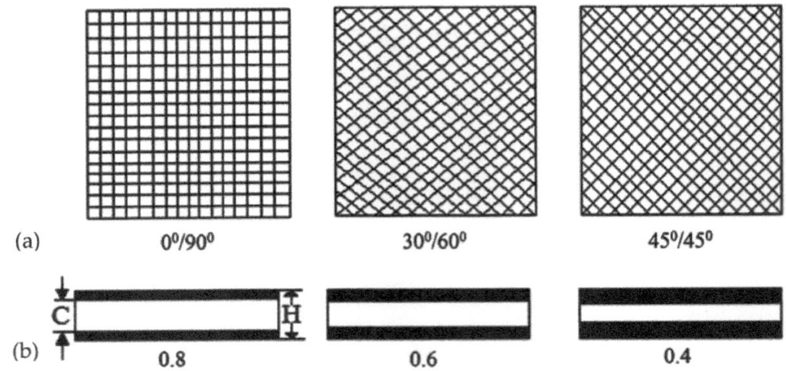

Figure 2. (a). Orientation of jute strands in the sandwich skins, (b). Variation of C/H ratio considered for analysis

5. Experimental details

The mechanical testing of sandwich composites to obtain parameters such as strength, stiffness etc. is a time consuming and often difficult process. It is, however, an essential process, and can be somewhat simplified by the testing of simple structures such as flat coupons. The data obtained from these tests can then be directly related with varying degrees of simplicity and accuracy to any structural shape. The test methods outlined in this section merely represent a small selection available to the composites scientist. Various FG sandwiches fabricated are characterized for three point bending condition. Influence of rubber up (rubber rich region towards the top) and ash up (ash rich region below the loading point) configurations are critically analyzed. Expected gradation in FG cores is presented in Figure 3 (rubber up and ash up).

The three point bending test is carried out in accordance with ASTM C 393 [42] using Instron universal testing machine of model 4206 with loading capacity ranging from 0.1 N to 150 kN. Figure 4 shows the sandwich sample mounted on flexural test set-up. The thickness to span ratio of the tested sandwich samples is 1:16. The crosshead displacement rate is maintained at 2 mm/min. The load deflection data is recorded at equal intervals up to a point at which the specimen shows the first sign of failure.

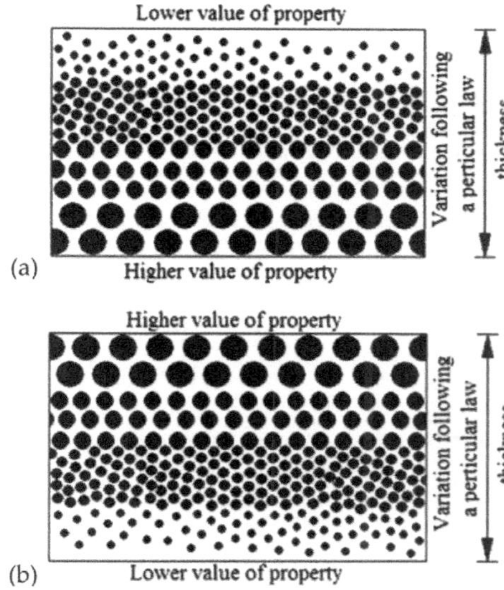

Figure 3. (a). Rubber Up condition in FG core, (b). Ash Up condition in FG core

Figure 4. Sandwich sample mounted on flexural test set-up

From load deflection data, bending modulus and strength are estimated using relations 1 and 2 respectively and the mean of five samples in each sandwich configuration is used for inference.

$$\text{Specific bending modulus} = \frac{\text{Flexural modulus}}{\text{Weight density}} = \frac{E_{bending}}{\rho \times g} \tag{1}$$

$$\text{Specific bending strength} = \frac{\text{Ultimate strength}}{\text{Weight density}} = \frac{\sigma_{u_{bending}}}{(\rho x g)} \tag{2}$$

where $\sigma_{u_{bending}} = \frac{6M}{BH^3}$ and $M = \frac{FL}{4}$

5.1. Details of finite element modelling

As outlined earlier, FE model helps to model the constituents of the FG composites and their sandwiches to study the interactions of these in load transfer and mechanisms influencing their failure. To understand and predict the effect of material as well as geometrical parameters on the mechanical behavior of FG fly ash filled rubber composites and their sandwiches finite element analysis can be a very effective technique. Towards this, a simple disctretized model is built in the software ANSYS® representing FG composites with properties varying from top layer to bottom representing gradation.

Static analysis is performed using FEM software ANSYS 5.4. In this analysis a two dimensional model of a FG system is constructed and meshed with 4-node PLANE42 element. Three different mesh sizes are tested with 4-node elements to check the convergence of the model, based on which medium mesh size (element edge length is taken as 0.5) is selected. Number of nodes and elements used in the analysis are 800 and 5000 respectively.

Finite element values are compared with experimental ones for bending behavior of FG sandwich. At the contact surfaces of the layers and between layers and faces of sandwich glue conditions are applied to eliminate relative movement of layers with respect of each other. Furthermore, nodes are merged at the interface allowing proper coupling between layers and interfaces. Figure 5 shows finite element mesh with boundary conditions as a typical case considered for three point bending analysis. Skins are being represented by top and bottom portions of the structure whereas in between are the four layers having graded properties.

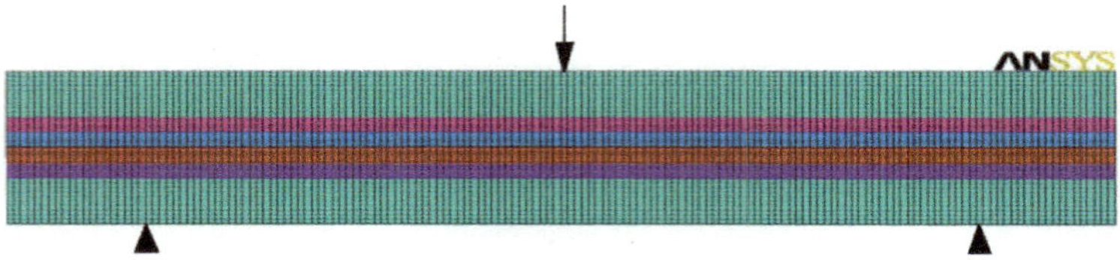

Figure 5. Finite element mesh with boundary condition for FG sandwich

While modeling gradation in ANSYS 5.4, the analogy of springs is used having differing stiffness ($K_1 < K_2 < K_3 < K_4$) from the top layer to bottom (Figure 6).

Sandwiches with FG core are modeled in FEA package ANSYS 5.4 [43] as emphasized before. Three different gradations of filler U (uniform), L (linear) and PL (piecewise linear) are considered during modeling of FG cores (Figure 7). Young's modulus and density of FG cores are determined for different weight fractions of fly ash from constituent properties are provided as input to FEA (Table 8).

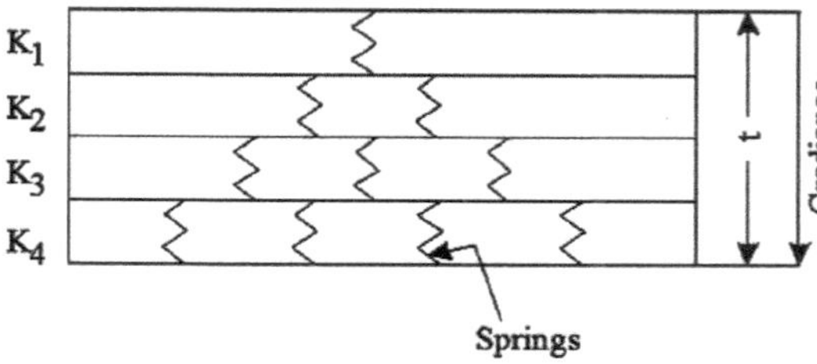

Figure 6. Spring analogy for gradation in modulus of core material

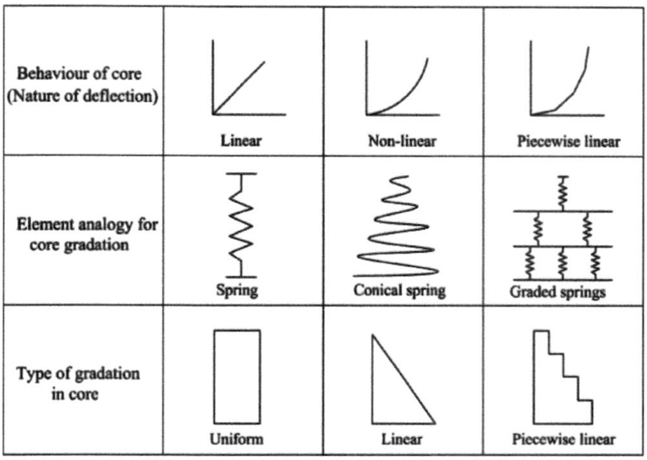

Figure 7. FG rubber core configurations used in FEA

Fly ash distributions taken into account for uniform configuration are 20%, 30% and 40% through the thickness. For these weight fractions Young's modulus is estimated using inverse rule of mixtures For skins, young's modulus is estimated by preparing five tensile samples of jute/epoxy with orientations of $0^0/90^0$, $30^0/60^0$ and $45^0/45^0$ which are subsequently tested as per ASTM D3039 [44] guidelines. Density of skins is determined experimentally using procedure outlined in ASTM D792 [45]. Table 8 presents properties of core and skin used in the FE analysis. Results of FE analysis are compared with experimental values.

FG Core							Element
Wt. % of fly ash	Young's modulus (GPa)			Density (Kg/m³)			
	U	L	PL*	U	L	PL*	
20%	0.7575	0.65 (upper) 0.75 (middle) 0.88 (bottom)	0.65 (L1)	1168.4	1163.9 (L1) 1167.5 (L2) 1172.5 (L3)	1162.8 (L1)	2D Plane 42
			0.71 (L2)			1165.2 (L2)	
			0.79 (L3)			1168.2 (L3)	
			0.88 (L4)			1173.5 (L4)	

FG Core							Element
Wt. % of fly ash	Young's modulus (GPa)			Density (Kg/m^3)			
	U	L	PL*	U	L	PL*	
30%	0.89	0.68 (upper)	0.68 (L1)	1330.2	1324.5 (L1)	1323.9 (L1)	
			0.79 (L2)		1331.1 (L2)	1328.4 (L2)	
		0.865 (middle)	0.94 (L3)		1336.9 (L3)	1334.6 (L3)	
		1.15 (bottom)	1.15 (L4)			1337.2 (L4)	
40%	1.1	0.71 (upper)	0.71 (L1)	1444.7	1435.2 (L1)	1434.9 (L1)	
			0.88 (L2)		1445.7 (L2)	1442.8 (L2)	
		1.015 (middle)	1.15 (L3)		1452.6 (L3)	1450.6 (L3)	
		1.66 (bottom)	1.66 (L4)			1455.2 (L4)	
Jute / Epoxy skin							
Orientation	Ex (GPa)		Ey (GPa)	Density (Kg/m^3)			
$0^0/90^0$	3.25		2.5	1468			
$30^0/60^0$	1.63		1.25	1451.2			
$45^0/45^0$	2.29		1.77	1444.3			

L-layer, *L1-top layer (rubber rich), L4-bottom layer (ash rich)

Table 8. Core and skin properties used in FEA

Bending tested samples are subjected to visual observation using regular photography technique for FG sandwich. These methods came in handy during the characterization of failures especially in impact failed samples.

6. Results and discussion

FG sandwiches are tested for Density, the results of which are presented in Table 9.

Sandwich code	Trial-1	Trial-2	Trial-3	Trial-4	Trial-5	Density (Kg/m^3)
$W_{20}R_{0.4}O_{0}$	1325.6	1328.9	1329.4	1332.8	1330.8	1329.5
$W_{20}R_{0.6}O_{30}$	1333.5	1334.8	1336.2	1336.4	1331.6	1334.5
$W_{20}R_{0.8}O_{45}$	1342.8	1350.7	1348.6	1345.5	1348.9	1347.3
$W_{30}R_{0.4}O_{30}$	1465.8	1464.6	1460.3	1462.1	1463.2	1463.2
$W_{30}R_{0.6}O_{45}$	1435.2	1435.9	1431.9	1432.8	1433.7	1433.9
$W_{30}R_{0.8}O_{0}$	1467.1	1466.9	1469.3	1470.5	1467.2	1468.2
$W_{40}R_{0.4}O_{45}$	1547.6	1549.8	1551.7	1550.6	1548.8	1549.7
$W_{40}R_{0.6}O_{0}$	1599.5	1598.8	1595.6	1594.4	1596.2	1596.9
$W_{40}R_{0.8}O_{30}$	1564.1	1561.8	1562.4	1560.9	1563.8	1562.6

Table 9. Density results of FG sandwiches

Experimental density values are subjected to statistical analysis (MINITAB 14) to propose regression equation which is presented in equation 3.

$$\begin{aligned}\text{Density}\left(\text{Kg}/\text{m}^3\right) &= 1099+\left[11.6 \times \text{Fly ash weight}\ \%\right]+ \\ &+\left[29.7 \times \text{C}/\text{H Ratio}\right]-[0.459 \times \text{Jute Orientation}]\end{aligned} \tag{3}$$

Equation 3 comes handy, which predicts density for large number of samples with varying combination of factors within the range of chosen levels without experimentation. Density increases with filler content as well as with C/H ratio (core to thickness ratio) being positive coefficients while shows a decreasing trend with increase in jute orientation. Obvious reason for this might be lower specific weight with increasing skin orientation.

Three point bending behavior of a FG sandwich composite is investigated under flexural loading condition. Results are analyzed for specific modulus and specific bending strength. Load deflection data is traced all along the path. The load and corresponding deflection data is noted at equal intervals up to a maximum load at which the specimen shows the first sign of failure (point 'A'). The load and deflections obtained during testing are plotted. A typical load deflection curve is shown in Figure 8.

Load-displacement consists of an initial linear part followed by a nonlinear portion (Figure 8). A nonlinear mechanics of materials analysis that accounts for the combined effect of the nonlinear behavior of the facings and core materials (material nonlinearity) and the large deflections of the beam (geometric nonlinearity) are observed. The nonlinear load-deflection behavior of the beams is attributed to the combined effect of material and geometric nonlinearity. The material nonlinearity of the sandwich beam is due to the nonlinear normal stress-strain behavior of the facing material and the FG core. For long beam spans, even though there is a geometric nonlinearity effect, the overall load-deflection curve of the beam does not deviate much from linearity.

For long beam spans the nonlinearity of the load-deflection curve is mainly due to the combined effect of the facings nonlinearity and the large deflections of the beam. Both effects, however, have a small contribution to the load-deflection behavior, which shows a small deviation from linearity. Some of the general observations made are listed below.

1. The load decreases sharply after the end of the elastic region due to failure initiation in sandwich composites (A to B).
2. All samples have shown small linear region (B to C) before skin failure in compressive side.
3. Variation in displacement value at which peak load is observed for various types of FG sandwiches is considerable.
4. The failure originates on the tensile side.

6.1. Specific bending modulus

From load deflection data the average specific modulus and strength for five samples (Table 10) are estimated using equations 1 and 2.

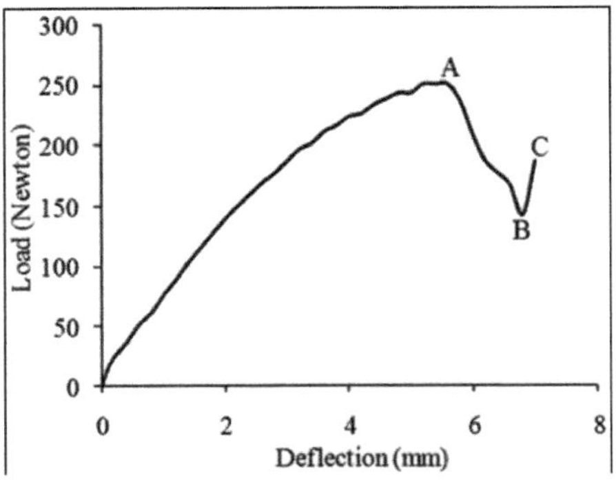

Figure 8. Load-deflection behavior under three point bend test for sandwich

Sample coding	Sp. flexural modulus (MPa/Nm^{-1})				Sp. flexural strength (MPa/Nm^{-1})			
	Rubber Up	Avg.	Ash Up	Avg.	Rubber Up	Avg.	Ash Up	Avg.
	3945.23		3410.9		132.7		103.55	
	3933.7		3404.2		128.1		105.32	
$W_{20}R_{0.4}O_{0}$	3961.5	3953.07	3419.16	3410.29	127.7	128.1	95.47	98.81
	3963.7		3416.4		128.1		99.1	
	3961.2		3400.8		123.9		90.61	
	5322.6		4540.15		85.3		72.3	
	5306.8		4545.39		88.1		70.7	
$W_{20}R_{0.6}O_{30}$	5321.3	5319.4	4539.4	4545.36	83.5	88.1	71.9	70.7
	5322.9		4544.95		92.4		70.7	
	5323.4		4556.9		91.2		67.9	
	7391.4		6150.4		54.59		45.23	
	7387.91		6160.73		54.5		50.4	
$W_{20}R_{0.8}O_{45}$	7377.4	7387.91	6155.14	6155.14	57.7	54.59	47.9	48.75
	7393.4		6155.14		58.1		46.8	
	7389.42		6154.28		48.06		53.42	
	2996.2		2390.31		141.4		115.43	
	3003.1		2398.92		141.4		116.23	
$W_{30}R_{0.4}O_{30}$	3004.5	3001.3	2398.92	2398.92	145.5	141.4	113.26	113.26
	3001.3		2398.92		149.1		111.59	
	3001.4		2407.53		129.6		109.79	
	4043.3		3533.59		94.76		80.17	
	4047.6		3528.61		99.14		81.34	
$W_{30}R_{0.6}O_{45}$	4042.4	4045.36	3531.75	3533.57	95.35	101.23	75.46	78.75
	4045.36		3523.16		106.4		79.1	
	4048.12		3550.73		110.5		77.68	

Sample coding	Sp. flexural modulus (MPa/Nm-1)				Sp. flexural strength (MPa/Nm-1)			
	Rubber Up	Avg.	Ash Up	Avg.	Rubber Up	Avg.	Ash Up	Avg.
$W_{30}R_{0.8}O_0$	6559.3	6562.65	6018.2	6018.2	148.7	153.1	120.1	119.3
	6562.8		6018.2		149.2		121.3	
	6570.4		6018.2		157.3		118.3	
	6560.55		6020.9		151.2		121.54	
	6560.21		6015.5		159.1		115.26	
$W_{40}R_{0.4}O_{45}$	2134.3	2138.92	1702	1692.71	149.3	151.4	110.34	117.17
	2138.69		1692.67		148.7		121.56	
	2141.92		1688.2		152.4		117.17	
	2139.26		1690.4		159.3		120.23	
	2140.42		1690.3		147.3		116.55	
$W_{40}R_{0.6}O_0$	4372.5	4365.98	4060.12	4065.98	188.98	192.21	159.21	154.45
	4370.28		4068.63		193.5		155.29	
	4365.39		4065.98		199.7		152.8	
	4360.87		4059.3		191.49		155.7	
	4360.86		4075.86		187.38		149.25	
$W_{40}R_{0.8}O_{30}$	6515.5	6518.2	6050.3	6062.65	155.23	159.53	121.44	125.45
	6520.7		6070.4		151.8		123.3	
	6521.4		6060.9		161.32		127.56	
	6518.2		6058.5		164.2		128.9	
	6515.2		6073.15		165.1		126.05	

Table 10. Specific bending modulus and strength for FG sandwich

It can be clearly seen from the table that, rubber up configuration registered higher results compared to ash up condition for both the properties in the range of 7 to 30%. Constrained straining and resisting forces set up in the FG core might be the reasons for such an observation in bending test as depicted in Figure 9.

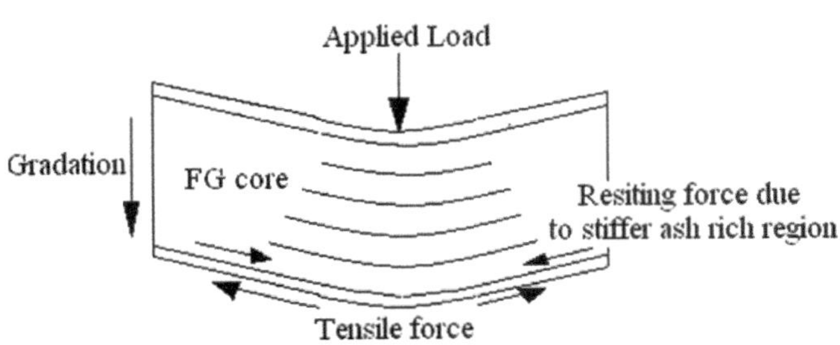

Figure 9. Loads acting on FG sandwich in bending test

Rubber up condition of FG core in sandwich represents ash rich region on tensile side. Crack initiation is observed to be from tensile region to compressive region in pre sent loading case. In rubber up condition, as stiffer zone is near tensile region, sandwich can take up higher loads resulting in better performance compared to homogenous cores and ash up condition in FG core. Thereby, such sandwiches are excellent examples of optimized designs.

Developed FG sandwiches can be used in practical cases wherein structures are continuously subjected to bending loads. Depending upon whether load is acting downwards or upwards sandwiches can be suitable placed with either rubber up or ash up configuration as regards to FG cores.

Figure 10 shows the signal to noise (SN) response plot for specific bending modulus with respect to the parameters under study. Response of SN ratio in Specific bending modulus for Rubber Up condition is presented in Table 11.

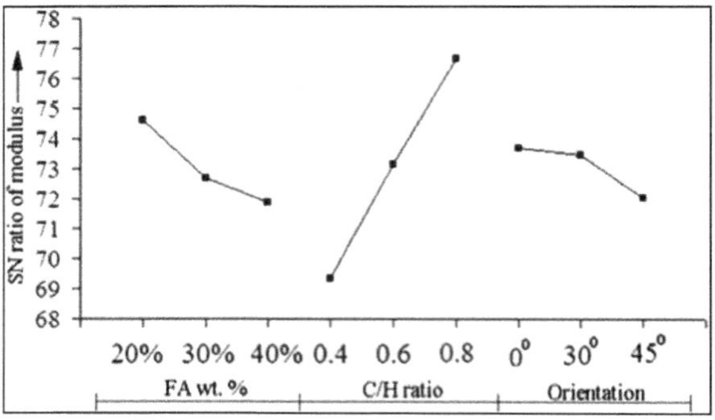

Figure 10. Variation of SN ratio in specific bending modulus (Rubber Up)

	Fly ash weight %	C/H ratio	Orientation
Level 1	74.61	69.36	73.69
Level 2	72.68	73.15	73.45
Level 3	71.90	76.66	72.04
Effect	2.71	7.30	1.66
Rank	2	1	3

Table 11. SN ratio table for specific bending modulus (Rubber Up)

From the data analysis, vide response Table 11, it is seen that C/H ratio and fly ash % exhibit greater influence compared to the orientation. It is further observed from the Table and Figure 10 that samples with fly ash content of 20%, C/H of 0.8 and an orientation of $0^0/90^0$ possess highest specific bending modulus. This could be due to higher C/H ratio implying larger rubber rich region imparting higher modulus to sandwich system.

6.2. Specific bending strength

Results of specific bending strength from Table 10 are statistically analyzed and are used to rank the variables as presented in Table 12.

	Fly ash weight %	C/H ratio	Orientation
Level 1	38.6	42.92	43.84
Level 2	42.27	41.56	41.99
Level 3	44.44	40.83	39.48
Effect	5.85	2.09	4.36
Rank	1	3	2

Table 12. SN ratio table for specific bending strength (Rubber Up)

From SN response Table, it can be seen that specific bending strength behavior is prominently governed by fly ash weight % followed by orientation and C/H ratio. Figure 11 presents SN plot for specific bending strength incase of rubber up condition.

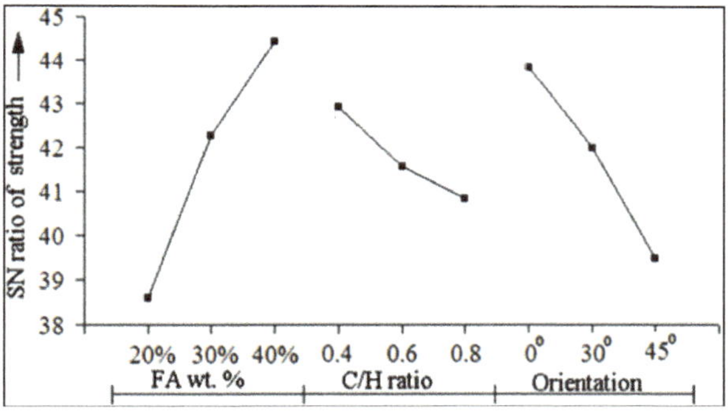

Figure 11. Plot of SN ratio in specific bending strength (Rubber Up)

From SN response plot shown in Figure 11, the best combination for specific strength is a sample with fly ash content of 40%, C/H of 0.4 and orientation of $0^0/90^0$. Reasons for this could be stiffening effect due to high modulus filler and larger skin-epoxy component for lower C/H ratios. Similar results are observed for ash up configuration. Even though $W_{20}R_{0.8}O_{45}$ and $W_{40}R_{0.6}O_0$ are showing higher values (Table 10) for modulus and strength respectively, inference on basis of these will not lead to an appropriate conclusion. The reason being these values are merely based on average of means. Inference on the grounds of SN analysis leads to a meaningful conclusion as it takes means and data spread into account. By the SN ratio analysis the best sandwich configurations are $W_{20}R_{0.8}O_0$ and $W_{40}R_{0.4}O_0$ for specific modulus and strength respectively. Similar observation is noted for ash up configuration. Regression equation is proposed based on the experimental data for specific bending properties are presented in equations 4-7.

$$\begin{aligned}\text{Specific Bending Modulus }(\text{Rubber Up}) = 1151 - [60.6 \times \text{Weight \% of Fly ash}] + \\ + [480 \times \text{C / H Ratio}] - [8.38 \times \text{Jute Orientation}]\end{aligned} \tag{4}$$

$$\begin{aligned}\text{Specific Bending Strength }(\text{Rubber Up}) = 70.4 + [3.87 \times \text{Weight \% of Fly ash}] - \\ - [44.7 \times \text{C / H Ratio}] - [1.19 \times \text{Jute Orientation}]\end{aligned} \tag{5}$$

$$\begin{aligned}\text{Specific Bending Modulus }(\text{Ash Up}) = 342 - [38.2 \times \text{Weight \% of Fly ash}] + \\ + [8945 \times \text{C / H Ratio}] - [14.2 \times \text{Jute Orientation}]\end{aligned} \tag{6}$$

$$\begin{aligned}\text{Specific Bending Strength }(\text{Ash Up}) = 54.2 + [2.98 \times \text{Weight \% of Fly ash}] - \\ - [29.8 \times \text{C / H Ratio}] - [0.912 \times \text{Jute Orientation}]\end{aligned} \tag{7}$$

6.3. Finite element analysis

Specific bending strength is estimated by simulating the sample and loading (Gupta et al. 2008) in FEA. Figure 12 represents the plot for bending stress in the sample for one typical loading case.

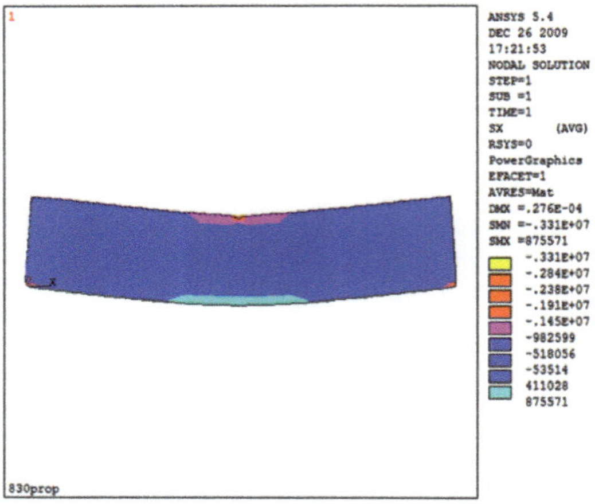

Figure 12. Bending stress in x-direction for typical case in FG sandwich

The breaking load taken from experiment is applied on FE model. For this applied load, maximum stress (von misses criteria) is recorded and finally specific strength is determined by taking the ratio of maximum stress to the weight of sample. The specific strength values obtained from FEA for three variations in gradation (Uniform-U, Linear-L and Piecewise linear-PL) and with experimental approach is presented in Table 13.

Sandwich configuration	FEA			Experimental	% Error with PL
	U	L	PL		
$W_{20}R_{0.4}O_{0}$	115.4	119.6	132.75	128.1	3.50
$W_{20}R_{0.6}O_{30}$	78.2	81.5	92.58	88.1	4.84
$W_{20}R_{0.8}O_{45}$	46.9	48.9	58.58	54.59	6.81
$W_{30}R_{0.4}O_{30}$	125.1	130.2	147.34	141.4	4.03
$W_{30}R_{0.6}O_{45}$	84.4	88.8	110.38	101.23	8.29
$W_{30}R_{0.8}O_{0}$	129.7	137.6	160.88	153.1	4.84
$W_{40}R_{0.4}O_{45}$	126.6	131.3	169.11	151.4	10.47
$W_{40}R_{0.6}O_{0}$	175.2	179.5	201.42	192.21	4.57
$W_{40}R_{0.8}O_{30}$	140.2	145.6	165.7	159.53	3.72

Table 13. Specific bending strength (MPa/Nm^{-1}) results for sandwich

It is significant to note that the experimental results for specific bending strength match well with FEA values especially for the ones with PL gradation. It is observed that bending strength obtained from FEA is slightly higher than experimental values. This could be due to inability of modeling inhomogenities creeping in during the processing of samples which may result in lowering specific strength.

6.4. Discussion on fractured samples

Within the elastic region of the load-displacement curve (Figure 8), where no damage is induced, the responses of all specimens to the applied loads are quite similar. This is visible in the form of nearly constant slope in the elastic region of the load-displacement curves. It is observed that the failure starts in the form of crack origination on the tensile side of the specimen as displacement increases. On further loading, the skin of the sandwich composite that is on the tensile side tends to fracture, causing the final failure of the specimen. However, it is not significant enough to lead to the final failure of the specimen. It is observed that the entire specimen fractures at a much later instant of skin fracture. Appearance of small linear region (B to C in Figure 8) at the end in the load-displacement curves is due to stiffening of FG core before final failure. During the loading process, deformation also takes place in the compression side of the specimen. Cracks initiate from the tensile side and propagate to the compressive side within the core in all sandwiches.

It is worth discussing the mode of failure. Sandwich samples tested under bending did not display the distinct separation into pieces at failure. The FG core being compliant is observed to be successfully absorbing media. Basically two types of failure mechanisms observed are skin cracking and delamination between skins and core. Figure 13 shows the failed sandwich specimens with their failure modes.

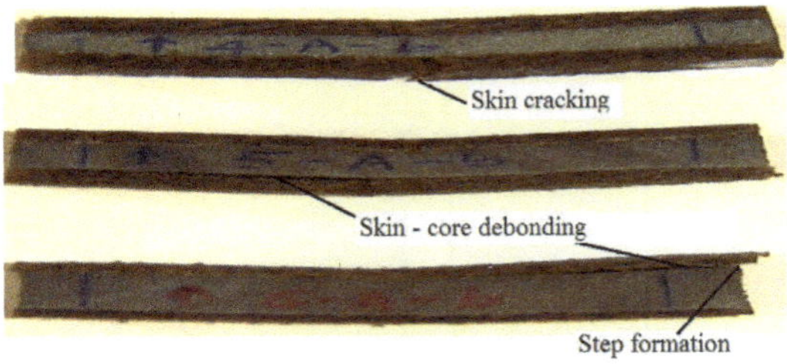

Figure 13. Sandwich failure modes under three point bending loads

The sandwich beams failed at the center of the two supporting rollers. In this portion of the beam, the shear force is zero and only the pure bending exists. Thus, the sandwich samples are capable of resisting higher bending moment. As the load on the specimen is increased, failures first start under the loads in tensile region and then they propagate towards the compressive zone through compliant FG core. All the samples failed under skin tension or compression and skin - core debonding. The sandwiches with higher C/H ratio have shown skin - core debonding. FG core takes up most of the load applied for higher C/H ratios (lesser skin thickness). Since core is made up of rubber composite being compliant in nature, relative movements are set up with respect to skin resulting in inter laminar shear stresses. As magnitude of these stresses crosses the adhesive strength delamination creeps in. Some sandwich samples are seen to be intact even after the first sign of failure. These samples exhibited a spring back effect. Samples bearing lower C/H ratio have failed mainly because of skin cracking along the jute orientation. Few samples failed due to shearing at skin-core interface displayed step formation.

7. Conclusions

This section highlights the significant conclusions drawn from the results presented earlier. Major inferences from both experimental and finite element investigations are discussed below.

Density of FG sandwiches increases with filler content and C/H ratio while decreases with jute orientation. An experimental investigation of sandwiches under bending loads for specific modulus and specific strength shows that C/H ratio and fly ash weight fraction are the influential factors respectively. Specific bending modulus in both cases (i.e. rubber up and ash up) the sample $W_{20}R_{0.8}O_{0}$ registered the higher value while $W_{40}R_{0.4}O_{0}$ shows higher value of specific strength. Rubber up configuration registered higher results compared to ash up condition for modulus and strength. The ash up condition recorded about 30% increase in strength. Increasing fly ash weight fraction rendered an increase in bending strength of about 29% for rubber up condition. Specific strength values estimated from FEA for bending loads match well with experimental results especially for piecewise gradation.

Author details

Mrityunjay R. Doddamani* and Satyabodh M Kulkarni
Mechanical Engineering, National Institute of Technology Karnataka, Surathkal, India

8. References

[1] Nicoleta Alina Apetre. Sandwich Panels with Functionally Graded Core. PhD thesis. Graduate school of the university of Florida; 2005.

[2] Vinson J. The Behavior of Sandwich Structures of Isotropic and Composite Materials. Pennsylvania: Technomic; 1999.

[3] Gdoutos EE, Daniel IM. Failure modes of composite sandwich Beams. Journal of Theoretical and Applied Mechanics 2008;35(1-3),105-118.

[4] Venkata Dinesh Muthyala. Composite sandwich structure with grid stiffened core. Master thesis. Graduate Faculty of the State University and Agricultural and Mechanical College, Louisiana, 2007.

[5] Zenkert D. Handbook of Sandwich Construction. London: EMAS, Chameleon Press; 1997.

[6] Abrate S. Impact on Composite Structures. Cambridge: Cambridge University Press; 1998.

[7] Paulino GH, Jin ZH. Correspondence Principle in Viscoelastic Functionally Graded Materials. Journal of Applied Mechanics 2001;68(1),129-132.

[8] Koizumi M. Concept of FGM. Ceramic Transactions 1993;34,3-10.

[9] Suresh S, Mortensen A. Fundamentals of Functionally Graded Materials. London: Institute of Materials; 1998.

[10] Miyamoto Y, Kaysser WA, Rabin BH, Kawasaki A, Ford RG. Functionally Graded Materials: Design, Processing and Applications. Dordrecht: Kluwer Academic; 1999.

[11] Cannillo V, Manfredini T, Siligardi C, Sola A. Preparation and experimental characterization of glass–alumina functionally graded materials. Journal of the European ceramic society 2006;26(6),993-1001.

[12] Al-Ajmi MA, Alhazza KA. 5th International Conference on Composite Science and Technology, ICCST/5 2005: conference proceeding. Vibration and damping analysis of

* Corresponding Author

sandwich beams with functionally graded viscoelastic core. Sharjah, United Arab Emirates, 125 - 128.

[13] Parameswaran V, Shukla A. Processing and Characterization of a Model Functionally Gradient Material. Journal of Materials Science 2000;35(1),21-29.

[14] Venkataraman S, Sankar BV. 42nd AIAA/ASME/ASCE/AHS/ASC Structures, Structural Dynamics and Materials Conference 2001: conference proceeding. Analysis of sandwich beams with functionally graded core. AIAA-2001-1281, Seattle, Washington.

[15] Kirigulige MS, Kitey R, Tippur HV. Dynamic fracture behaviour of model sandwich structures with functionally graded core: a feasibility study. Composites Science and Technology 2005;65(7-8),1052-1068.

[16] Pollien A, Conde Y, Pambaguian L, Mortensen A. Graded open-cell aluminium foam core sandwich beams. Materials Science and Engineering: A 2005;404(1-2),9-18.

[17] Gupta N. A functionally graded syntactic foam material for high energy absorption under compression. Materials Letters 2007;61(4-5),979-982.

[18] Krajcinovic D. Sandwich Beam Analysis. Journal of Applied Mechanics 1971;38(1),773-778.

[19] Krajcinovic D. Sandwich Beams with Arbitrary Boundary Conditions. Journal of Applied Mechanics 1975;42(1),873-880.

[20] DiTaranto RA. Static Analysis of a Laminated Beam. Journal of Engineering for Industry 1973;95(2),755-761.

[21] Johnson AF, Sims GD. Mechanical Properties and Design of Sandwich Materials. Composites 1986;17(4),321-328.

[22] Teti R, Caprino G. First International Conference on Sandwich Construction 1989: conference proceeding. Mechanical Behavior of Structural Sandwiches. Stockholm, Sweden, 53-67.

[23] Lingaiah K, Suryanarayana BG. Strength and Stiffness of Sandwich Beams in Bending. Experimental Mechanics 1991;31(1),1-7.

[24] Gupta N, Woldesenbet E. Microscopic Studies of Syntactic Foams Tested Under Three-Point Bending Conditions. American Society of Mechanical Engineers 2002;1,147-152.

[25] Maharsia R, Gupta N, Jerro HD. Investigation of flexural strength properties of rubber and nanoclay reinforced hybrid syntactic foams. Materials Science and Engineering: A 2006;417(1-2),249-258.

[26] Kishore, Ravi, Sankaran S. Short-Beam Three-Point Bend Test Study in Syntactic Foam. Part III: Effects of Interface Modification on Strength and Fractographic Features. Journal of Applied Polymer Science 2005;98(2),687-693.

[27] Avila AF. Failure mode investigation of sandwich beams with functionally graded core. Composite Structures 2007;81(3),323-330.

[28] Karthikeyan CS, Sankaran S, Kishore. Influence of chopped strand fibres on the flexural behavior of a syntactic foam core system. Polymer International 2000;49(2),158-162.

[29] Karthikeyan CS, Chedarampet S, Sankaran S, Kishore. Flexural Behaviour of Fibre-Reinforced Syntactic Foams. Macromolecular Materials and Engineering 2005;290(1),60-65.

[30] Karthikeyan CS, Sankaran S, Kishore. Investigation of bending modulus of fiber-reinforced syntactic foams for sandwich and structural applications. Polymers for Advanced Technologies 2007;18(3),254-256.

[31] Gupta N, Woldesenbet E. Characterization of Flexural Properties of Syntactic Foam Core Sandwich Composites and Effect of Density Variation. Journal of Composite Materials 2005;39(23),2197-2212.

[32] Montgomery DC. Design and analysis of experiments. New York: Wiley; 2001.

[33] Ferrigno TH., Katz HS., Milewski JV., editor. Handbook of fillers and reinforcements for plastics. New York: Van Nostrand Reinhold; 1978, 66-71.

[34] Kulkarni SM, Anuradha D, Murthy CRL, Kishore. Analysis of filler-fibre interaction in fly ash filled short fibre-epoxy composites using ultrasonic NDE. Bulletin of Materials Science 2002;25(2),137-140.

[35] Pedlow JW., Torrey S., editor. Cenospheres, in Coal ash utilization Fly ash, Bottom ash and Slag. New Jersy: Noyes; 1978, 353-362.

[36] Mohapatra R, Rajagopala Rao. Some aspects of characterisation, utilisation and environmental effects of fly ash. Journal of Chemical Technology and Biotechnology 2001;76(1),9-26.

[37] Kishore, Kulkarni SM, Sharathchandra S, Sunil D. On the use of an instrumented set-up to characterize the impact behavior of an epoxy system containing varying fly ash content. Polymer testing 2002;21(7),763-771.

[38] Mohan R, Kishore, Shridhar MK, Rao RMVGK. Compressive strength of jute-glass hybrid fibre composites. Journal of Materials Science Letters 1982;2(2),99-102.

[39] Shah AN, Lakkad, SC. Mechanical properties of jute-reinforced plastics. Fibre Science and Technology 1981;15(1),41-46.

[40] Gowda TM, Naidu ACB, Rajput Chhaya. Some mechanical properties of untreated jute fabric reinforced polyester composites. Composites Part A: Applied Science and Manufacturing 1999;30(3),277-284.

[41] Mohan, Rengarajan, Kishore. Jute-Glass sandwich composites. Journal of Reinforced plastics and composites 1985;4(2),186-194.

[42] ASTM C393. Standard Test Method for Flexural Properties of Flat Sandwich Construction. ASTM International, PA, USA, 2000.

[43] Gupta N, Sandeep Kumar Gupta, Mueller BJ. Analysis of a functionally graded particulate composite under flexural loading conditions. Materials Science and Engineering: A 2008;485(1-2),439-447.

[44] ASTM D3039. Standard Test Method for Tensile properties of polymer matrix composite materials. ASTM International, PA, USA, 2008.

[45] ASTM D792. Standard test methods for density and specific gravity (Relative density) of plastics by displacement. ASTM International, PA, USA, 2008.

Section 3

Applications of FEA in "Wave Propagation and Failure-Analysis"

Chapter 8

Modeling Dynamics Behaviour of Materials: Theoretical Framework and Applications

Giovanni Leonardi and Michele Buonsanti

Additional information is available at the end of the chapter

1. Introduction

The study of the dynamic behaviour of materials falls in a multidisciplinary area where many different disciplines converge. The definition of the state of the solid body subject to various actions is very different from the conditions of almost static load, or single dynamic load.

Complex dynamic actions (i. e. explosion, travelling waves, etc.) request an approach where both inertia and kinetics of the material are fundamental elements to describe the variable answer in terms of stress and strain.

The topics covered in the chapter are the following:

A first section, where particular attention about the shock waves-induced phase transformations and chemical changes will be given. A modelling coupled multifield processes will be introduced in the multiphase solids case through constitutive assumption, energy balance and mass transfer and a reaction-diffusion model.

A second part in which some applications of finite element analysis to multi-physics dynamic problems is presented and discussed.

2. Waves equations

When an elastic media is subject, over one or more points, to fast actions then media acceleration results. The strain field resulting is carried out within the media by elastic waves, and so the new and variable stress field should be equilibrated [1, 2].

Let us call u the displacement field, λ and μ the Lamè elastic constants, ρ the media density then, in the isotropic, homogeneous and elastic media we have the follow motion equation:

$$\rho \ddot{u} = \mu \nabla^2 u + (\lambda + \mu)\nabla(\nabla u) \tag{1}$$

In the Eq. (1) the symbol ∇ represents the nabla differential operator. Under the volume forces equal to zero one possible solution to the Eq. (1) has the form:

$$u(x,t) = u(t - nx / c) \tag{2}$$

where n is a constant unit vector and c represents the velocity. Representing the Eq. (2) the solution to the Eq. (1) we observe that the Eq. (2) is a plane wave equation scattering in n direction, with c velocity. When a direction is fixed, for example x_1, then we have:

$$u(x,t) = u[t - (x_1 / c)] \tag{3}$$

Generally, an elastic wave can be of two kinds, longitudinal (volume wave) or transversal (shear wave), and respective velocities go with the function:

$$\alpha = sqr[(\lambda + 2\mu) / \rho]; \qquad \beta = sqr(\mu / \rho) \tag{4}$$

As a principle, we have an elastic wave's emission when a fast, and localized variation on the body force exists.

In this case, the Eq. (1) appears as a Green tensor, that is a second order tensor time dependent, G_{ij} $(x, t; \varepsilon, \tau)$. Neglecting isotropic source, the dynamics source gives out longitudinal and transversal waves with various amplitude according to the direction.

The wave shape represents the signal shape reproduced by the source, in other words the temporal course of the source namely, the $F(t)$ function.

Since the Green tensor calculation appears with difficulty, through known references, it becomes possible to proceed by the Helmholtz potential method, and therefore to derive, for convolution, the Green tensor final form [3]:

$$\begin{aligned} G_{ij}(x,t;0,0) &= (1/4\pi\rho)(3\gamma_i\gamma_j - \delta_{ij})r^{-3}t[H(t - r^{\alpha-1}) - H(t - r\beta^{-1})] + \\ &+ [(1/4\pi\rho\alpha^2)\gamma_i\gamma_j\delta(t - r\alpha^{-1})] + [(1/4\pi\rho\beta^2)(\delta_{ij} - \gamma_i\gamma_j)\delta r^{-1}(t - r\beta^{-1})] \end{aligned} \tag{5}$$

Where γ_i and γ_j are the director cosine of x, δ the x varied position and δ_{ij} the Kronecker delta.

The Eq. (5) is composed by 3 terms, all depending on the distance. We have the first one, called close field, while the other two called away fields. We observe not separable waves in the close field while, in the away field longitudinal and transversal waves appear distinct. All of this allows, in the next modelling to consider only the close field and then longitudinal and transversal motions together.

Here we consider plane waves travelling in an elastic half-space and, without loss of generality, we affirm that the wave normal lies in the vertical plane of the half-space. Referring to the infinite space case, we assume that the particle motion, due to dilatational effects, belongs to the wave normal direction and therefore lies in the vertical plane. Otherwise, the particle motion, due to shear effects, present components in either vertical or horizontal plane. Introducing the functions H and Φ, called the Helmholtz potential functions, the governing equations related on this approach follow:

$$\begin{aligned} u_x &= (\partial\Phi / \partial x) + (\partial H_z / \partial y); \quad u_y = (\partial\Phi / \partial y) + (\partial H_z / \partial x); \\ u_y &= (\partial H_z \partial y) + (\partial H_y / \partial x); \quad (\partial H_x / \partial x) + (\partial H_y / \partial x) = 0. \end{aligned} \tag{6}$$

The stress-displacement relations are given by:

$$\begin{aligned} \sigma_{xx} &= (\lambda + 2\mu)[(\partial^2\Phi / \partial x^2) + (\partial^2\Phi / \partial y^2)] - 2\mu[(\partial^2\Phi / \partial y^2) - (\partial^2\Phi / \partial x^2) - \partial^2 H_x / \partial y\partial x)] \\ \sigma_{yy} &= (\lambda + 2\mu)[(\partial^2\Phi / \partial x^2) + (\partial^2\Phi / \partial y^2)] - 2\mu[(\partial^2\Phi / \partial y^2) - (\partial^2\Phi / \partial y^2) - \partial^2 H_x / \partial y\partial x)] \\ \tau_{xy} &= \mu[2(\partial^2\Phi / \partial x\partial y) + (\partial^2 H_z / \partial y^2) - (\partial^2 H_z / \partial x^2)] \\ \tau_{yz} &= \mu[2(\partial^2 H_x / \partial^2 y) + (\partial^2 H_y / \partial y\partial x)] \end{aligned} \tag{7}$$

Moreover at Eq. (7) the boundary conditions should be summed:

$$\sigma_{yy} = \tau_{yx} = \tau_{yz} = 0; \qquad y = 0 \tag{8}$$

In according with [5], we see that the problem above defined can be uncoupled and therefore resolve the motion problem into two parts, namely the first one is plane strain, such that $u_z = 0$, u_x, $u_y \neq 0$. The second one is the secondary wave motion where only $u_z \neq 0$.

From now on, we wish to study the interaction of elastic waves with discontinuities or boundaries of more complex shape than that of the half-space framework.

Particularly, we focus the attention over the scattering of compression waves against absorbed obstacles [3, 5], as well as inclusion in elastic half-space. The propagation and reflection of waves, generated by dynamical forcing over the external surface, against inner surfaces or discontinuity [4-6] has, also, great interest in seismology, structural foundations since the vibratory phenomenon represents a very important further load condition for global stability and strength [7-9].

Moreover this building framework picks up more general problems, for example voids, flaws or stress raise in half-space constituent materials. The approach adopted follows the stated assumptions and hypothesis , that is expansion of the wave fields in series.

3. Shock waves

The impact between two solid elements represents the simplified condition for the generation of shock waves. In the specific case of parallel impact the two surfaces enter in contact simultaneously and all the points of the two surfaces enter in contact at the same time.

The true profile of a shock wave is complex. In the following Figure 1, it is possible to observe the difference between the ideal and the true profile where, for the latter, it is clear the dependence form the characteristics of the material and the pressure applied at contact.

An impulsive stress on contact has an initial, middle and final pressure value. Initially it's a shock wave (discontinuity in compression); the mean reaction is characterized from a slow variation of pressure and the final from a dissolution which tends to the undisturbed state.

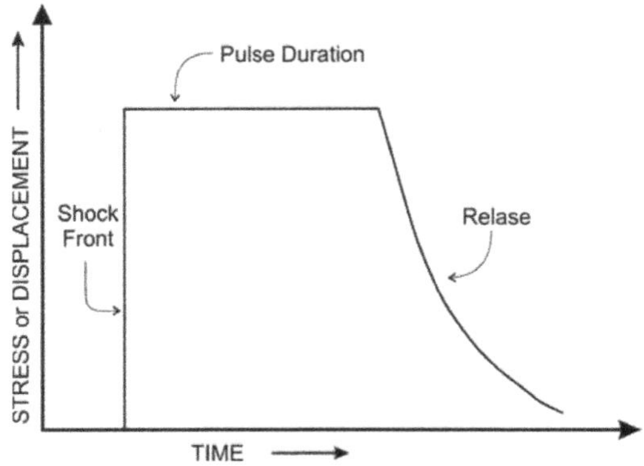

Figure 1. *(a)* idealized and *(b)* generic realistic shock wave profile (from Meyers [10]).

In a previous paper the authors has investigated the waves generation after the impact on a granular plate [9]. The study has been developed, initially from a microscopic point of view and subsequently on macroscale.

The effects are strictly linked with material degradation associated with damage evolution. In accordance to [11] the shock waves can induce phase transitions in the solid, (Figure 2), then transitions form elastic to plastic response (in our case plasticization of the mixture binding component).

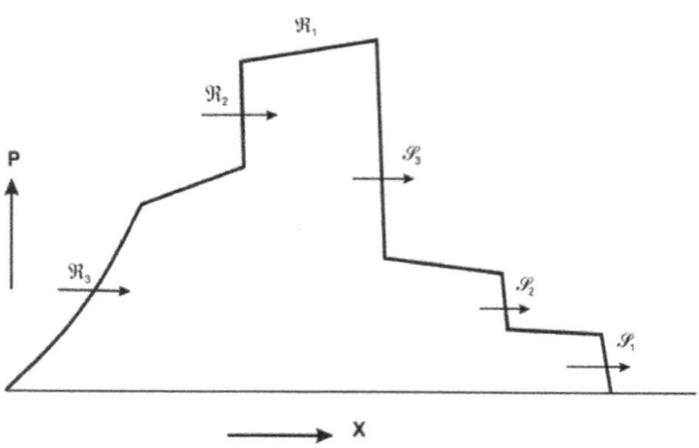

Figure 2. Pressure distribution in a pulse propagating through a material undergoing a phase transformation and a transition from elastic to inelastic behaviour (from Meyers [10]).

On a theoretical point of view, we classify the problem as the propagation of a shock wave, where a uniform contact pressure is applied on a plane solid surface in an elastic semi space. Given the geometrical origin $x = 0$ and the beginning of the phenomenon at time $t = 0$, after a laps of time t the shock front divides the space in two regions, one undisturbed, the other compressed and accelerated. Therefore the flow equation is reduced to the jump condition:

$$\Delta P = \frac{(v_s - v_o)(v_p - v_s)}{V_s} \tag{9}$$

with v_s the wave propagation speed, v_o and v_p are the speeds of the particles respectively behind and in front of the shock front, and, V_s is the specific volume of the medium.

The clear result is the introduction of a pressure step which travels across the medium, with changes of shape which depend on the mechanical proprieties of the element.

In the case of impact the contact time tends to zero, therefore $t_i = t_p$ where t_i is the impact time and t_p is the plasticization time. In theory, the problem can be represented as two successive phases.

First phase: transversal speed at the centre of the body remains constant. This phase is necessary to absorb the remaining kinetic energy in the body.

Second phase: a concentrated plasticization begins which starts to expand from the core to the external part of the body. The time it takes is given by the expression:

$$t = \mu v^o r^2 / 6M^o \tag{10}$$

Where μ is the Lame's material constant and the associated maximum permanent transversal displacement in the contact zone can be approximated as:

$$\omega \cong 3p_0^2 r^2 / 4\mu p_c \tag{11}$$

In regards to the mechanical proprieties of the medium subject to the impact actions in the case, the response of solids made of asphalt mixtures can be divided, accordingly to [12, 13] in three groups: elastic visco-elastic and visco-plastic. In the one-dimensional case we have:

$$E_{ii}^{e} = \epsilon^{-1} T_{ii} \tag{12}$$

$$E_{ii}^{ve} = A(T_{ii}) t^{\alpha} \tag{13}$$

$$E_{ii}^{vp} = B(T_{ii}) f(N) t^{\beta} \tag{14}$$

where $f(N)$, $A(T)$, $B(T)$ are functions of the stress in the viscose phase. A, B, α, β and ε are constants determined at constant temperature.

In the multidimensional case the equations above become:

$$[E]^e = [K] \cdot [T] \tag{15}$$

$$[E]^{ve} = At^{\alpha}[H] \cdot [T^h] \tag{16}$$

where K is the deformability matrix, $H = \epsilon K$ and $T^h = [T_{ijk}]T$

In regard to the visco-elastic part, it has to be specified that the critical points which arise in this phase with the load time can be resolved using the Perzyna theory [13]. For an associated visco-plastic flow we have:

$$[E]^{vp} = \gamma \cdot \phi(t) \cdot \partial F / \partial t \tag{17}$$

where:

- γ is a fluidity parameter associated to the loading times and the number of loading cycles;
- ϕ(F) is the viscose flow function;
- F is the plasticity function $F(T, k)$ with k the hardening parameter.

Passing to the numerical implementation, in the case of reduced load intervals, an iterative procedure, as a Newton-Raphson, can be applied.

$$A[T]^n = [K][A[E]]^n - A\alpha t^{\alpha-1}[K][T^h]^n - A[E^{vp}]^n \tag{18}$$

and, after the rightful developing, stress and strain in approximated as:

$$[T]^{n-1} = [T]^n + A[T]^n \tag{19}$$

$$[E^{vp}]^{n+1} = [E^{vp}] + A[E^{vp}]^n \tag{20}$$

Therefore it follows the link between micro-scale effects and material behaviours at macro-scale.

So we focus the micromechanics of the damage processes because the nonlinear response of typical engineering materials is almost entirely dependent on the primary change in the concentration, distribution, orientation and defects in its structural composition.

The relation between the continuum damage mechanics and the fracture mechanics is very complicated, in essence, a question of scale. The important role of scale can be clarified by an energetic point of view.

In view of an approximated continuum theory with the physical foundation of micromechanical models, a promising strategy would consist of combining the best features of both models. In this approach we consider only the first layer of the pavement package because, at micro-scale, damage distribution at the edge of the body, where surface degradation is of importance, is expected to be significantly different from the damage distribution far from the edge in the body.

We follow the volume element theory RVE, it's possible to represent a non-homogeneous solid with periodic microstructure. Particularly in the transition toward the micro-scale our RVE can be represented by more granular elements joint by means of an asphalt mixture, so considerations are applied on the contact area among two granular elements. In this manner the homogenization problems can be satisfied.

Following Sneddon's solution [14] type we model the physics of impact by means of a rigid frictionless asymmetric concentrated impact, with generic concave profile described by the function $f(r)$. We find respectively, the σ_z pressure distribution under the concentrated impact and the displacement on the surface.

4. Shell structures and blast loading

To reproduce a possible genuine model becomes fundamental to describe the single load conditions since blast action, fundamentally, can be decomposed in thermal and shock wave's loads. Here we develop the theoretical assumptions in both cases just starting with some structural considerations about the thick shell behaviour.

From a structural point of view the tunnel can be considered as well as a half thick-walled cylinder subject to internal and external pressures. So we consider a half cylinder of inner radius a and outer radius b and subject to an internal pressure p_a and an external pressure p_b. We choose, as the closest to real behaviour, the plane stresses condition so that the ends of the cylinder be free to expand.

Assuming the z-axis as the revolution axis, the deformation becomes symmetrical respect to the z-axis. Consequently it's convenient to use cylindrical coordinates r, θ, z.

According to [16-18] the plane stresses conditions involve $\sigma_z = 0$ and $\tau_{rz} = 0$ and the equilibrium conditions, without body forces become.

$$\frac{\partial \sigma_r}{\partial r} + \frac{\sigma_r - \sigma_\theta}{r} = 0 \tag{21}$$

while the deformation field E as the components in the form:

$$E_\theta = \frac{1}{r}\frac{\partial u_\theta}{\partial \theta} + \frac{u_r}{r}; \quad E_r = \frac{\partial u_r}{\partial r}; \quad E_r = \frac{\partial u_z}{\partial z} \tag{22}$$

where the function $u(r, \theta, z)$ represents the displacements field over the shell.

Introducing the Lame's constitutive equations (with ν and ε, respectively, Poisson's and Young's modulus) after some simple calculations we get the basic equations governing the thick-walled half-cylinder:

$$u_r = \frac{1-\nu}{\epsilon}\frac{p_a a^2 - p_b b^2}{b^2 - a^2} r + \frac{1+\nu}{\epsilon}\frac{a^2 b^2}{r}\frac{p_a - p_b}{b^2 - a^2}$$

$$\sigma_r = \frac{p_a a^2 - p_b b^2}{b^2 - a^2} - \frac{b^2 a^2}{r^2}\frac{p_a - p_b}{b^2 - a^2}$$

$$\sigma_\theta = \frac{p_a a^2 - p_b b^2}{b^2 - a^2} + \frac{b^2 a^2}{r^2} \frac{p_a - p_b}{b^2 - a^2} \tag{23}$$

Under these conditions, we recall either of the specific conditions of the internal and external pressure loads. In the first case (internal pressure) the above equations becomes:

$$\begin{aligned} \sigma_r &= \frac{p_a a^2}{b^2 - a^2}\left(1 - \frac{b^2}{r^2}\right) \\ \sigma_\theta &== \frac{p_a a^2}{b^2 - a^2}\left(1 + \frac{b^2}{r^2}\right) \end{aligned} \tag{24}$$

From the equations above, a consideration can be drawn about the circumferential stress (σ_θ tensile stress), which is at its greatest on the inner surface and is always greater than p_a. In the second case (external pressure) the general equations assume the form:

$$\begin{aligned} \sigma_r &= -\frac{p_b}{b^2 - a^2}\left(1 - \frac{a^2}{r^2}\right) \\ \sigma_\theta &= \frac{p_b b^2}{b^2 - a^2}\left(1 + \frac{a^2}{r^2}\right) \end{aligned} \tag{25}$$

The stress paths, when no inner holes were present, are uniformly distributed in the cylinder. From now we will be able to describe the coupling actions over the thick-walled half-cylinder shell and for this we run recalling some basic thermo-elasticity assumption. There is a large literature over the question but we prefer to follow [18-20].

We focus the consistence of thermal stresses induced in thick-walled half-cylinder when the temperature field is symmetrical about the z-axis. In this case we suppose the temperature T as radius function only and independent from z then plane strain $E_z = 0$. With analogous considerations as above, the basic equations, for the coupled problem, can be written as.

$$\begin{aligned} u_r &= \frac{(1+\nu)\alpha}{(1-\nu)r}\left\{\int_a^r Trdr + \frac{(1-2\nu)r^2 + a^2}{b^2 - a^2}\int_a^b Trdr\right\} \\ \sigma_r &= \frac{\alpha\epsilon}{(1-\nu)}\frac{1}{r^2}\left\{\frac{r^2 - a^2}{b^2 - a^2}\int_a^b Trdr - \int_a^r Trdr\right\} \\ \sigma_\theta &= \frac{\alpha\epsilon}{(1-\nu)}\frac{1}{r^2}\left\{\frac{r^2 - a^2}{b^2 - a^2}\int_a^b Trdr + \int_a^r Trdr - Tr^2\right\} \\ \sigma_z &= \frac{\alpha\epsilon}{(1-\nu)}\left\{\frac{2}{b^2 - a^2}\int_a^b Trdr - Tr\right\} \end{aligned} \tag{26}$$

If the temperature T is positive and if the external temperature is equal to zero then the radial stress is always compressive, like other stresses in the inner surface.

After numerous disasters in the building and structures, the fire-structure question was developed for many researchers, which has reproduced a large and specific literature. For instance it is our opinion, referring at some as important in [16-19]. From now we will deepen the other coupled action namely the structural effects after the burst. According to [10] the interaction of a detonating explosive with a material in contact with it or in close proximity is extremely complex, since it evolves detonation waves, shock waves, expanding gases, and their interrelationships.

The question was developed, principally, by military requirement which study has developed the computational apparatus, for instance the Gurney equation [25-26].

It's our interest some basic assumption linked to the effective problem that requires us namely, only actions from shock waves. So, we affirm the following basic assumptions:

a. A shock is a discontinuous surface and has no apparent thickness.
b. The shear modulus is assumed to be zero and so it responds to the wave as a fluid, and the theory can be restricted to higher pressures.
c. Body forces and heat conduction at the shock front are negligible.
d. There is no elastic-plastic behaviour.
e. Material do not undergo phase transformations.

Now we will consider the dynamic behaviour of thick-wall cylindrical shell under internal pressure produced by shock wave.

Let p_c be the collapse pressure, then the shell is subject to a symmetrical internal pressure pulse, in the interval time $0 \leq t \leq \tau$, while $p = 0$ when $t \geq \tau$. Again we assume a perfectly rigid plastic material behaviour.

Supposing the pressure load symmetric, then the yielding is controlled by force in the shell middle plane. So, let N_θ be the generalized membrane forces, at the yielding point we have $N_\theta = N_c$ (with N_c the fully plastic membrane forces). Neglecting the elastic effects, the dynamics response consist of two phases motion with $N_\theta = N_\phi = N_c$.

For major clarify we consider, as the second phase, the time as $\tau \leq t \leq t^*$ where t^* is the response duration time. Let ω be the transverse displacement of the shell middle plane and let v° be the spherically symmetric outwards impulsive velocity, we then find the radial displacement

$$\omega = N_c t^2 / \mu r - v^o t \tag{27}$$

After some calculations we have the associated permanent radial displacement field over the shell.

$$\omega_f = -\mu r v^o / 4N_c \tag{28}$$

5. Applications examples

In the following, we will give two examples of finite element applications to dynamic analysis in particular, interaction problems of structures systems subject to explosion waves and impact loads are presented [9] [15].

5.1. Impact loads on flexible pavement

In this example we assessed the effects of a heavy impact caused by aircraft landing gear wheels on a flexible airport pavement.

Flexible pavements are usually idealized as closed systems consisting of several layers; so the surface, base, sub-base and sub-grade material were modelled using 3-D finite elements. While an elastic constitutive model was assumed for the granular layers and the base course, a time hardening creep model was incorporated to simulate the viscoelastic behaviour of the HMA surface layer

The aircraft considered in the model was the Airbus 321 [26]. The most common way of applying wheel loads in a finite element analysis is to apply pressure load to a circular or rectangular equivalent contact area with uniform tyre pressure [27]. To investigate the impact simulation in exceptional condition, the dynamic parameters of an "hard" landing, that caused the broken of some gear components, were considered [28]. Starting from this, considering the damping effect of the gear system, it is possible to calculate the acceleration graph during the landing (Figure 3).

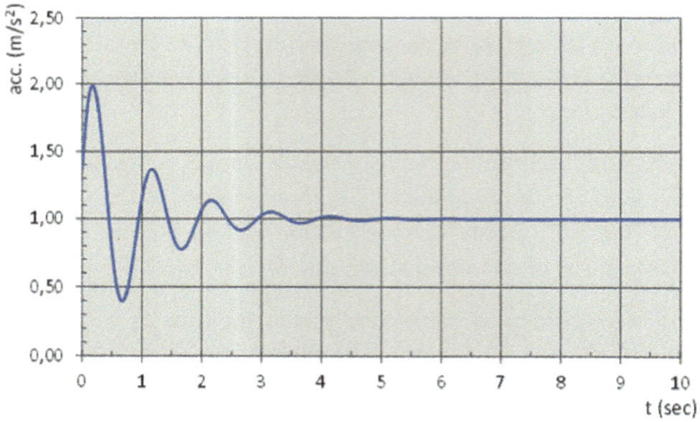

Figure 3. Acceleration graph.

As shown in Figure 3 the peak acceleration value, during the hard landing, is 1.99 m/s^2. This value of acceleration was used in the finite element model to calculate the maximum wheel load.

The finite element model has the following dimensions: 10 m in x and y directions and 2.5 m in the z- direction. The three-dimensional view of finite element model is shown in Figure 4.

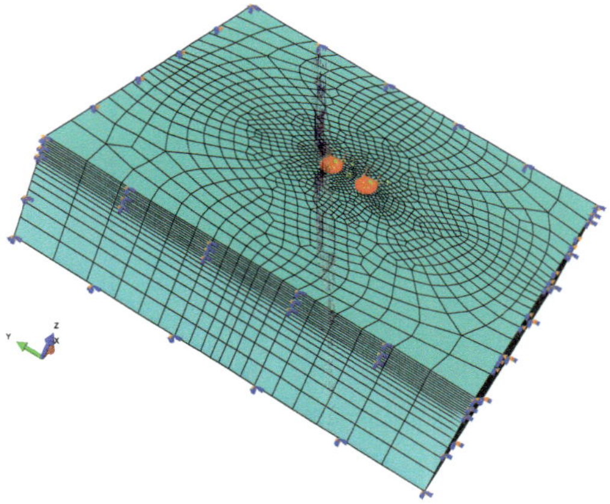

Figure 4. Three-dimensional view of the finite element model.

The degree of mesh refinement is the most important factor in estimating an accurate stress field in the pavement: the finest mesh is required near the loads to capture the stress and strain gradients. The mesh presented has 126245 nodes and 29900 quadratic hexahedral elements of type C3D20R (continuum 3-dimensional 20 node elements with reduced integration). Quadratic elements yield better solution than linear interpolation elements [29].

The loads (vertical and horizontal) were uniformly applied to the element, which was created to be the same size as the wheel imprint of an Airbus A321.

In this example the surface was considered to be free from any discontinuities (with no cracks) or unevenness, and the interface between layers was considered to be fully bonded (with no gaps).

The model was constrained at the bottom (encastre: U1 = U2 = U3 = UR1 = UR2 = UR3 = 0); X-Symm (U1 = UR2 = UR3 = 0) on the sides parallel to y-axis; and Y-Symm (U2 = UR1 = UR3 = 0) on the sides parallel to x-axis.

The results of the non-linear FE analysis are illustrated in the following figures. Figure 5 shows the Mises stress distribution for the considered FE model at the landing aircraft impact instant and Figure 6 presents the results of pavement surface deflection along transversal direction.

Finally, in the graph of Figure 7 are plotted the predicted transversal surface deflection profiles along the transverse center line.

This example shows how finite element analysis of pavement structures, if validated, can be extremely useful, because it can be used directly to estimate pavement response parameters without resorting to potentially costly field experiments.

If accurate correlations between the theoretically-calculated and the field-measured response parameters can be obtained, then the finite element model can be used to simulate pavement response utilizing measurements from strain gages. In particular, the proposed model has clearly confirmed the need and importance of 3-Dimensional finite element analyses on flexible pavements to consider the behaviour of the structure under high stress.

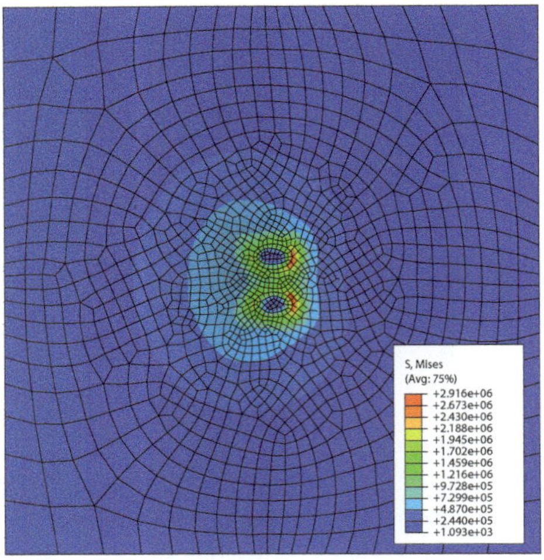

Figure 5. Mises stress at the instant of impact.

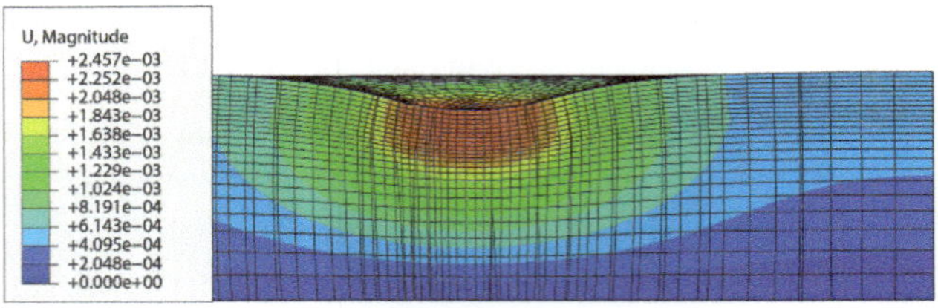

Figure 6. Displacement contours at the instant of impact.

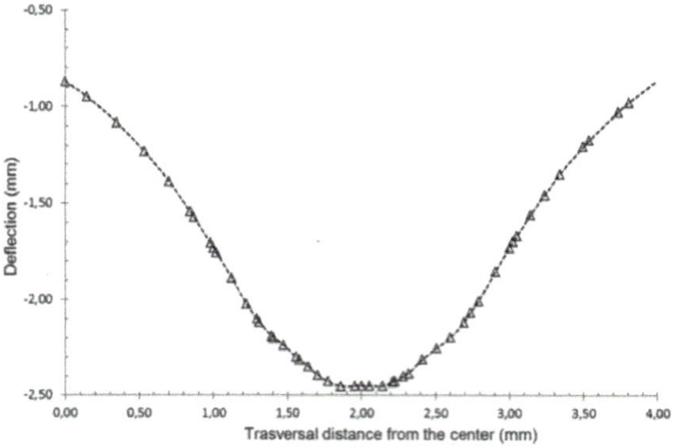

Figure 7. Predicted deflection profiles (y-direction).

5.2. Confined explosions

In the following application a 3-D simulation of tunnel structures under Blast loading is proposed.

The Finite Element model was based on a single track railway tunnel system consisting of concrete tunnel tube with the section dimensions reported in Figure 8. The tunnel was about 10 m below the ground surface. The model extended 150 m in the longitudinal direction of the tunnel, while the length and height of the model were of 26.8 m. The finite Element model was fixed at the base and roller boundaries were imposed to the four side. The modelled tunnel structure is surrounded by soil and this load represents the starting state of stress. Drucker-Prager elasto-plastic model was used to model the soil. For the characterization of the reinforced concrete of the tunnel structure it was considered a C50/60 class concrete having thermal characteristics according with the indications of the Eurocode 2 [30].

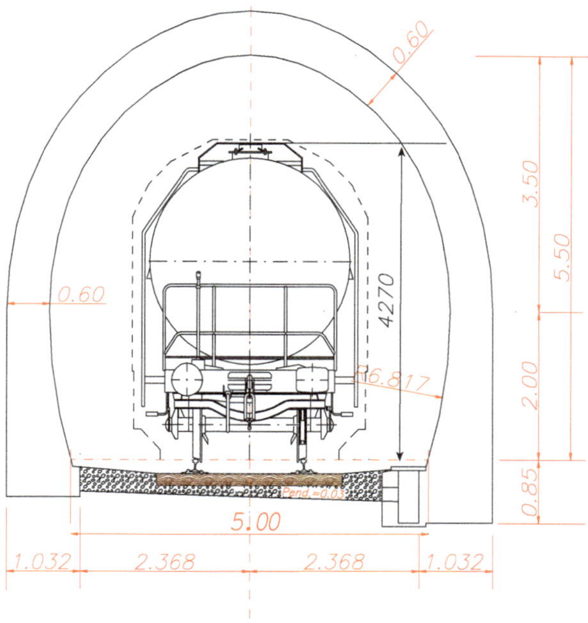

Figure 8. Rail tunnel section.

A fundamental aspect in the study of fire resistance in underground structures is the definition beforehand of the fire scenario taken in the analysis, therefore choosing the best fit standard curve. A standard curve is the cellulose curve defined in several standards, e.g. ISO 834 [23]. Specific temperature curves have been developed in some countries to simulate hydrocarbon fires in tunnels. Examples of such curves are the RABT/ZTV Tunnel Curve in Germany [31] and the Rijkswaterstaat Tunnel Curve (RWS curve) in The Netherlands (based on laboratory scale tunnel tests performed by TNO in 1979 [32]). In the considered model the HC curve was used to simulate the fire action.

The blast overpressure was generated from an instantaneous release of 50 m^3 LPG rail tanker at 326K. The pressure-time curve was assumed to be of triangular shape, the duration of which was obtained from CONWEB reflected pressure diagram [33]. To calculate the decay of blast overpressure during the longitudinal direction of the tunnel the Energy Concentration Factor (ECF) method was used [34].

During the propagation of the blast wave over the first 75 m from the BLAVE to the tunnel opening, the blast overpressure falls from 1700 kPa (vapor pressure at 326 K) down to approximately 86 kPa. This decay is solely through the intense energy dissipation in the strong leading shock of the blast wave.

The 3-Dimensional model is representative of a tunnel section 300 meters long. This model was implemented by quadratic tetrahedral type elements [31] obtaining 95003 elements and 147528 nodes as shown in Figure 9.

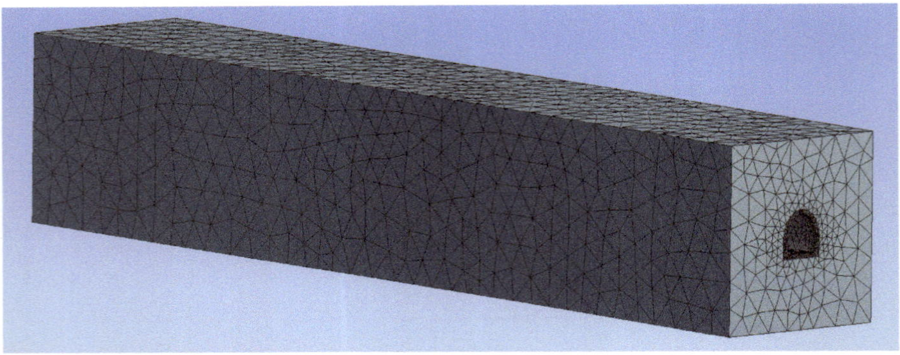

Figure 9. Meshed model.

The analysis was carried out in two steps [35]. The first step obtained the initial stress state caused by soil load and fire and the second step analysed the dynamic response under blast loading. Consequently the following load conditions were considered in the FE analysis:

1. from time $t = 0$ to time $t = 120$ min the tunnel was subjected to the surrounding soil load and to the fire thermal stress;
2. at the instant $t = 2$ sec the structure was subjected to the blast over pressure.

Therefore, on the base of this analysis, the distribution of the temperature inside of the structure is known. In Figure 10 the temperature distribution is showed.

Subsequently, the mechanical behaviour of the models was analysed introducing also to thermal stress, the explosion load. Figure 11 shows the deformation and the Mises stress of tunnel section, from meddle, where explosion is localized, to the tunnel opening.

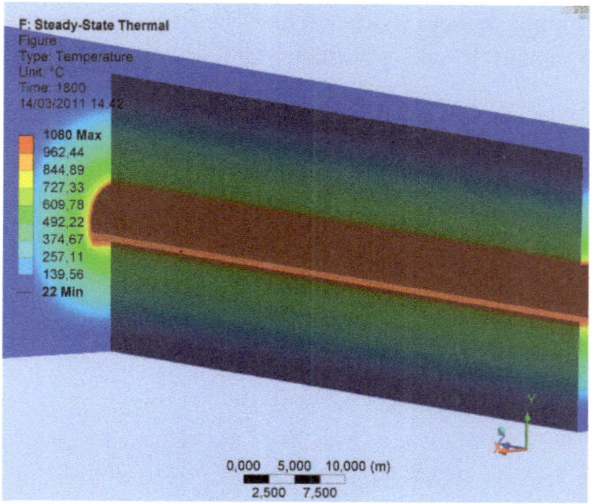

Figure 10. Temperature distribution (°C) $t = 1800$ sec.

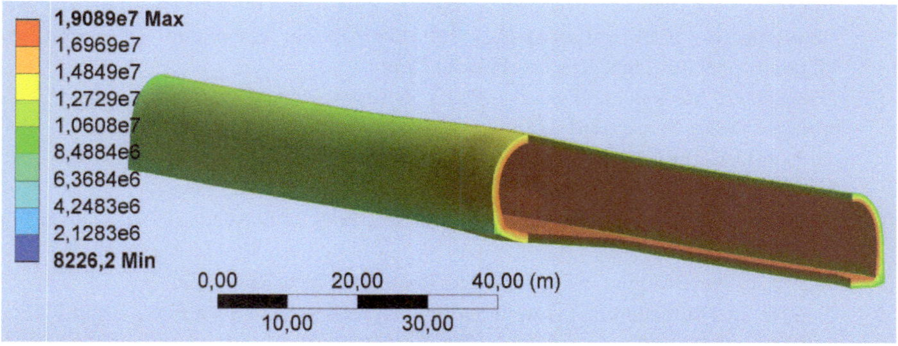

Figure 11. Mises stress (Pa) of the tunnel at the explosion instant.

6. Conclusions

The topics developed in this chapter belong to multi-physics problems and consequently represent a great computational weight on the results. Again, further complexities arise in the hypothesis of the mechanical process being dynamic.

In the almost static case the strain, in any instant of time, is in a situation of almost equilibrium with the loads; instead in the dynamic case the stress state is variable in space, therefore there are portions of the solid under stress against others in almost absence of stress. In other words the stress travels inside the solid as a stress-wave and it becomes a fundamental parameter for the description of the behaviour of the material. The dynamic processes in materials involve different scientific disciplines and areas, as materials science, shock physics/chemistry, mechanics combustion, applied mathematics and large scale computation.

Certainly, in developing this approach we have had the opportunity to deepen analysis about strength of materials and structures, and damage and fracture at micro and macroscale.

In the second section of the chapter, two numerical simulations relatively to one impact against air field pavements and one explosion in tunnel structures have been presented. Both simulations assume the problems as multi-field, and the results are quantitatively adequate.

Author details

Giovanni Leonardi* and Michele Buonsanti
Faculty of Engineering, University of Reggio Calabria,
Department of Mechanics and Materials – MECMAT, Reggio Calabria, Italy

7. References

[1] Buonsanti M., Cirianni F., Leonardi G. Study of the barriers for the mitigation of railway vibrations. In: ICSV16 Proceedings 16th International Conference of Sound and Vibrations, 5-9 July 2009, Kracòw, Poland; 2006.
[2] Truesdell C. Mechanics of Solids Waves in Elastic and Viscoelastic Solids, vol. IV, Springer-Verlag Berlin; 1984.
[3] Hudson J. A. The Excitation and Propagation of Elastic Waves, Cambridge University Press, Cambridge; 1988.
[4] Love A. E. H. A Treatise on the Mathematical Theory of Elasticity, IV Ed., Dover Pbs. New York; 1972.
[5] Graff K. F. Wave Motion in Elastic Solids, Dover Pbs. New York; 1991.
[6] Wood R. D. Diffraction of Transient Horizontal Shear Waves by a Finite Rigid Ribbon. International Journal of Engineering Science. 1970; 8: 857-874.
[7] Wood R. D., Pao Y.H. Diffractions and Horizontal Shear Waves by a Parabolic Cylinder and Dynamic Stress Concentration. Journal Applied Mechanics. 1966; 33: 785-792.
[8] Woods R. D. Screening of Surface Waves in Soils, Journal of the Soil Mechanics and Foundations Division. ASCE. 1968; 94: 951-979.
[9] Buonsanti M., Cirianni F., Leonardi G., Scopelliti F. Impact on Granular Plate. In: De Roeck G., Degrande G., Lombaert G., Muller G. (eds.) Eurodyn2011: proceedings of the 8th International Conference on Structural Dynamics, EURODYN2011, Leuven, Belgium, 2011.
[10] Meyers M. Dynamic behavior of materials. Wiley-Interscience; 1994.
[11] Duvall G., Graham R. Phase transitions under shock-wave loading, Reviews of Modern Physics. 1977; 49:523-579.

* Corresponding Author

[12] Lu Y., Wright P. Numerical approach of visco-elastoplastic analysis for asphalt mixtures, Computers & Structures. 1998; 69: 139-147.
[13] Krishnan J., Rajagopal K. On the mechanical behavior of asphalt, Mechanics of materials. 2005; 37: 1085-1100.
[14] Little R. W., Keer L. Elasticity. Journal of Applied Mechanics. 1974; 41: 7.
[15] Buonsanti M, Leonardi G, Scopelliti F. 3-D Simulation of shock waves generated by dense explosive in shell structures. Procedia Engineering. 2011; 10: 1550-5.
[16] Timoshenko S, Woinowsky-Krieger S, Woinowsky S. Theory of plates and shells: McGraw-hill New York; 1959. C. Truesdell, Mechanics of solids: Springer Verlag, 1973.
[17] Sadd M. Elasticity Theory, Applications, and Numerics. 2005. Burlington, MA: Elsevier Butterworth-Heinemann..
[18] Srinath LS. Advanced mechanics of solids: Tata McGraw-Hill Publishing Company Limited; 2009.
[19] Doghri I. Mechanics of deformable solids: linear, nonlinear, analytical, and computational aspects: Springer Verlag; 2000.
[20] Crozier DA, Sanjayan JG. Tests of load-bearing slender reinforced concrete walls in fire. ACI Structural Journal. 2000; 97(2).
[21] O'meagher A, Bennetts I. Modelling of concrete walls in fire. Fire safety journal. 1991; 17(4): 315-35.
[22] Pesavento F, Gawin D, Majorana CE, Witek A, Schrefler B. Modelling of thermal damaging of concrete structures during fire. In: VII International Conference Computational Plasticity, 2003.
[23] Schrefler B, Brunello P, Gawin D, Majorana C, Pesavento F. Concrete at high temperature with application to tunnel fire. Computational Mechanics. 2002; 29(1): 43-51.
[24] Gurney RW. The Initial Velocities of Fragments from Bombs, Shell and Grenades. DTIC Document, 1943.
[25] Kennedy J, editor. Behavior and Utilization of Explosives in Engineering Design. 1972.
[26] AIRBUS. Airplane Characteristics A321. 1995.
[27] Huang Y. Pavement analysis and design: Prentice Hall; 1993.
[28] AAIB. AAIB Bulletin: 6/2009 EW/C2008/07/02. 2009.
[29] Kuo CM, Hall KT, Darter MI. Three-dimensional finite element model for analysis of concrete pavement support. Transportation Research Record. 1995; 1505:119-27.
[30] Committee E. Eurocode2: Design of concrete structures-Part 1-2: General rules-Structural fire design. ENV 1992-1-2, 1995.
[31] Forschungsgesellschaft für Straßen - und Verkehrswesen. Richtlinien für Ausstattung und Betrieb von Tunneln (RABT). Ausgabe1985.
[32] Instituut TNO voor Bouwmaterialen en Bouwconstructies. Rapport betreffende de beproeving van het gedrag van twee isolatiematerialenter bescherming van tunnels tegen brand. Rapport B-80-33. Delft, The Netherlands 1980.
[33] Hyde D. CONWEP, Conventional Weapons Effects Program. US Army Engineer Waterways Experiment Station, Vicksburg, MS. 1992.

[34] Silvestrini M, Genova B, Leon Trujillo F. Energy concentration factor. A simple concept for the prediction of blast propagation in partially confined geometries. Journal of Loss Prevention in the Process Industries. 2009; 22(4): 449-54.
[35] Buonsanti M, Leonardi G, 3-D Simulation of tunnel structures under Blast loading. Archive of Mechanics and Engineering, in press, 2012.

Chapter 9

Perfectly Matched Layer for Finite Element Analysis of Elastic Waves in Solids

Koji Hasegawa and Takao Shimada

Additional information is available at the end of the chapter

1. Introduction

Numerical analysis of scattering and propagation of elastic waves in solids gives insight into physical phenomena under operation of ultrasonic devices such as electromechanical filters and resonators, nondestructive testing with ultrasonic waves and seismic prospecting. To take anisotropy of solids and complex structures of composite solids into account, commercial simulator based on the finite element method are available. Numerical models for finite element analysis (FEA) must be bounded and infinite half spaces of models should be replaced with finite domains and absorbing boundary conditions.

Perfectly matched layer (PMLs) is one of popular absorbing boundary conditions for truncating the computational domain of open regions without reflection of oblique incident waves. In 1994, Berenger invented a PML for electromagnetic waves in the finite difference time domain (FD-TD) method by a splitting field method.[1] Because fields in Berenger's PML do not satisfy the Maxwell's equations, two concepts have been introduced for implementation in the finite element method (FEM) of electromagnetic wave problems: the analytic continuation or the complex coordinate stretching[2, 3](CCS) and anisotropic PMLs.[4] Nowadays PMLs for electromagnetic waves are widely used in the FD-TD method and the FEM.

Extension of PMLs to elastic waves in isotropic solids in the Cartesian coordinate first appeared in 1996.[5, 6] In the cylindrical and spherical coordinates, PMLs were presented by using splitting field method in isotropic solids in 1999[7] and by using analytic continuation in anisotropic solids in 2002.[8] Recently validity and usefulness of PMLs derived from the analytic continuation in piezoelectric solids was demonstrated. [9–11] Hastings et. al.[5] reported better performance of PMLs by the FD-TD method than the second-order absorbing boundary condition (ABC) of Peng and Toksöz: in the range of the incident angle from 0° to 80°, reflection powers of S- or P-wave, which is excited by a pure S- or P-wave line source and propagating in a two-dimensional infinite isotropic solid modeled by a rectangular solid with its opposite sides attached sponge mediums and other sides imposed ABC or loaded PMLs, from the PML side are suppressed below -45 and -80 dB with 8 and 16 grid spaces of the PML region, respectively. On the other hand, reflection power level at the computational domain edge imposed ABC is in the range of -90 dB to -10 dB. This implies that PMLs yield more superior approximation of perfect matching than the ABC with thickening PML and increasing the number of grid spaces.

We recommend that readers who are unfamiliar with PMLs consult Basu and Chopra[12] about explanations and finite element (FE) implementation of PMLs for time-harmonic elastodynamics, Michler et al.[13] about derivation of material constants of PML in FE method by analytic continuation, and Taflove and Hagness[14] about PMLs for electromagnetic waves in FD-TD method.

Although PML is one of attractive artificial materials, two questions of PMLs derived from the analytic continuation are left: why are the particle displacements in the complex coordinate identical to those in the real coordinate and why must we multiply stress tensors by the Jacobian of the coordinate transformation?

For replying to the questions, we will examine a derivation of PMLs for elastic waves in the Cartesian, the cylindrical and the spherical coordinates from the differential form on manifolds. Our results reveal that the components of stress tensors and the particle displacement vectors in the analytic continuation are not transformed to the real space.[15] In addition, the rule for determining PML parameters in the Cartesian coordinate holds in the cylindrical and spherical coordinates.[16]

Mathematical models of PMLs, which are given by differential equations and boundary conditions, are exactly perfect matching medium. In numerical models, however, discretizing PMLs changes phase velocities of propagating waves and generates reflection waves from the PML region.[17] Furthermore, approximation of infinite regions with finite thick layers also generates reflection waves from the PML terminal.[1, 17, 18]

Estimating matching performance and optimizing parameters of PMLs in a numerical domain are required before solving problems. Chew and Jin investigated dependence of PML's performance on attenuation parameters of FE analysis of electromagnetic wave problems.[18] For FD-TD method Collino and Monk also carried out such an investigation.[19] Recently, Bermúdez et al. investigated absorbing functions for time harmonic Helmholtz equations in the Cartesian and cylindrical coordinates under the condition of ignoring reflection caused by FE-discretization and showed the advantages of non-integrable absorbing functions over conventional functions of power series.[20, 21]

Most of these investigations of optimizing attenuation parameters of PML employed numerical analysis of scattering problems in the two dimensions such as plane or cylindrical wave scattering problems. For tackling optimization problem of PML parameters, plane wave scattering problem is appropriate because required resource of computation is small. For FEA, Chew and Jin[18] modeled scattering of plane waves as electromagnetic field analysis in the thick layer in the one dimension. But this model has not been applied to elastic wave scattering.

In this chapter, we also examine PML performance of FE-models in the frequency range with scattering problems of elastic waves in an isotropic solid as field analysis in the thick layer in the one dimension. To the best of our knowledge, quantification of reflection power generated by FE-discretization has not attracted attention. Recently, for electromagnetic waves, we reported that the reflection power caused by discretization can be computed by the equivalent transmission line with its impedance and propagation constant determined by discretized wave numbers.[15] Because, for elastic waves, the transfer matrix is popular, we explain the reflection from PMLs by the transfer matrix of elastic waves, and confirm that numerical results of FE-models may be predicted by replacement of propagation constants of elastic waves in PML with discretized wave numbers.

2. Derivation of perfectly matched layers for elastic waves by using complex coordinate stretching and differential form

2.1. Differential form

A particle displacement vector $\mathbf{u}$, particle velocity vector $\mathbf{v}$, density of momenta $\mathbf{P}$, stress tensor $\bar{\bar{T}}$ and displacement gradient tensor $\bar{\bar{F}}$ are given as follows:

$$\mathbf{u} = u^i \frac{\partial}{\partial x^i}, \tag{1}$$

$$\mathbf{v} = v^i \frac{\partial}{\partial x^i}, \tag{2}$$

$$\mathbf{P} = \frac{1}{3!} P^i_{\alpha\beta\gamma} \frac{\partial}{\partial x^i} \otimes dx^\alpha \wedge dx^\beta \wedge dx^\gamma, \tag{3}$$

$$\bar{\bar{T}} = \frac{1}{2} T^i_{\alpha\beta} \frac{\partial}{\partial x^i} \otimes dx^\alpha \wedge dx^\beta, \tag{4}$$

$$\bar{\bar{F}} = F^i_\alpha \frac{\partial}{\partial x^i} \otimes dx^\alpha = d\mathbf{u}, \tag{5}$$

where $\frac{\partial}{\partial x^i}$ and $dx^i (i = 0,1,2)$ are the contravariant and covariant basis vectors, $\otimes$ and $\wedge$ represent the tensor product and the cross product, respectively. d is the exterior differential operator. Newton's equation of motion is

$$d\bar{\bar{T}} = \frac{\partial \mathbf{P}}{\partial t}. \tag{6}$$

Changing the coordinate gives relations of tensor components: for a tensor with a tensor type of the contravariant of rank 1 and the covariant of rank q, $V = V^i_{X\alpha_1\cdots\alpha_q} \frac{\partial}{\partial X^i} \otimes dX^{\alpha_1} \wedge \cdots \wedge dX^{\alpha_q} = V^k_{x\beta_1\cdots\beta_q} \frac{\partial}{\partial x^k} \otimes dx^{\beta_1} \wedge \cdots \wedge dx^{\beta_q}$, the relation of tensor components is

$$V^i_{X\alpha_1\cdots\alpha_q} = \frac{\partial X^i}{\partial x^k} \frac{\partial x^{\beta_1}}{\partial X^{\alpha_1}} \cdots \frac{\partial x^{\beta_q}}{\partial X^{\alpha_q}} V^k_{x\beta_1\cdots\beta_q}. \tag{7}$$

Using CCS[2, 3, 8] given by $X^i = \int^{x^i} \tilde{s}_i(\tau)d\tau = \int^{x^i} \tilde{s}_{iR}(\tau) + j\tilde{s}_{iI}(\tau)d\tau$ with the two real functions $\tilde{s}_{iR}(\tau)$ and $\tilde{s}_{iI}(\tau)$, we have the relation

$$V^i_{X\alpha_1\cdots\alpha_q} = \tilde{s}_i(x^i)\left[\tilde{s}_{\alpha_1}(x^{\alpha_1})\cdots\tilde{s}_{\alpha_q}(x^{\alpha_q})\right]^{-1} V^i_{x\alpha_1\cdots\alpha_q}. \tag{8}$$

Here j is the imaginary unit.

2.2. PMLs in the Cartesian, the cylindrical and the spherical coordinates

In the complex coordinate stretching (CCS), we consider that the real coordinate (x^0, x^1, x^2) is (x, y, z), (r, θ, z) or (r, θ, ϕ) for the Cartesian, the cylindrical or the spherical coordinate, respectively. Assuming that the same constitutive equations in the real Cartesian, cylindrical and spherical coordinate exist in the complex coordinate, (X^0, X^1, X^2), we have

$$\mathbf{P}^c = \rho \mathbf{v}^c, \tag{9}$$

$$\begin{aligned} T^c_{ij} &= C_{ijkl} S^c_{kl} \\ &= C_{ijkl}(F^c_{kl} + F^c_{lk})/2 \\ &= C_{ijkl} F^c_{kl}. \end{aligned} \tag{10}$$

Here, the superscript c denotes the value in the complex coordinate and the mass density ρ and the stiffness C_{ijkl} $(i, j, k, l = X^0, X^1, X^2)$ are the values corresponding to original material parameters, mass density and stiffness constants, of its PML in the Cartesian, the cylindrical and the spherical coordinates. Using eq. (8) to eqs.(2)-(5), and recalling that the base differentials $d\xi, d\eta$ and $d\zeta$ of the general orthogonal coordinate system (ξ, η, ζ) are dual to the unit vectors $\frac{\hat{\xi}}{h_\xi}, \frac{\hat{\eta}}{h_\eta}$ and $\frac{\hat{\zeta}}{h_\zeta}$, we have

$$v^c_i = s_i v_i \text{ (no summation)}, \tag{11}$$

$$P^c_i = \frac{s_i}{s_0 s_1 s_2} P_i \text{ (no summation)}, \tag{12}$$

$$T^c_{ij} = \frac{s_i s_j}{s_0 s_1 s_2} T_{ij} \text{ (no summation)}, \tag{13}$$

$$F^c_{ij} = \frac{s_i}{s_j} F_{ij} \text{ (no summation)}. \tag{14}$$

Here $s_i = \frac{h^c_i}{h^r_i}\tilde{s}_i$ with h^r_i and h^c_i being scale factors of general orthogonal coordinate systems (x^0, x^1, x^2) and (X^0, X^1, X^2), respectively. Note that the scale factors h_i are given by follows: in the cylindrical coordinate (r, θ, z) $h_0 = 1, h_1 = r, h_2 = 1$, and in the spherical coordinate (r, θ, ϕ) $h_0 = 1, h_1 = r, h_2 = r\sin\theta$. In addition, in the Cartesian coordinate, $h_0 = h_1 = h_2 = 1$.

The quotient rule and eqs. (9)- (14) yield PML material constants: the mass density ρ^{PML} is

$$\rho^{\text{PML}} = s_0 s_1 s_2 \rho \tag{15}$$

and the stiffness is

$$C_{ijkl}^{\mathrm{PML}} = \frac{s_0 s_1 s_2 s_k}{s_i s_j s_l} C_{ijkl} \text{ (no summation).} \tag{16}$$

Here, $s_0 = \tilde{s}_0$, $s_1 = \tilde{s}_1 \frac{R}{r}$, $s_2 = \tilde{s}_2$ in the cylindrical coordinate system (r, θ, z) with its complex coordinate (R, Θ, Z), and $s_0 = \tilde{s}_0$, $s_1 = \tilde{s}_1 \frac{R}{r}$, $s_2 = \tilde{s}_2 \frac{R \sin\Theta}{r \sin\theta}$ in the spherical coordinate system (r, θ, ϕ) with its complex coordinate (R, Θ, Φ). In addition, $s_i = \tilde{s}_i$ in the Cartesian coordinate system.

Eqs. (15) and (16) show that PML parameters for elastic waves in solids in the cylindrical and spherical coordinates may be calculated by the same procedure in the Cartesian coordinate.

2.3. Derivation of PML constants by the analytic continuation

For simplicity, we present a procedure of deriving material constants in only cylindrical coordinates by the analytic continuation[8] below. Note that in spherical coordinates the same procedure may be applied. We recommend that the reader who is interesting in the procedure consult Zheng and Huang.[8]

First we consider Newton's equation of motion. In a cylindrical coordinate (r, θ, z) , the governing equations are

$$-\rho\omega^2 u_r = \frac{1}{r}\frac{\partial}{\partial r}(rT_{rr}) + \frac{1}{r}(\frac{\partial T_{r\theta}}{\partial \theta} + \hat{r} \bullet \frac{\partial \hat{\theta}}{\partial \theta} T_{\theta\theta}) + \frac{\partial T_{rz}}{\partial z}, \tag{17}$$

$$-\rho\omega^2 u_\theta = \frac{1}{r}\frac{\partial}{\partial r}(rT_{\theta r}) + \frac{1}{r}(\hat{\theta} \bullet \frac{\partial \hat{r}}{\partial \theta} T_{r\theta} + \frac{\partial T_{\theta\theta}}{\partial \theta}) + \frac{\partial T_{\theta z}}{\partial z}, \tag{18}$$

$$-\rho\omega^2 u_z = \frac{1}{r}\frac{\partial}{\partial r}(rT_{zr}) + \frac{1}{r}\frac{\partial T_{z\theta}}{\partial \theta} + \frac{\partial T_{zz}}{\partial z}. \tag{19}$$

Here, we use phasor notation. The time dependences of the fields are $\exp(j\omega t)$ where ω is angular frequency. Applying CCS with a complex coordinate (R, Θ, Z), multiplying the CCS equations by $s_0 s_1 s_2$ and using the assumption of $u_i^c = u_i$, $\hat{R} = \hat{r}$, $\hat{\Theta} = \hat{\theta}$ and $\hat{Z} = \hat{z}$, we get the governing equations in the PML:

$$-\rho^{\mathrm{PMLA}}\omega^2 u_r = \frac{1}{r}\frac{\partial}{\partial r}(rs_1 s_2 T_{rr}^c) + \frac{1}{r}(\frac{\partial(s_0 s_2 T_{r\theta}^c)}{\partial \theta} + \hat{r} \bullet \frac{\partial \hat{\theta}}{\partial \theta}(s_0 s_2 T_{\theta\theta}^c)) + \frac{\partial(s_0 s_1 T_{rz}^c)}{\partial z}, \tag{20}$$

$$-\rho^{\mathrm{PMLA}}\omega^2 u_\theta = \frac{1}{r}\frac{\partial}{\partial r}(rs_1 s_2 T_{\theta r}^c) + \frac{1}{r}(\hat{\theta} \bullet \frac{\partial \hat{r}}{\partial \theta}(s_0 s_2 T_{r\theta}^c) + \frac{\partial(s_0 s_2 T_{\theta\theta}^c)}{\partial \theta}) + \frac{\partial(s_0 s_1 T_{\theta z}^c)}{\partial z}, \tag{21}$$

$$-\rho^{\mathrm{PMLA}}\omega^2 u_z = \frac{1}{r}\frac{\partial}{\partial r}(rs_1 s_2 T_{zr}^c) + \frac{1}{r}\frac{\partial(s_0 s_2 T_{z\theta}^c)}{\partial \theta} + \frac{\partial(s_0 s_1 T_{zz}^c)}{\partial z}. \tag{22}$$

Here, the mass density of PML is defined as $\rho^{\mathrm{PMLA}} = s_0 s_1 s_2 \rho$. When we rewrite components of stress tensors in the PMLs as $T_{ij}^{\mathrm{PMLA}} = \frac{s_0 s_1 s_2}{s_j} T_{ij}^c$, we may identify eqs.(20)~(22) to eqs.(17)~(19).

Next we consider the displacement gradient $\nabla\mathbf{u}^c = [\Gamma^c_{kl}]$ in the complex coordinate (R,Θ,Z). Using definition $d\mathbf{u} = \nabla\mathbf{u} \bullet (\hat{x}^i h_i dx^i)$ and the assumption of $u^c_i = u_i$, $\hat{R} = \hat{r}$, $\hat{\Theta} = \hat{\theta}$, $\hat{Z} = \hat{z}$, and applying CCS with a complex coordinate (R,Θ,Z), we have the relation:

$$\nabla\mathbf{u}^c = \begin{bmatrix} \frac{\partial u^c_R}{\partial R} & \frac{1}{R}\left(\frac{\partial u^c_R}{\partial\Theta} + \hat{R}\cdot\frac{\partial\hat{\Theta}}{\partial\Theta}u^c_\Theta\right) & \frac{\partial u^c_R}{\partial Z} \\ \frac{\partial u^c_\Theta}{\partial R} & \frac{1}{R}\left(\hat{\Theta}\cdot\frac{\partial\hat{R}}{\partial\Theta}u^c_R + \frac{\partial u^c_\Theta}{\partial\Theta}\right) & \frac{\partial u^c_\Theta}{\partial Z} \\ \frac{\partial u^c_Z}{\partial R} & \frac{1}{R}\frac{\partial u^c_Z}{\partial\Theta} & \frac{\partial u^c_Z}{\partial Z} \end{bmatrix}$$

$$= \begin{bmatrix} \frac{1}{s_0}\frac{\partial u_r}{\partial r} & \frac{1}{s_1}\frac{1}{r}\left(\frac{\partial u_r}{\partial\theta} + \hat{r}\cdot\frac{\partial\hat{\theta}}{\partial\theta}u_\theta\right) & \frac{1}{s_2}\frac{\partial u_r}{\partial z} \\ \frac{1}{s_0}\frac{\partial u_\theta}{\partial r} & \frac{1}{s_1}\frac{1}{r}\left(\hat{\theta}\cdot\frac{\partial\hat{r}}{\partial\theta}u_r + \frac{\partial u_\theta}{\partial\theta}\right) & \frac{1}{s_2}\frac{\partial u_\theta}{\partial z} \\ \frac{1}{s_0}\frac{\partial u_z}{\partial r} & \frac{1}{s_1}\frac{1}{r}\frac{\partial u_z}{\partial\theta} & \frac{1}{s_2}\frac{\partial u_z}{\partial z} \end{bmatrix}. \quad (23)$$

Hence we have $\Gamma^c_{kl} = \frac{1}{s_l}\Gamma_{kl}$.

Using the quotient rule and the constitutive equation, $T^c_{ij} = C_{ijkl}\Gamma^c_{kl}$, we get the constitutive equation of the PML in the real coordinate (r,θ,z): $T^{\text{PMLA}}_{ij} = \frac{s_0s_1s_2}{s_js_l}C_{ijkl}\Gamma_{kl}$. Therefore, we may define the stiffness of the PML derived by the analytic continuation: $C^{\text{PMLA}}_{ijkl} = \frac{s_0s_1s_2}{s_js_l}C_{ijkl}$.

2.4. Comparison with PML material constants derived from differential forms and the analytic continuation

By the analytic continuation, Zheng and Huang[8] derived the mass density and stiffness of PML in the cylindrical and spherical coordinates: $\rho^{\text{PMLA}} = s_0s_1s_2\rho$ and $C^{\text{PMLA}}_{ijkl} = \frac{s_0s_1s_2}{s_js_l}C_{ijkl}$. The mass density agree with our result, eq. (15), because multiplying the stress tensors by the Jacobian of the coordinate transformation, $s_0s_1s_2$, adjusts the mass density. We note that the form of eq. (15) is also derived from eq. (6) with the tensor type of mass density being covariant of rank 3, i.e. 3-form. The stiffness is different from eq. (16) because in the analytic continuation, the manipulation of the coordinate transformation corresponding to the part of stress tensor and the particle displacement vector, contravariant of rank 1, is excluded. This fact can be confirmed by the derivation procedure presented in the previous section for the cylindrical coordinate:we put $T^{\text{PMLA}}_{ij} = \frac{s_0s_1s_2}{s_j}T^c_{ij}$ and use the assumption $u_i = u^c_i$.

To show a difference between PML material constants, we consider an isotropic solid with following stiffness constants in the Cartesian coordinate (x_0,x_1,x_2): $C_{ijkl} = \lambda\delta_{ij}\delta_{kl} + \mu(\delta_{ik}\delta_{jl} + \delta_{il}\delta_{jk})$. Here, λ and μ are the Lamé constants of an isotropic solid, the subscripts i,j,k and l denote the x_i-, x_j- , x_k- and x_l-axis, respectively, and δ_{ij} is the Kronecker delta. Components of stiffness tensors derived from the differential form and analytic continuation, C^{PML}_{ijkl} and C^{PMLA}_{ijkl}, respectively, are given by

$$C^{\text{PML}}_{ijkl} = \left[\left(\frac{\lambda}{s_i^2}\delta_{ij}\delta_{kl} + \frac{\mu}{s_j^2}\delta_{ik}\delta_{jl} + \frac{\mu}{s_i^2}\delta_{il}\delta_{jk}\right|s_0s_1s_2 \quad \text{(no summation)}, \quad (24)$$

$i\ j$	$\frac{T_{ij}}{s_0 s_1 s_2}$	
	C_{ijkl}^{PML}:differential form	C_{ijkl}^{PMLA}:analytic continuation
0 0	$\frac{1}{s_0^2}[(\lambda+2\mu)u_{0,0}+\lambda(u_{1,1}+u_{2,2})]$	$(\lambda+2\mu)\frac{u_{0,0}}{s_0^2}+\frac{\lambda}{s_0}(\frac{u_{1,1}}{s_1}+\frac{u_{2,2}}{s_2})$
1 1	$\frac{1}{s_1^2}[(\lambda+2\mu)u_{1,1}+\lambda(u_{2,2}+u_{0,0})]$	$(\lambda+2\mu)\frac{u_{1,1}}{s_1^2}+\frac{\lambda}{s_1}(\frac{u_{2,2}}{s_2}+\frac{u_{0,0}}{s_0})$
2 2	$\frac{1}{s_2^2}[(\lambda+2\mu)u_{2,2}+\lambda(u_{0,0}+u_{1,1})]$	$(\lambda+2\mu)\frac{u_{2,2}}{s_2^2}+\frac{\lambda}{s_2}(\frac{u_{0,0}}{s_0}+\frac{u_{1,1}}{s_1})$
1 2	$\mu\left(\frac{u_{1,2}}{s_2^2}+\frac{u_{2,1}}{s_1^2}\right)$	$\frac{\mu}{s_2}\left(\frac{u_{1,2}}{s_2}+\frac{u_{2,1}}{s_1}\right)$
2 1	$\mu\left(\frac{u_{1,2}}{s_2^2}+\frac{u_{2,1}}{s_1^2}\right)$	$\frac{\mu}{s_1}\left(\frac{u_{1,2}}{s_2}+\frac{u_{2,1}}{s_1}\right)$
2 0	$\mu\left(\frac{u_{2,0}}{s_0^2}+\frac{u_{0,2}}{s_2^2}\right)$	$\frac{\mu}{s_0}\left(\frac{u_{2,0}}{s_0}+\frac{u_{0,2}}{s_2}\right)$
0 2	$\mu\left(\frac{u_{2,0}}{s_0^2}+\frac{u_{0,2}}{s_2^2}\right)$	$\frac{\mu}{s_2}\left(\frac{u_{2,0}}{s_0}+\frac{u_{0,2}}{s_2}\right)$
0 1	$\mu\left(\frac{u_{0,1}}{s_1^2}+\frac{u_{1,0}}{s_0^2}\right)$	$\frac{\mu}{s_1}\left(\frac{u_{0,1}}{s_1}+\frac{u_{1,0}}{s_0}\right)$
1 0	$\mu\left(\frac{u_{0,1}}{s_1^2}+\frac{u_{1,0}}{s_0^2}\right)$	$\frac{\mu}{s_0}\left(\frac{u_{0,1}}{s_1}+\frac{u_{1,0}}{s_0}\right)$

Table 1. Components of a stress tensor in a PML material of an isotropic solid in the Cartesian coordinate.

$$C_{ijkl}^{\mathrm{PMLA}} = \left[\frac{\lambda}{s_i s_k}\delta_{ij}\delta_{kl}+\frac{\mu}{s_j^2}\delta_{ik}\delta_{jl}+\frac{\mu}{s_i s_j}\delta_{il}\delta_{jk}\right]s_0 s_1 s_2 \quad \text{(no summation)}. \tag{25}$$

Table 1 shows all components of the stress tensor computed with C_{ijkl}^{PML} and C_{ijkl}^{PMLA}. C_{ijkl}^{PMLA} gives $\tilde{T}_{ij} \neq \tilde{T}_{ji}$ $(i \neq j)$ and we predict that rotational forces may be observed. With C_{ijkl}^{PML}, however, we have a symmetric stress tensor, $T_{ij} = T_{ji}$ $(i \neq j)$.

3. Reflection from PMLs discretized for finite element models in the frequency domain

We consider a plane elastic wave propagating in a half infinite isotropic solid attached with its PML backed with a vacuum region as shown in Fig.1. Here θ is the incident angle, θ_p and θ_s are propagation angles of P-waves and SV- or SH-waves, L is thickness of the PML, $\mathbf{k}_{\mathrm{i}}$ and $\mathbf{k}_{r,m}$ $(m = 0,1,2)$ are wave vectors of the incident wave and reflected P-, SV- and SH-waves, respectively.

We use the phasor notation and assume that the time dependences of all fields are $\exp(\mathrm{j}\omega t)$, where j is the imaginary unit and ω is the angular frequency.

When the stiffness component of the isotropic solid C_{ijkl} $(\mathrm{i,j,k,l} = x,y,z)$ is given by $C_{\mathrm{ijkl}} = \lambda\delta_{\mathrm{ij}}\delta_{\mathrm{kl}} + \mu(\delta_{\mathrm{ik}}\delta_{\mathrm{jl}} + \delta_{\mathrm{jl}}\delta_{\mathrm{ik}})$ where λ and μ are the Lamé constants and δ_{ij} is the Kronecker delta, the stiffness component of its PML $C_{\mathrm{ijkl}}^{\mathrm{PML}}$ is

$$C_{\mathrm{ijkl}}^{\mathrm{PML}} = (\frac{\lambda}{s_{\mathrm{i}}^2}\delta_{\mathrm{ij}}\delta_{\mathrm{kl}} + \frac{\mu}{s_{\mathrm{j}}^2}\delta_{\mathrm{ik}}\delta_{\mathrm{jl}} + \frac{\mu}{s_{\mathrm{i}}^2}\delta_{\mathrm{il}}\delta_{\mathrm{jk}})s_x s_y s_z. \tag{26}$$

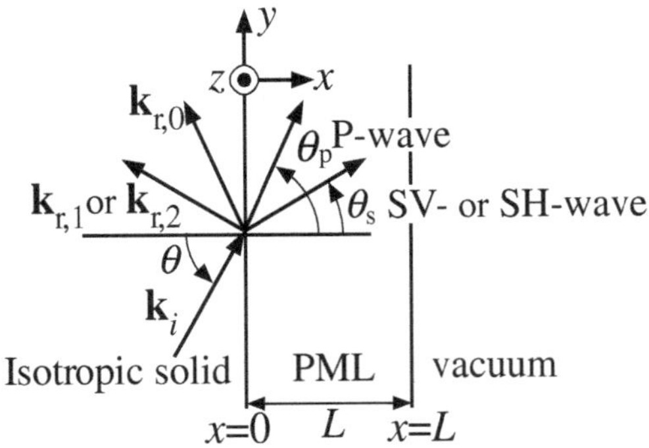

Figure 1. Reflection by the plane boundary between an isotropic solid and its PML.

Here s_i $(i = x, y, z)$ is a coordinate stretching factor of i-direction.[2] The mass density of the PML $\rho^{\rm PML}$ is given by

$$\rho^{\rm PML} = s_x s_y s_z \rho. \tag{27}$$

Here ρ is the mass density of the isotropic solid. For examining absorbing performance of PMLs in the x direction, taking an assumption of considering fields being consisted by plane waves propagating on the x-y plane, we have a differential equation in one variable x: from Newton's equation of motion and constitutive equation we get the differential equation in the PML

$$C_{\rm ijkl}^{\rm PML} \frac{\partial}{\partial x_j} \left(\frac{\partial u_k}{\partial x_l} \right) = -\omega^2 \rho^{\rm PML} u_i \tag{28}$$

where u_i is the component of the particle displacement in the i-direction $(i = x, y, z)$.

In this case, we may choose the coordinate stretching factor as follows:

$$\begin{aligned} s_x &= 1 - {\rm j} s_{xI}(x), \\ s_y &= s_z = 1. \end{aligned} \tag{29}$$

Here $s_{xI}(x)$ is the imaginary part of s_x and therefore a real function, which controls absorbing performance of propagating waves in PMLs.

Boundary conditions at the interface of isotropic solid and PML, $x = 0$, are the nonslip condition and the continuous condition of the normal component of the stress:[15]

$$u_i(-0) = s_i u_i(+0), \tag{30}$$

$$T_{ix}(-0) = \frac{s_i s_x}{s_x s_y s_z} T_{ix}(+0), \quad i = x, y, z. \tag{31}$$

At the terminal of PML, $x = L$, the boundary condition is

$$\frac{s_i s_x}{s_x s_y s_z} T_{ix}(L) = 0, \quad i = x, y, z. \tag{32}$$

3.1. Numerical procedure

3.1.1. *Finite element analysis*

Because finite element formulation of a thick plate with line elements as shown in Fig. 2 is well known and we use COMSOL MultiPhysics for FEA, we explain the Robin condition at $x = 0$ and a formula for reflection coefficients. In the half isotropic solid, the field distribution,

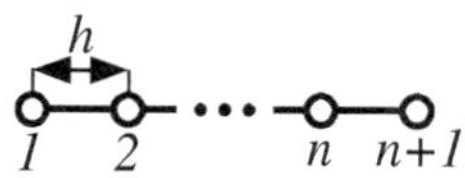

Figure 2. Line element with (n+1)-nodes.

components of the particle displacement and stress, can be expressed by superposition of incident and reflected plane waves:

$$\mathbf{u}(\mathbf{r}) = \sum_l R_l \mathbf{u}_l(\mathbf{r}) e^{-j\mathbf{k}_{r,l}\cdot\mathbf{r}} + \mathbf{u}_i e^{-j\mathbf{k}_i\cdot\mathbf{r}}. \tag{33}$$

Where R_l, $\mathbf{k}_{r,l}$, $\mathbf{u}_l(l = 0,1,2)$ and $\mathbf{u}_i$ are reflection constants, the wave vector, the particle displacement vectors of reflection P-, SV- and SH-waves and the incident wave respectively, which are given by the solutions of the Christoffel equation for the isotropic solid. When $\mathbf{r} = \mathbf{0}$, we have

$$[\mathbf{u}(-\mathbf{0})] = [L] \begin{bmatrix} R_0 \\ R_1 \\ R_2 \end{bmatrix} + [\mathbf{u}_i(-\mathbf{0})], \tag{34}$$

$$[L] = \begin{bmatrix} \hat{x}\cdot\mathbf{u}_0 & \hat{x}\cdot\mathbf{u}_1 & \hat{x}\cdot\mathbf{u}_2 \\ \hat{y}\cdot\mathbf{u}_0 & \hat{y}\cdot\mathbf{u}_1 & \hat{y}\cdot\mathbf{u}_2 \\ \hat{z}\cdot\mathbf{u}_0 & \hat{z}\cdot\mathbf{u}_1 & \hat{z}\cdot\mathbf{u}_2 \end{bmatrix}, \tag{35}$$

$$[\mathbf{u}(x)] = [\hat{x}\cdot\mathbf{u}(x) \;\; \hat{y}\cdot\mathbf{u}(x) \;\; \hat{z}\cdot\mathbf{u}(x)\,]^T \tag{36}$$

where the superscript T denotes transpose and $\hat{i}(i = x, y, z)$ is the unit vector of the i-direction. Derivative of (33) with respect to x is

$$\frac{\partial}{\partial x}\mathbf{u} = \sum_l (-j\mathbf{k}_{r,l}\cdot\hat{x}) R_l \mathbf{u}_l e^{-j\mathbf{k}_{r,l}\cdot\mathbf{r}} + (-j\mathbf{k}_i\cdot\hat{x})\mathbf{u}_i e^{-j\mathbf{k}_i\cdot\mathbf{r}}, \tag{37}$$

and for $\mathbf{r} = \mathbf{0}$ we have

$$\begin{bmatrix} R_0 \\ R_1 \\ R_2 \end{bmatrix} = [K]^{-1} \left(j\mathbf{k}_i\cdot\hat{x} \begin{bmatrix} \hat{x}\cdot\mathbf{u}_i(-\mathbf{0}) \\ \hat{y}\cdot\mathbf{u}_i(-\mathbf{0}) \\ \hat{z}\cdot\mathbf{u}_i(-\mathbf{0}) \end{bmatrix} + \begin{bmatrix} \frac{\partial\hat{x}\cdot\mathbf{u}}{\partial x}\big|_{x=-0} \\ \frac{\partial\hat{y}\cdot\mathbf{u}}{\partial x}\big|_{x=-0} \\ \frac{\partial\hat{z}\cdot\mathbf{u}}{\partial x}\big|_{x=-0} \end{bmatrix} \right). \tag{38}$$

Here

$$[K] = -\mathrm{j}\begin{bmatrix} \mathbf{k}_{r,0}\cdot\hat{x}\hat{x}\cdot\mathbf{u}_0 & \mathbf{k}_{r,1}\cdot\hat{x}\hat{x}\cdot\mathbf{u}_1 & \mathbf{k}_{r,2}\cdot\hat{x}\hat{x}\cdot\mathbf{u}_2 \\ \mathbf{k}_{r,0}\cdot\hat{x}\hat{y}\cdot\mathbf{u}_0 & \mathbf{k}_{r,1}\cdot\hat{x}\hat{y}\cdot\mathbf{u}_1 & \mathbf{k}_{r,2}\cdot\hat{x}\hat{y}\cdot\mathbf{u}_2 \\ \mathbf{k}_{r,0}\cdot\hat{x}\hat{z}\cdot\mathbf{u}_0 & \mathbf{k}_{r,1}\cdot\hat{x}\hat{z}\cdot\mathbf{u}_1 & \mathbf{k}_{r,2}\cdot\hat{x}\hat{z}\cdot\mathbf{u}_2 \end{bmatrix}. \tag{39}$$

Using eqs.(34) and (38), we have

$$\begin{aligned}[\mathbf{u}(-0)] = {} & [L][K]^{-1}\begin{bmatrix} \frac{\partial\hat{x}\cdot\mathbf{u}}{\partial x}|_{x=-0} \\ \frac{\partial\hat{y}\cdot\mathbf{u}}{\partial x}|_{x=-0} \\ \frac{\partial\hat{z}\cdot\mathbf{u}}{\partial x}|_{x=-0} \end{bmatrix} \\ & + \left([I] + \mathrm{j}\mathbf{k}_i\cdot\hat{x}[L][K]^{-1}\right)\begin{bmatrix} \hat{x}\cdot\mathbf{u}_i(-\mathbf{0}) \\ \hat{y}\cdot\mathbf{u}_i(-\mathbf{0}) \\ \hat{z}\cdot\mathbf{u}_i(-\mathbf{0}) \end{bmatrix}. \end{aligned} \tag{40}$$

Equation (31) yields

$$\left.\frac{\partial[\mathbf{u}]}{\partial x}\right|_{x=-0} = [C_{i1}]^{-1}[C_{t1}]\left.\frac{\partial[\mathbf{u}]}{\partial x}\right|_{x=+0}, \tag{41}$$

$$[C_{i1}] = \begin{bmatrix} \lambda+2\mu & 0 & 0 \\ 0 & \mu & 0 \\ 0 & 0 & \mu \end{bmatrix}, \tag{42}$$

$$[C_{t1}] = \begin{bmatrix} \lambda+2\mu & 0 & 0 \\ 0 & \mu & 0 \\ 0 & 0 & \mu \end{bmatrix}, \tag{43}$$

$$[C_{t1}] = \begin{bmatrix} \lambda+2\mu & 0 & 0 \\ 0 & \frac{s_y}{s_x}\mu & 0 \\ 0 & 0 & \frac{s_z}{s_x}\mu \end{bmatrix}. \tag{44}$$

Substituting eqs. (30) and (41) into eq. (40), we get the Robin condition:

$$\begin{aligned} & [L][K]^{-1}[C_{i1}]^{-1}[C_{t1}]\left.\frac{\partial[\mathbf{u}]}{\partial x}\right|_{x=+0} - [s][\mathbf{u}(+\mathbf{0})] \\ & = -\left([I] + \mathrm{j}k_{ix}[L][K]^{-1}\right)[\mathbf{u}_i(-\mathbf{0})]. \end{aligned} \tag{45}$$

After we solve the distributions of the particle displacements in the PML by COMSOL MultiPhysics, the reflection coefficients $R_l(l = 0,1,2)$ are computed with the following

Order of finite element	Discretized wave number $\tilde{\beta}_{\mathrm{PML}}$
1	$\frac{1}{h}\cos^{-1}\left[\frac{6-2(\beta_{\mathrm{PML}}h)^2}{6+(\beta_{\mathrm{PML}}h)^2}\right]$
2	$\frac{1}{2h}\cos^{-1}\left[\frac{15-26(\beta_{\mathrm{PML}}h)^2+3(\beta_{\mathrm{PML}}h)^4}{15+4(\beta_{\mathrm{PML}}h)^2+(\beta_{\mathrm{PML}}h)^4}\right]$
3	$\frac{1}{3h}\cos^{-1}\left[\frac{2800-11520(\beta_{\mathrm{PML}}h)^2+4860(\beta_{\mathrm{PML}}h)^4-324(\beta_{\mathrm{PML}}h)^6}{2800+1080(\beta_{\mathrm{PML}}h)^2+270(\beta_{\mathrm{PML}}h)^4+81(\beta_{\mathrm{PML}}h)^6}\right]$
4	$\frac{1}{4h}\cos^{-1}\left[\frac{19845-148680(\beta_{\mathrm{PML}}h)^2+134064(\beta_{\mathrm{PML}}h)^4-28800(\beta_{\mathrm{PML}}h)^6+1280(k_{\mathrm{PML}}h)^8}{19845+10080(\beta_{\mathrm{PML}}h)^2+3024(\beta_{\mathrm{PML}}h)^4+768(\beta_{\mathrm{PML}}h)^6+256(\beta_{\mathrm{PML}}h)^8}\right]$

Table 2. Discretized wave number in PML.

relation derived from eq. (34):

$$\begin{bmatrix} R_0 \\ R_1 \\ R_2 \end{bmatrix} = [L]^{-1}\left(\begin{bmatrix} s_x & 0 & 0 \\ 0 & s_y & 0 \\ 0 & 0 & s_z \end{bmatrix}[\mathbf{u}(+\mathbf{0})] - [\mathbf{u}_i(-\mathbf{0})]\right). \tag{46}$$

Here $\mathbf{u}(+0)$ and $\mathbf{u}_i(-0)$ are particle displacements at PML's incident side given by FEA solution and known incident field vector of displacements.

3.1.2. *Discretized wave number*

The finite element approximation of the propagating elastic fields changes the propagation constant given by the Christoffel equation, which is called the intrinsic wave number β_{PML}, to discretized wave number $\tilde{\beta}_{\mathrm{PML}}$. Table 2 shows discretized wave numbers for nodal finite elements with the polynomial interpolate function as shown in Fig.2 after Scott[22]. Here β and h are the x-component of the intrinsic wave number of P-, SV- or SH-wave propagating in the PML and equal interval between nodes, respectively. Figure 3 shows the difference of the discretized wave number and the intrinsic wave number as the function of the x-axis propagation constant.

3.1.3. *Transfer matrix analysis*

Because the structure shown in Fig.1 is a layered structure where the propagation constants in the isotropic solid and its PML are given as the intrinsic wave numbers and discretized wave numbers respectively, we can compute the reflection coefficient by the transfer matrix.

In this section, we consider the fields composed of P- and SV-waves propagating on the x-y plane with the same y-component k_y of the P- and SV-wave numbers only since SH-waves are not coupled with P- or SV-waves and SH-wave scattering problem is straightforward. Assuming that field distributions do not vary in the z-direction, we have the particle displacements in the solid:[15]

$$u_{x,m} = \mathrm{e}^{-\mathrm{j}k_y y}\left(\sum_{i=1}^{4} A_{i,m}\frac{f_{x,i}}{s_x}\mathrm{e}^{-\mathrm{j}k_{x,i}s_x x}\right), \tag{47}$$

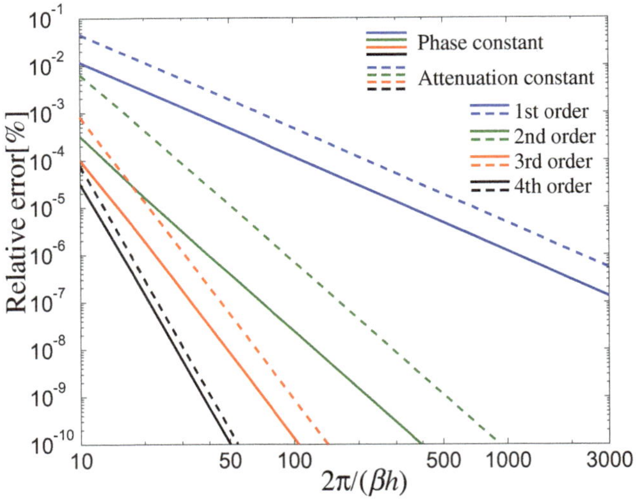

Figure 3. Phase error and attenuation error as a function of $2\pi/(\beta h)$ for 1st-, 2nd-, 3rd-, and 4th order elements.

$$u_{y,m} = \mathrm{e}^{-\mathrm{j}k_y y} \left(\sum_{i=1}^{4} A_{i,m} f_{y,i} \mathrm{e}^{-\mathrm{j}k_{x,i}s_x x} \right). \tag{48}$$

Here, $k_{x,i}$ is the x-component of the intrinsic wave number for the isotropic solid and the discretized wave number for the PML, the subscripts $i = 1$ and $i = 3$ denote P-waves propagating to $+x$- and $-x$-direction respectively, $i = 2$ and $i = 4$ denote SV-waves propagating to $+x$- and $-x$-direction respectively, and $A_{i,m} (i = 1, 2, 3, 4,\ m = 0, 1)$ is the amplitude at $x = 0$ in the isotropic solid ($m = 0$) or PML ($m = 1$). $f_{x,i}$ and $f_{y,i}$ are shown in Table 3. Here θ_p and θ_s are angles between the x-direction and the wave vectors of P-waves or SV-waves as shown in Fig. 1. In the isotropic region, we set $s_x = 1$.

i	1	2	3	4
$f_{x,i}$	$\cos\theta_p$	$-\sin\theta_s$	$-\cos\theta_p$	$\sin\theta_s$
$f_{y,i}$	$\sin\theta_p$	$\cos\theta_s$	$\sin\theta_p$	$\cos\theta_s$

Table 3. Displacement directions of P- and SV-waves, $f_{x,i}$ and $f_{y,i}$.

Using the boundary conditions at $x = 0$ and $x = L$, and eliminating $A_{i,1}$, we get the relation

$$\begin{bmatrix} A_{3,0} \\ A_{4,0} \end{bmatrix} = \begin{bmatrix} X_{31} & X_{32} \\ X_{41} & X_{42} \end{bmatrix} \begin{bmatrix} X_{11} & X_{12} \\ X_{21} & X_{22} \end{bmatrix}^{-1} \begin{bmatrix} A_{1,0} \\ A_{2,0} \end{bmatrix} \tag{49}$$

where $[X]$ is the square matrix with four columns and rows given by

$$[X] = [Y_0]^{-1}[s][Y_1][T(L)][Y_1]^{-1}[s]^{-1}. \tag{50}$$

Here,

$$[s] = \begin{bmatrix} s_x & 0 & 0 & 0 \\ 0 & s_y & 0 & 0 \\ 0 & 0 & \dfrac{s_x}{s_y s_z} & 0 \\ 0 & 0 & 0 & \dfrac{1}{s_z} \end{bmatrix}, \tag{51}$$

$$[Y_0] = \begin{bmatrix} Y_{11,0} & Y_{12,0} & Y_{13,0} & Y_{14,0} \\ Y_{21,0} & Y_{22,0} & Y_{23,0} & Y_{24,0} \\ Y_{31,0} & Y_{32,0} & Y_{33,0} & Y_{34,0} \\ Y_{41,0} & Y_{42,0} & Y_{43,0} & Y_{44,0} \end{bmatrix}, \tag{52}$$

$$Y_{1i,0} = f_{x,i}, \tag{53}$$

$$Y_{2i,0} = f_{y,i}, \tag{54}$$

$$Y_{3i,0} = -\mathrm{j}\left(k_{x,i}(\lambda + 2\mu) f_{x,i} + k_y \lambda f_{y,i}\right), \tag{55}$$

$$Y_{4i,0} = -\mathrm{j}\left(k_y \mu f_{x,i} + k_{x,i} \mu f_{y,i}\right), \tag{56}$$

$$[Y_1] = \begin{bmatrix} Y_{11,1} & Y_{12,1} & Y_{13,1} & Y_{14,1} \\ Y_{21,1} & Y_{22,1} & Y_{23,1} & Y_{24,1} \\ Y_{31,1} & Y_{32,1} & Y_{33,1} & Y_{34,1} \\ Y_{41,1} & Y_{42,1} & Y_{43,1} & Y_{44,1} \end{bmatrix}, \tag{57}$$

$$Y_{1i,1} = f_{x,i}/s_x, \tag{58}$$

$$Y_{2i,1} = f_{y,i}, \tag{59}$$

$$Y_{3i,1} = -\mathrm{j}\left(k_{x,i} s_x \frac{1}{s_x}(\lambda + 2\mu)\frac{f_{x,i}}{s_x} + k_y \frac{1}{s_x}\lambda f_{y,i}\right), \tag{60}$$

$$Y_{4i,1} = -\mathrm{j}\left(k_y s_x \mu \frac{f_{x,i}}{s_x} + k_{x,i} s_x \frac{1}{s_x}\mu f_{y,i}\right), \tag{61}$$

$$[T(x)] = \begin{bmatrix} \mathrm{e}^{-\mathrm{j}k_{x,1}s_x x} & 0 & 0 & 0 \\ 0 & \mathrm{e}^{-\mathrm{j}k_{x,2}s_x x} & 0 & 0 \\ 0 & 0 & \mathrm{e}^{-\mathrm{j}k_{x,3}s_x x} & 0 \\ 0 & 0 & 0 & \mathrm{e}^{-\mathrm{j}k_{x,4}s_x x} \end{bmatrix}. \tag{62}$$

The reflection coefficients at the boundary $x = 0$, we obtain $(A_{3,0}/A_{1,0}$ and $A_{4,0}/A_{1,0})$ with $A_{2,0} = 0$ when the incident wave is the P-wave and in the case of SV-wave incidence we obtain $(A_{3,0}/A_{2,0}$ and $A_{4,0}/A_{2,0})$ with $A_{1,0} = 0$.

3.2. Computed results

Figure 4 shows the computed results of the reflection coefficient dependence on $2\pi/(\beta h)$ in the case of the SH-wave incidence with incident angle $\theta = 0$, the attenuation coefficient $s_{xI}(x) = 0.1$ and normalized thickness $k_s L = 24\pi$. Here k_s is the intrinsic wave number of the SH-wave in the isotropic solid. Decreasing the interval between adjacent nodal points h, the reflection coefficient approaches the value estimated by the truncation effect which caused by the reflection at the PML end terminal and can be estimated by attenuated waves in the PML, $20\log_{10}(\exp(2k_s L s_{2I})) = 20 \times 4.8\pi \log_{10} \mathrm{e} = 131\mathrm{dB}$. A higher order element causes lower reflection because of a better approximation of the intrinsic wave number. Figure 5 shows dependence on the incident angle. Smaller intervals of finite element nodes, $k_s h = 0.1\pi$, gives a better approximation than $k_s h = 0.2\pi$. Increasing the incident angle, β decreases and the approximation of the intrinsic wave number with discretized wave number becomes better. Hence, the reflection by FE-discretization decreases. However, incident angle becomes larger than the angle such as about 63.5 degrees for 1st order element, reflection increases because decreasing β yields decreasing of wave attenuation in PMLs and the reflection by the truncation effect increases. Figures 4 and 5 show that the results of the transfer matrix agree well those of FEA and we confirm that the reflection of the FE-model of the PML may be explained with the discretized wave number and the truncation effect.

We consider an isotropic solid and its PML with the Poisson ratio σ=0.3 in this section.

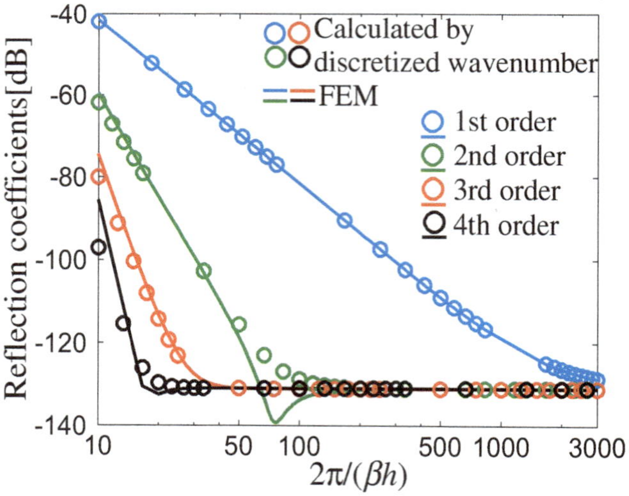

Figure 4. Dependence of SH-wave perpendicular incidence on $2\pi/(\beta h)$ for $k_s L = 24\pi$ and $s_{xI}(x) = 0.1$.

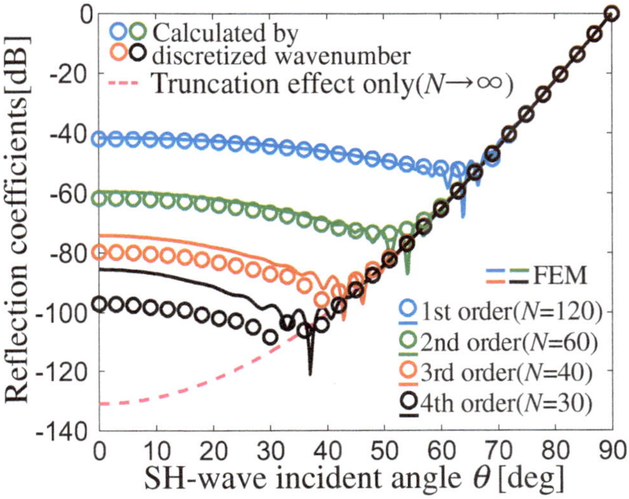

(a) $k_s h = 0.2\pi$.

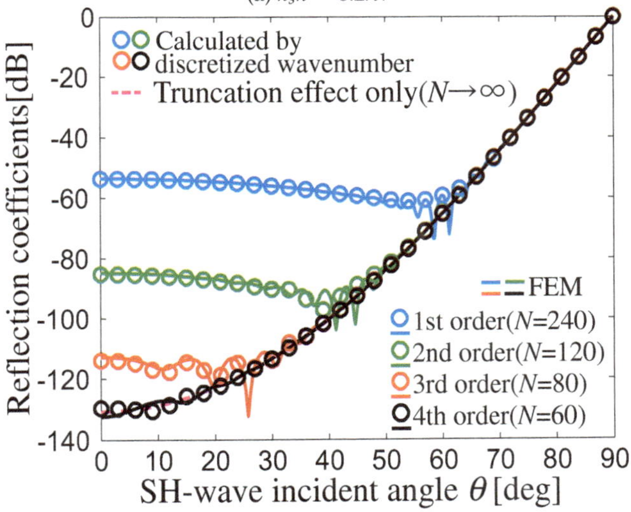

(b) $k_s h = 0.1\pi$.

Figure 5. Dependence of reflection coefficients of the SH-wave perpendicular incidence with the attenuation coefficient $s_{xI}(x) = 0.1$ and normalized thickness $k_s L = 24\pi$ on $2\pi/(\beta h)$. Here N is the number of FEs.

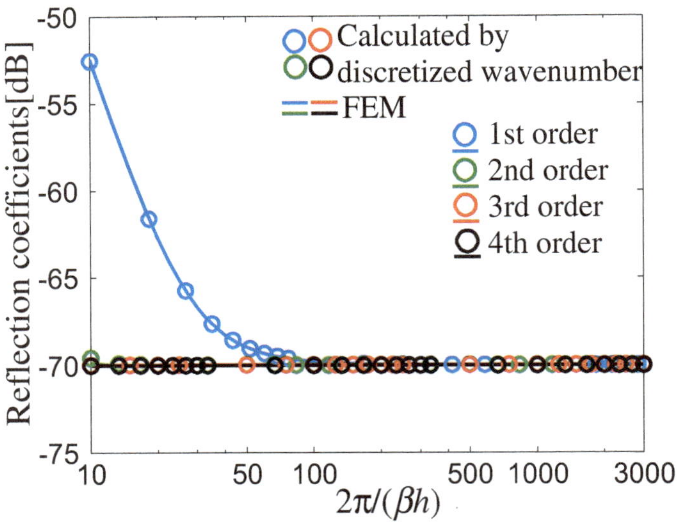

Figure 6. Dependence of reflection coefficients on $2\pi/(\beta h)$ in the case of perpendicular P-wave incident, $k_s L = 24\pi$ and $s_{xI}(x) = 0.1$.

Next, we consider P-wave or SV-wave scattering problems. Figure 6 shows the computed result of the reflection coefficient dependence on $2\pi/(\beta h)$ in case of the P-wave perpendicular incidence with the attenuation coefficient $s_{xI}(x) = 0.1$ and normalized thickness $k_s L = 24\pi$. Because the result of the SV-wave perpendicular incidence is the same result of the SH-vave perpendicular incidence owing to a symmetry of the problem, Fig. 4 also shows the result of SV-wave incidence with the attenuation coefficient $s_{xI}(x) = 0.1$ and normalized thickness $k_s L = 24\pi$. Both cases also approaches the value estimated by the truncation effect, -70.0dB and -131dB. Note that the wave number of the P-wave is $\sqrt{\mu/(\lambda + 2\mu)} = \sqrt{2/7}$ times of the SV-wave wave number. Dependencies of P- and SV-wave reflections on P- and SV-wave incident angle are shown in Figs. 7 and 8, respectively. Reflection coefficients of incident waves computed by the transfer matrix except the range that is larger than the critical angle, about 32.3 degrees, of SV-wave incidence are good agreement with the results of FEA. However, reflection coefficients of converted waves from incident waves are smaller for the SV-wave excited by the incident P-wave and larger for the P-wave than the results of FEA. We still can not explain this discrepancy.

Increasing s_{xI}, the reflection coefficient of P-wave in case of the P-wave incidence decreases in the incident angle range that is larger than 59 degrees. In the lower range, the reflection does not decrease because FE-discretization effect dominates the reflection. In the case of SV-wave incidence, we confirm that the P-wave converted from the SV-wave is amplified in the incident angle range that is lager than the critical angle when P-wave's wave number is zero in the isotropic region because of PML's intrinsic characteristics for non-propagating waves.

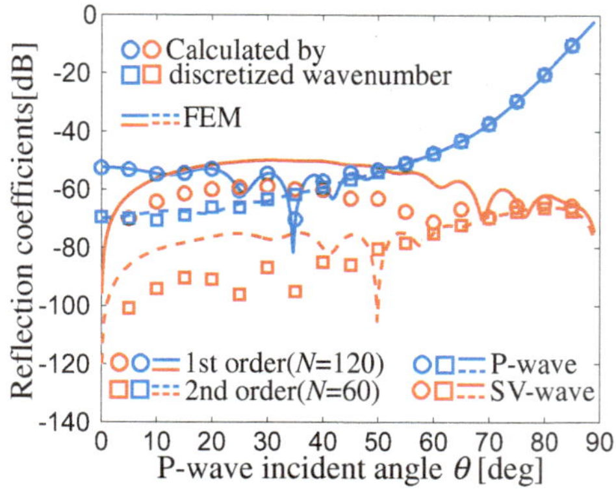

(a) $s_{xI}(x) = 0.1$.

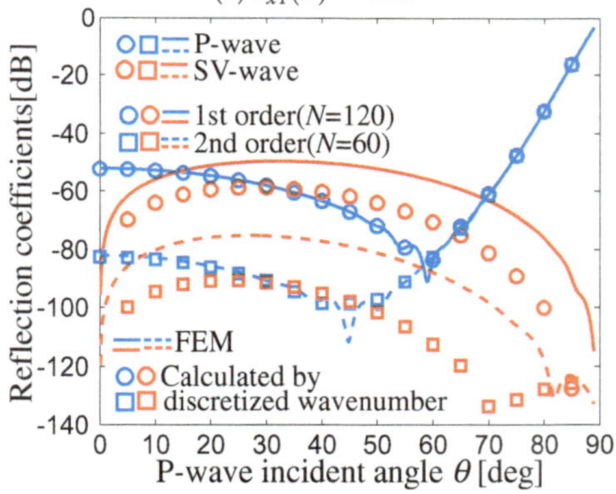

(b) $s_{xI}(x) = 0.2$.

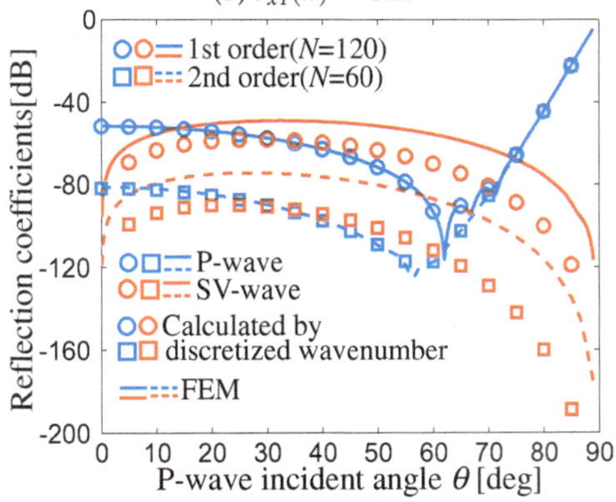

(c) $s_{xI}(x) = 0.3$.

Figure 7. Dependence on P-wave incident angle θ_i for $k_s h = 0.2\pi$ and $k_s L = 24\pi$.

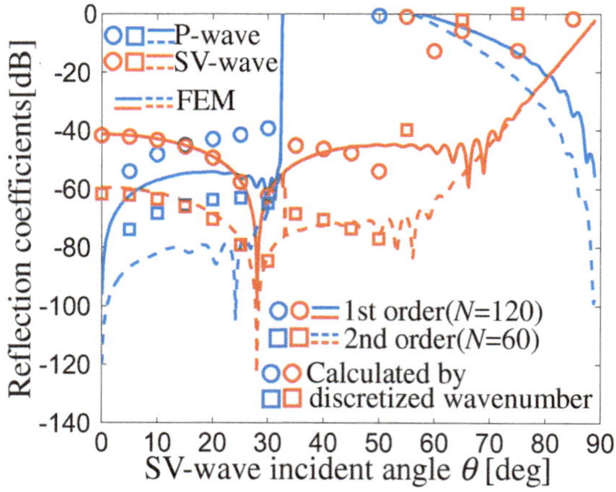

(a) $s_{xI}(x) = 0.1$.

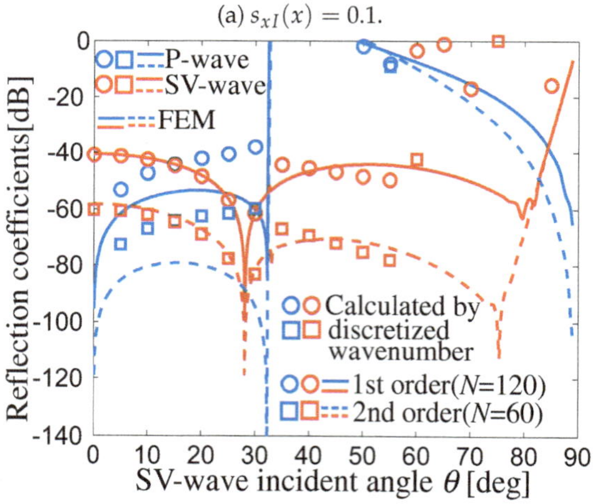

(b) $s_{xI}(x) = 0.3$.

Figure 8. Dependence on SV-wave incident angle θ_i for $k_s h = 0.2\pi$ and $k_s L = 24\pi$.

4. Conclusions

In this chapter, first, PMLs in the Cartesian, the cylindrical and the spherical coordinates for elastic waves in solids were derived from differential forms on manifolds. Our results show that PML parameters in any orthogonal coordinate system for elastic waves in solids may be determined by the same procedure in the Cartesian coordinates. Next, scattering of elastic waves in an isotropic solid was analyzed by field analysis in the thick layer in the one dimension. Numerical results show that the reflection from PMLs by the transfer matrix of elastic waves approximates the numerical results of FE-models successfully. We concluded that the reflection by FE discritization may be explained by FE-approximation of the intrinsic wave number.

Author details

Koji Hasegawa
Division of Information and Electronic Engineering, Graduate School of Engineering, Muroran Institute of Technology, Muroran, Japan

Takao Shimada
Department of Electrical and Electronic Engineering, Tsuyama National College of Technology, Tsuyama, Japan

5. References

[1] J.P. Berenger (1994) A perfectly matched layer for the absorption of electromagnetic waves. J. Comput. Phys. 114 : 185-200.
[2] W.C.Chew and W.H. Weedon (1994) A 3D perfectly matched medium for modified Maxwell's equations with stretched coordinates. Microwave Opt. Tech. Lett. 7 : 599-604.
[3] F.L.Teixeira and W.C.Chew (2000) Complex space approach to perfectly matched layers: a review and some new developments. Int. J. Numer. Model. 13 : 441-455.
[4] Z.S.Sacks, D.M.Kingsland, R. Lee and J.F.Lee (1995) A perfectly matched anisotropic absorber for use as an absorbing boundary condition. IEEE Trans. Antennas Propag. 43: 1460-1463.
[5] F.D.Hastings, J.B.Schneider and S.L.Broschat (1996) Application of the perfectly matched layer (PML) absorbing boundary condition to elastic wave propagation. J Acoust. Soc. Am. 100: 3061-3069.
[6] W.C.Chew and Q.H.Liu (1996) Perfectly matched layers for elastodynamics: a new absorbing bounary condition. J. Comput. Acoust. 4 : 341-359.
[7] Q.H.Liu (1999) Perfectly matched layers for elastic waves in cylindrical and spherical coordinates. J. Acoust. Soc. Am. 105: 2075-2084.
[8] Y.Zheng and X.Huang (2002) Anisotropic perfectly matched layers for elastic waves in cartesian and curvilinear coordinates. MIT Earth Resources Laboratory Industry Consortium Meeting.
[9] S.Ballandras,D.Gachon, J.Masson and W.Daniau (2007) Development of absorbing conditions for the analysis of finite dimension elastic wave-guides. Proc. Int. Freq. Contr. Symp.: 729-732.
[10] M.Mayer, S.Zaglmayr, K.Wagner and J. Schöberl (2007) Perfectly matched layer finite element simulation of parasitic acoustic wave radiation in microacoustic devices. Proc. IEEE Ultrason. Symp.: 702-706.
[11] Y.Li, O.B.Matar, V.Preobrazhensky and P. Pernod (2008) Convolution-Perfectly Matched Layer(C-PML) absorbing boundary condition for wave propagation in piezoelectric solid. Proc. IEEE Ultrason. Symp.: 1568-1571.
[12] U.Basu and A.K.Chopra (2003) Perfectly matched layers for time-harmonic elastodynamics of unbounded domains: theory and finite-element implementation. Comput. Methods Appl. Mech. Eng. 192 : 1337-1375.

[13] C.Michler, L.Demkowicz, J.Kurtz and D.Pardo (2007) Improving the performance of perfectly matched layers by means of hp-adaptivity. Num. Methods Partial Diffe. Equa. 23 : 832-858.

[14] A.Taflove and S.C.Hagness (2005) *Computational electrodynamics*(Artech House, Boston) 3rd ed. Ch7, p.273.

[15] T.Shimada and K.Hasegawa: (2010) Perfectly matched layers for elastic waves propagating in anisotropic solids. IEICE Trans. J93-C : 215-223 [in Japanese].

[16] T. Shimada and K. Hasegawa: (2010) Perfectly matched layers in the cylindrical and spherical coordinates. Jpn. J. Appl. Phys. 49 : 07HB08.

[17] T.Shimada, K.Hasegawa, and S.Sato (2009) Absorbing characteristics of electromagnetic plane waves in perfectly matched layers discretized by finite element method. Keisan Suri Kogaku Ronbunshu 9 : 04-091211 [in Japanese].

[18] W.C.Chew and J.M.Jin (1996) Perfectly matched layer in the discretized space: an analysis and optimaization. Electromagnetics 16 : 325-340.

[19] F.Collino and B.P.Monk: (1998) Optimizing the perfectly matched layer. Comput. Methods. Appl. Mech. Eng. 164 : 157-171.

[20] A.Berumúdez, L.Hervella-Nieto, A.prieto, and R.Rodrïguez: (2007) An optimal perfectly matched layer with unbounded absorbing function for time-harmonic acoustic scattering problems. J. Compu. Phys. 223 : 469-488.

[21] A.Berumúdez, L.Hervella-Nieto, A.prieto, and R.Rodrïguez (2010) Perfectly matched layers for time-haromonic second order elliptic problems. Arch. Comput. Methods Eng. 17 : 77-107.

[22] W.R. Scott, Jr. (1994) Errors due to spatial discretization and numerical precision in the finite-element method. IEEE Trans. Antennas Propagat. 42: 1565-1570.

Chapter 10

Finite Element Analysis Applications in Failure Analysis: Case Studies

Ahmad Ivan Karayan, Deni Ferdian, Sri Harjanto,
Dwi Marta Nurjaya, Ahmad Ashari and Homero Castaneda

Additional information is available at the end of the chapter

1. Introduction

The use of the finite element in engineering applications has grown rapidly in recent years. Finite element analysis (FEA) is based on numerical computation that calculates all parameters and boundaries given. Supported with powerful computer processors and continuous software development, the finite element method is rapidly advancing. The use of finite element analysis is not limited to the engineering field, as there are also medical and geospatial applications. The early development of the finite element can be traced back to the work of Courant [1], followed by the work of Martin [2], which applied the solutions for structural analyses at Boeing Company in the 1950s. Further work by Argyris, Clough, Turner, and Zienkiewicz developed the governing mathematical equation for the finite element method [3].

The numerical simulation introduced in the 70s for the stress model on concrete as published by Hillerborg [4] is a clear example of the FEA concept. Huiskes *et al.* [5] also stated that the finite element has been used in a structural stress analysis of human bones for biomechanic applications. Zienkiewicz *et al.* [6] applied the finite element method to the linear and non-linear problems encountered during the analysis of a reactor vessel. Gallagher [7] studied brittle material design through use of the finite element method, which incorporated thermal and elastic analysis aspects of the overall design. Miller *et al.* [8] used the finite element method during the study of crack stability of a turbine blade and proposed a hypothesis based on material strength characteristics, plastic zone size/history, and residual plastic strains.

In the mid 80s, Oritz *et al.* [9] proposed a method that aimed to enhance the performance of general classes of elements that undergo strain localization. An overview of the application of the finite element in machining from the 70-90s was well documented by Mackerle [10].

The application of the method to localize fractures was studied by Broberg [11], and further work was continued by Borst *et al.* [12]. Nowadays, finite element analysis is widely used by the engineering field in fluid dynamic and electrical applications.

In general, the finite element analysis is widely used in the pre-production manufacturing process to determine the most cost-effective decision based on the analysis. Finite element simulations allow comparison between different "designs." Finite element analysis can simulate operational and environmental conditions and formulate modifications without creating a physical prototype.

In order to adequately determine the root cause of material failure, two outcomes are required: the answer and an explanation. The failure of a mechanical component is usually associated with materials, the environment, a third party, or human error. An investigation through metallurgical failure analysis is usually conducted to reveal the root cause and the failure mechanism. Work by Griffith [13] on fracture mechanics created a breakthrough in the understanding of the material fracture mechanism. In certain cases, a conventional failure analysis approach is not enough to reveal the failure, and therefore, a more comprehensive analysis through the finite element is needed. Prawoto *et al.* [14] explained that the use of the finite element is an effective approach when the causes of failure are determined using qualitative metallographic and fractographic testing. The finite element requires an understanding of how a component works and will support the correct information and data. The quality of the data provided is a key element for the successful outcome of the simulation.

2. The finite element in failure analysis

The use of finite element analysis in mechanical applications has heavily increased in recent years. The continuous efforts to improve calculations and analyses so that models accurately incorporate actual conditions have been rewarded as a consequence of computer tool evolution. The finite element analysis has become an important tool for improving the design quality in numerous applications. The finite element analysis is a computer-based technique to solve problems using numerical solutions. The analysis includes a method based on creating a geometrical model of the structure that is divided into individual nodes or elements.

Finite element modeling provides different insights into the engineering analysis that cannot be obtained with the classical failure analysis method. Classical commercial software such as ABAQUS® and ANSYS® has been widely used to analyze failures or defects and reconstruct possible root causes. The images and animation produced by the software help to give a better understanding by visualizing the root cause behind the failure event. Previously, the development of the finite element evolved slowly due to a lack of tools to solve mathematical equations, and therefore, the method remained dormant until the computer era.

The early finite element software was commissioned by NASA in mid 1960s, which introduced NASTRAN® as an application that helped to design more efficient structures for vehicles that were developed by them [15]. The ANSYS® software, which was released in

1970, was run on a massive mainframe computer that was less powerful than the personal computers of today [16]. The early software was only capable of simulating the 2D beam model, but it eventually progressed towards the modern 3D solid model of today as computer hardware developed.

The applications of finite element analysis in engineering failure analysis are in continuous evolution as more factors related to the failure event are taken into account. For example, in the case of turbine blades involved in jet engine failure, engineers should incorporate the thermal effect and load effect received by the blade in the simulation, not to mention other possibilities from foreign particles that could initiate the failure. Therefore, FEA calculations now incorporate a combination of multiple physical environments. Current research related to the application of the finite element in failure analysis has been published in peer-reviewed journals, and can be seen in publications such as *Engineering Failure Analysis Journal®* and *Failure Analysis and Prevention Journal®*.

In general, a finite element method consists of three phases: (1) pre-processing, where the analyst creates the finite element mesh and applies certain parameters or boundaries to the model, (2) solver/solution, where the program runs the governing mathematical equation that was created by the model, and (3) post-processing, where the result is evaluated and validated for further interpretation.

Some limitations must be recognized in the finite element method. The ability to define and analyze the system and the model will determine the quality of the simulation results. The evaluations of any failure event need a comprehensive approach to reveal the root cause of the failure. FEA is a tool where multidisciplinary fields are combined to help find or support the solution with better accuracy for a final conclusion. Below are three case studies that illustrate how finite element analysis can be utilized in failure analysis.

3. FEA in mechanical failure

The use of FEA for mechanical failure has become a necessary tool that is easily obtained, as the commercial software is readily available for users. FEA is widely used in failure analysis assessment when analyzing and characterizing the quantitative and qualitative approaches used to determine the root cause of an event leading to a failure. The software is given historical background information related to the failure along with other boundary conditions. Below is an example of a mechanical failure analysis assessment that used finite element analysis software to verify the findings regarding the failure event.

In the first case, the finite element method was used to analyze the stress distribution of a failed 28MW horizontal hydroturbine shaft [17]. The data corresponded to the fractography and metallographic observations. The finite element analysis was performed for normal conditions as well as the type of high load conditions that would be experienced during the start-up period.

The shaft was constructed by the joining of a forged and cast part by slag welding. A crack was observed at the radius area of the casted part. The part was given an estimated 200,000 hours of operating life, and failure was detected after 163,411 hours in service. A visual examination of the fracture surface showed a fatigue pattern with obvious ratchet marks. Further observation of the fracture surface showed visible distorted fatigue lines around numerous gas holes and areas of increased porosity.

An original document indicated that the shaft was heat-treated to complete an austenization process. However, metallographic examination showed cast ferrite-pearlite with an undissolved dendrite structure, which might indicate that an improper heat treatment process occurred. In addition, a large non-metallic inclusion was also observed in the cast part.

A linear FEA was used to determine the stress state of the turbine shaft and shaft flange. The commercial ANSYS® software was used for the finite element modeling. The model represents a discrete continuum by an 8-node finite element with three degree of freedom that comprises 49,430 nodes and 47,547 elements. All of the boundary conditions were incorporated into the model.

A numerical calculation to determine the shaft flange stress states was performed for two characteristic load cases, where one load was taken from the manufacturer's documentation and the second load was the static load that occurs during start-up. Calculation of the finite element for both load conditions showed that the maximum stress was at the crack initiation site at the shaft flange.

The data collected from the chemical composition test and mechanical test showed that the material did not comply with the minimum standard required. Therefore, the crack location where the failure occurred was more susceptible to stress. Finite element analysis showed a high distribution of stress at the failed area. The finite element analysis revealed that the obtained tensile stress value on the shaft flange transition radius due to the load in case 2 was higher than recommended, and was characterized by the stress intensity factor at the crack tip being higher than the material threshold.

It was concluded that corrosion fatigue was the cause of the shaft failure [16]. The root cause for this case was improper corrosion protection at the failed area and a lack of periodical inspection, both of which were necessary due to the high stress on the region.

4. FEA in corrosion failure

Karayan *et al.* [18] studied the failure of a seawater inlet pipe. The failure was first characterized by a small leak at approximately the 4-8 o'clock position. A schematic drawing of the inlet pipe showing the backing bar near the leak location is shown in Figure 1. This backing bar was installed on the welded surface. The visual examination of this failed pipe is shown in Figure 2. In order to find out the root cause of failure, a number of laboratory tests were performed. The results showed that the failure was caused by cavitation, as evident by the presence of a crater-like surface near the backing bar (Fig. 3). These localized craters seemed unusual since they were only noticed near the backing bar. An additional tool such as finite element analysis was used to determine why this was the case. A finite

Figure 1. Big and small leak near the backing bar at about 6 o'clock position viewed from the inner side of seawater inlet pipe.

Figure 2. Big and small leak near the backing bar at about 6 o'clock position viewed from the inner side of seawater inlet pipe.

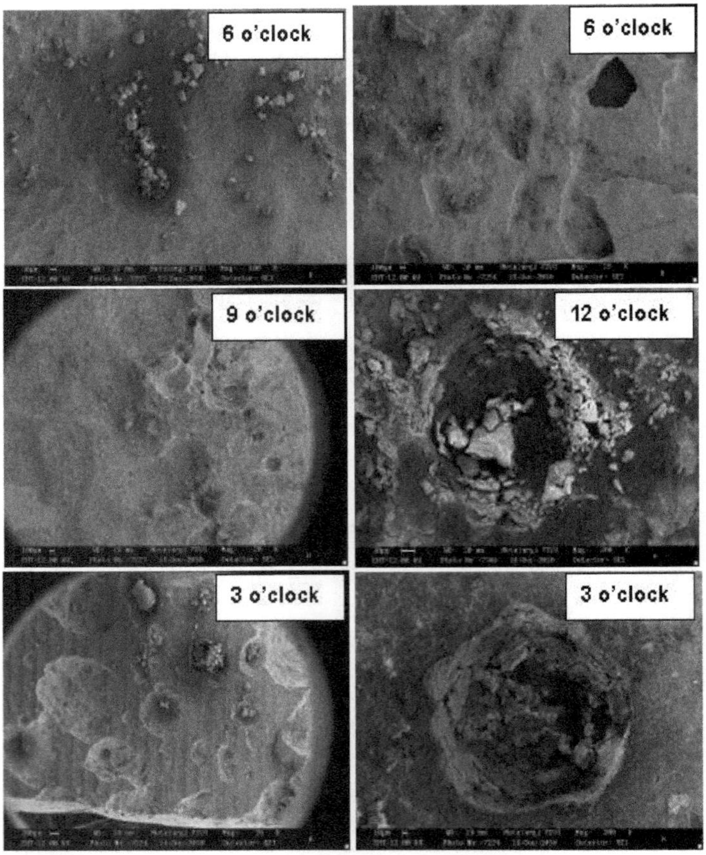

Figure 3. Surface morphologies of brown crater – like surface taken from 6 – 3 o'clock position.

element analysis was executed based on the pipe dimension and actual fluid conditions such as velocity, pressure, temperature, and implicit parameters. Because there were no data for the initial height of the unwanted backing bar, the authors assumed that the initial height was the highest backing bar found on the specimen. Interestingly, the failure location predicted by the finite element analysis matched up with the actual evidence (Fig. 4). It precisely showed that the failure could be located around the backing bar where the eddy zone was formed in this area. This suggests that sometimes the results obtained from laboratory tests cannot explain why a failure occurs, although evidence indicates the existence of a certain problem. In this case, a finite element analysis is the only tool that can help a failure analyst find the root cause of failure. As can be seen from Figure 4, the leaks and crater-like surfaces found in the area near the backing bar were attributed to the formation of eddy zones in this area. The length of the eddy zone predicted that the area that might suffer from a flow-induced attack.

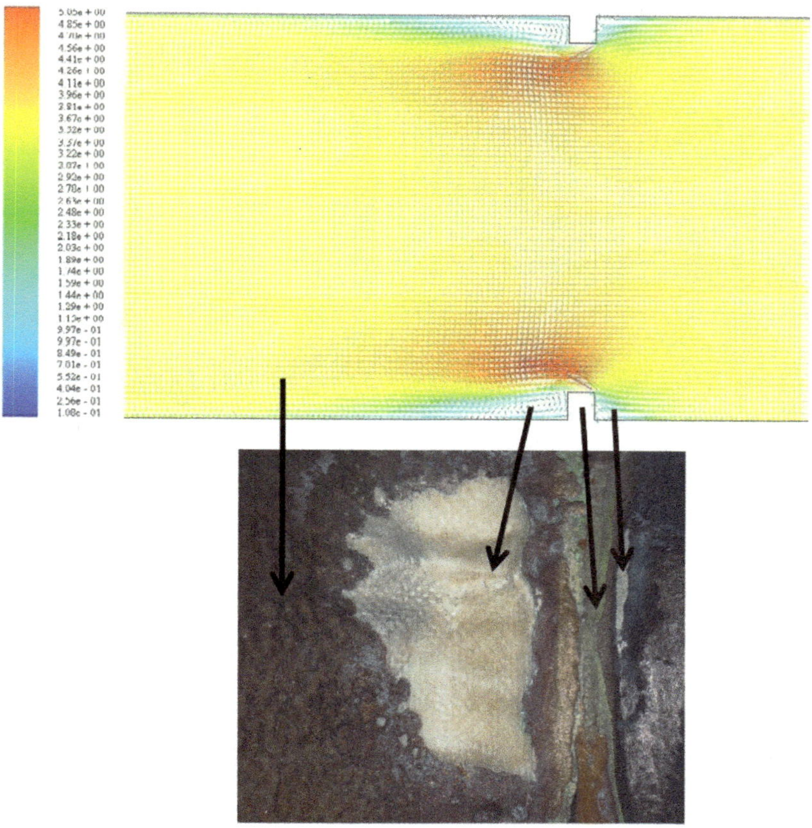

Figure 4. Finite element analysis of the inner pipe showing the orifice effect and the eddy zone near the backing bar correlated with the actual leaks on the inlet pipe.

The last case is a trunkline that burst during service [19]. This incident produced a significant impact on the gas production of the company and also on the environment. The failure was characterized by a mesa-like attack and wall thinning at the 5-7 o'clock position on the inner surface. The location of failure is shown in Figure 5. A reddish brown corrosion product was noticed on the surface, but there with no indication of the occurrence of an H_2S attack found in the material. This trunkline, carrying the gas with a total pressure of 905 psi, spanned across the jungle in the descending position. The analysis of the gas composition is listed in Table 1. The topography and characteristics of the soil in which the trunkline was located and also the material specification data is listed in Table 2.

Figure 5. The specimen sent to the laboratory for analysis is indicated by arrow

Component		Mole %
H_2	Hydrogen	0.0000
O_2	Oxygen	0.0084
N_2	Nitrogen	3.6036
CO_2	Carbon Dioxide	1.9406
H_2S	Hydrogen Sulfide	0.0000
C_1	Methane	82.7838
C_2	Ethane	6.7222
C_3	Propane	3.0077
iC_4	Iso-Butane	0.6144
nC_4	n-Butane	0.6610
iC_5	Iso-Pentane	0.2384
C_5	n-Pentane	0.1418
C_6	Hexanes	0.1352
C_7	Heptanes	0.0982
C_8	Octanes	0.0306
C_9	Nonanes	0.0072
C_{10}	Decanes	0.0069
C_{11+}	Undecanes plus	0.0000

Table 1. Result of gas analysis in trunkline

Item	Description	Result
Property	Q	± 3 MM
	WC	-
	GOR	-
	P (Psi)	750
	T (F)	140
Fluid Composition	H_2S	0.00
	CO_2 (%)	1.9406
	SRB content	-
	Chloride (%)	0.7635
	Water (%)	0.0898
Pipe Arrangement	Laydown/buried/support w/ith trestle	Laydown
	Seam position	-
Corrosion Form	Uniform/localized/pitting/etc	pitting/localized
	external/internal/both	Internal
	On seam/not	-
	On 6 o'clock/others	6 o'clock
Environment	Elevation	Descending
	Any river/road crossing	No
	Soil pH	-
	Any trees/bushes	Yes
Document	Work over on wells using the line	No job in last one year
	WT inspection	not yet

Table 2. Fluid properties and failed trunkline condition.

A visual examination showed that the burst area was located at the 6 o'clock position of the (a) downstream part and the (b) upstream part (Fig. 6). Some points on the inner surface of the downstream part at the 5-7 o'clock position showed surface degradation with wall thinning and pits (Fig. 7). Some points on the inner surface of the upstream part at the 5-7 o'clock position also showed surface degradation with wall thinning and pits (Fig. 8). A uniform attack at the burst area from the 5-7 o'clock position is shown in Figure 9.

(a)

(b)

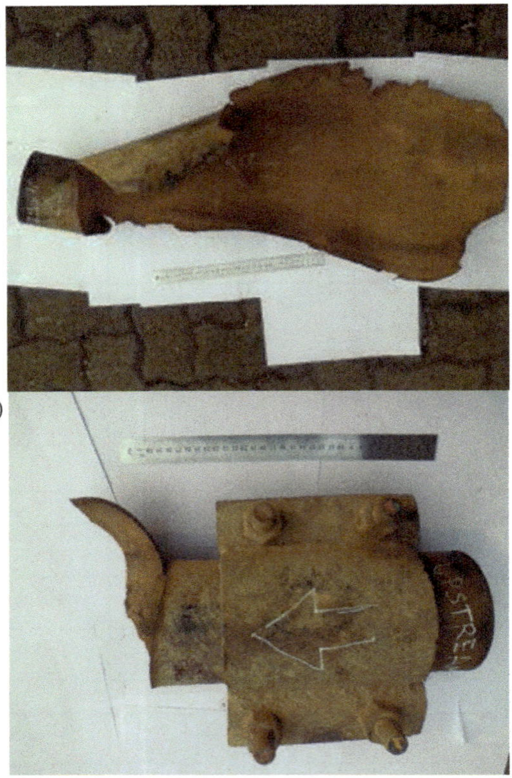

Figure 6. (a) Downstream part and (b) upstream part showing the burst located at 6 o'clock position.

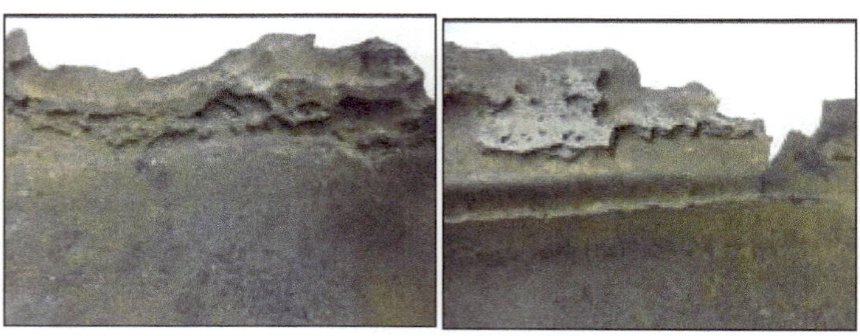

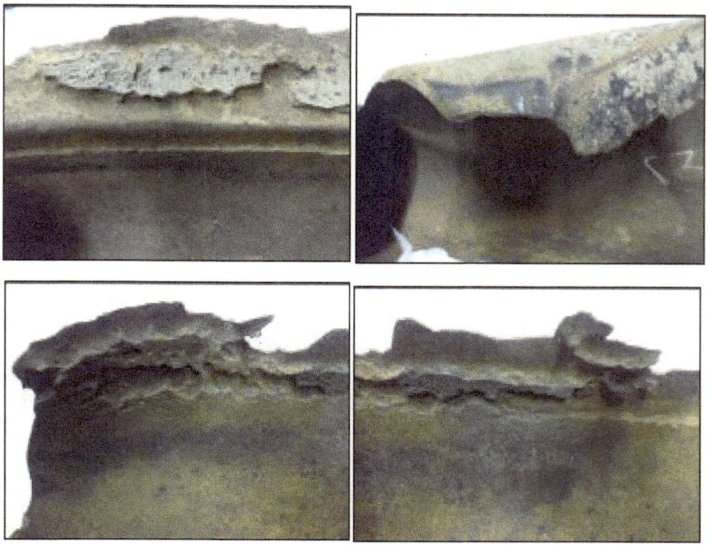

Figure 7. Wall thinning and pits at 5 – 7 o'clock position of downstream part.

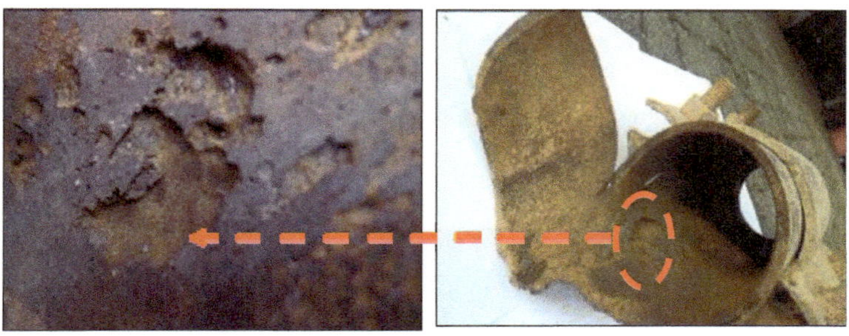

Figure 8. Wall thinning and pits at 5 – 7 o'clock position of upstream part.

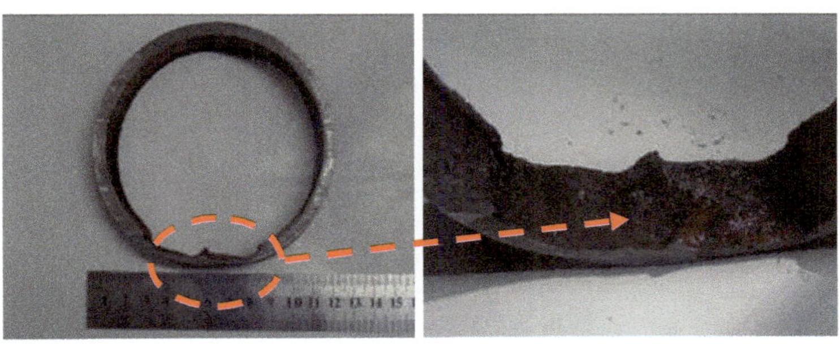

Figure 9. Uniform attack at the burst area from 5-7 o'clock position.

The metallographic preparation and macroetching were performed on the perimeter of the trunkline, and the results showed that the trunkline was made of seamless pipe (Figs. 10 and 11). The thinning area shown in Figure 9 was an ERW (electric resistance welding)-free area (Fig. 10). This information indicates that the failure could not be attributed to the ERW pipe issue.

Figure 10. Results of macroetching showing the absence of ERW.

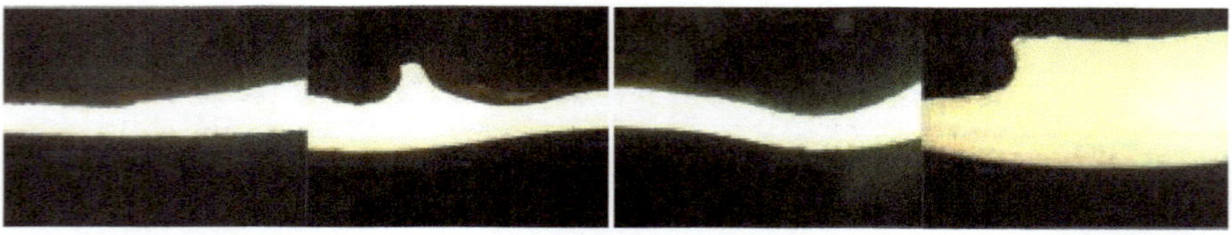

Figure 11. Macrograph of 5-7 o'clock position showing the absence of ERW.

The microstructure of the trunkline was taken from the cross-section, and the results are displayed in Figure 12. As can be seen, the trunkline is composed of ferrite (light phase) and pearlite (dark phase) with equiaxed grains, which is typical of seamless pipe. The chemical composition of the trunkline was tested using an optical emission spectrometer and the results (Table 3) show that this trunkline is composed of an API 5L X60 steel [20]. Due to the insufficient geometry of the trunkline, the mechanical property of the trunkline was only examined by hardness testing. In order to verify that the specification of this material was API 5L, the resultant hardness values were converted to tensile strength (Table 4).

Figure 12. Microstructure of trunkline showing a typical seamless microstructure with equiaxed grains. The dark phase is pearlite and the light one is ferrite.

Material	C (%)	Mn (%)	P (%)	S (%)	Ti (%)
API 5L X60 PSL 1 [20]	0.22	1.40	0.030 (max) <	0.030 (max)	0.040
Failed Trunkline	0.165	0.594	0.003	< 0.003	< 0.002

Table 3. Chemical composition of trunkline

Material	*Average HRB*	*Approxmate UTS Based on Conversion (MPa)*
Failed trunkline	82	524
API 5L X60 [20]	-	517 (min)

Table 4. Converted tensile strength in comparison with API 5LX60 specification.

The finite element analysis was executed around the overfill, and the results (Fig. 13) showed that the area at which the pipe burst had a dead or eddy zone due to the excessive overfill (Figs. 14 and 15). Because there was no information about the initial overfill height, we assumed an overfill height of 1 cm for the finite element simulation. This selection was based on the fact that there was one point at the remaining overfill that had a height of 1 cm. The other areas were mostly degraded, with heights of less than 0.5 cm. Severe corrosion was noticed on every peak of overfill.

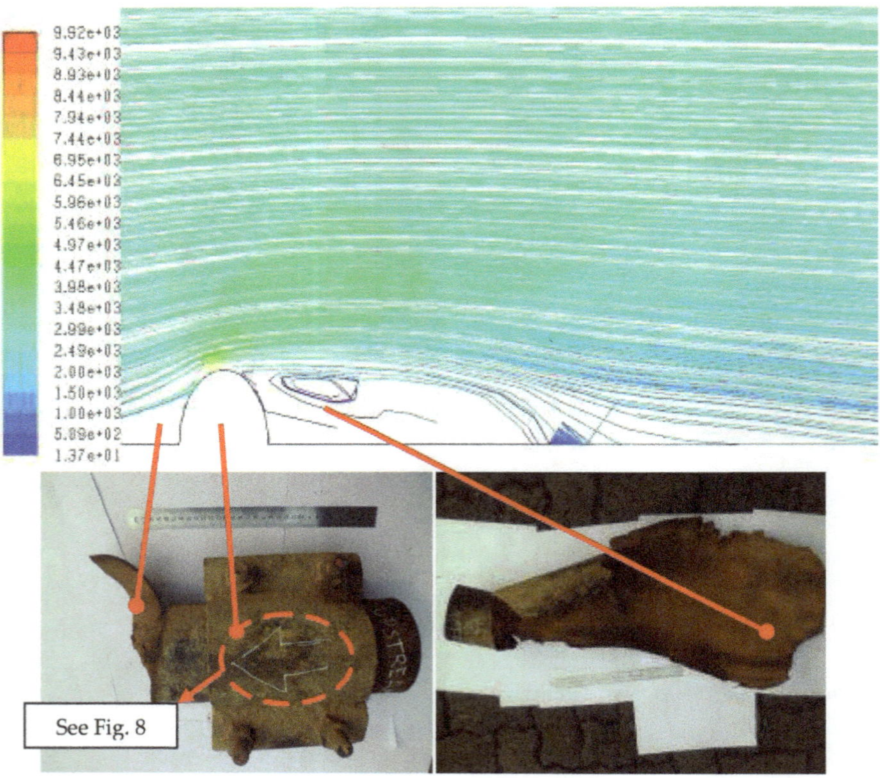

Figure 13. Finite element analysis showing the eddy zone due to overfill in comparison with the actual failed area of pipe.

As can be seen in Table 1, the only corrosive gas in the fluids flowing in this failed trunkline was CO_2. Carbon dioxide systems are one of the most common environments in the oil and gas field industry where corrosion occurs. In a relatively slow reaction, carbon dioxide forms a weak acid known as carbonic acid (H_2CO_3) in water, but the corrosion rate of CO_2 is greater than that of carbonic acid. Cathodic depolarization may occur, and other attack mechanisms may also occur. The presence of salts is relatively unimportant in sweet (CO_2) service [21], and thus, the presence of chloride in this system (0.7635%) did not significantly contribute to the failure.

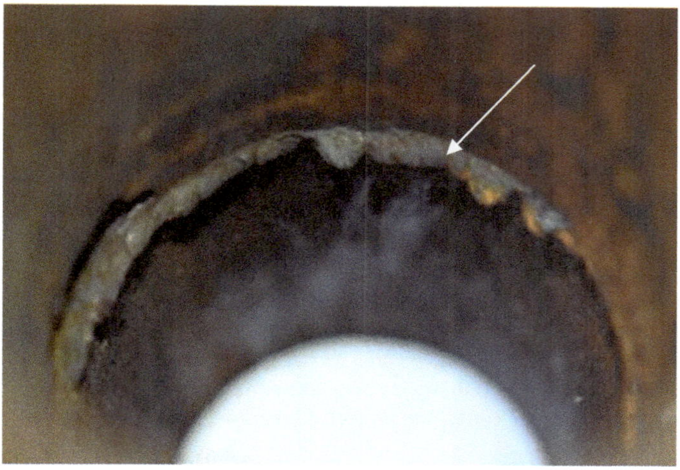

Figure 14. The inner side of circle area in Fig. 11 showing an excessive overfill (arrow) in the failed pipe at 12 o'clock position.

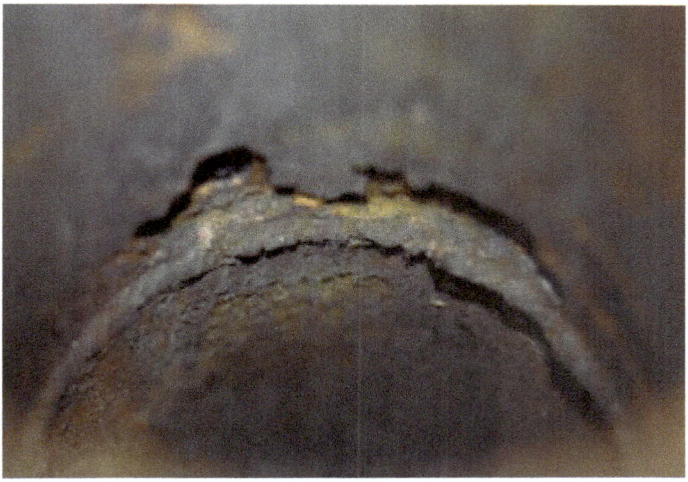

Figure 15. An excessive overfill and severe surface degradation beside overfill at 6 o'clock position.

The sweet environment of this system with a CO_2 partial pressure of 17.56 psig may influence the presence of corrosion. The relationship between corrosion tendency and CO_2 partial pressure in the sweet environment with pH 7 or less has been reported elsewhere [22]. When the CO_2 partial pressure is less than 7 psig, the system is non-corrosive. When it is somewhere between 7-30 psig, corrosion in the system may be present. Lastly, when it is higher than 30 psig, the system is corrosive [22]. Temperature and flow regime are closely linked because CO_2 corrosion is dynamic and very sensitive to electrochemical and physical imbalances (especially fluctuating pressure, temperature, and volume). Steady-state (P,T,V,) conditions tend to promote protective film compaction, and therefore, passivation and low corrosion rates. Lower temperatures <120°F (approximately 50°C) tend to promote patchy corrosion with softer multi-layered iron carbonate (siderite) scales that provide some barrier protection up to 140-160°F (60-70°C). Above these temperatures, damaging localized corrosion is observed as films lose stability and spall off, giving rise to galvanic mesa attack [23]. The failed pipe we studied with an operating temperature of 140°F might have formed a protective film. However, the phenomenon of film removal and its effect on the failure of this trunkline was not evident in our laboratory test. Our finite element analysis in the failed area in Figure 13 shows the unstable and chaotic flow in the failed area (called the dead or eddy zone). As illustrated, the eddy zone was triggered by the excessive overfill. This suggests that the pipe surface in this eddy area was severely attacked by flow. The presence of a passive carbonate layer could not protect the surface from this type of flow-induced attack that led to a mesa attack.

5. Conclusions

The three case studies discussed in this chapter have clearly shown us that finite element analysis (FEA) is an excellent and powerful tool that can be employed in failure analysis. Finite element analysis provides a failure analyst with more quantitative and qualitative information about the causes of failure. Although visual examinations and laboratory tests may not be able to determine a failure mechanism, the results of finite element analysis will support all data obtained from these tests. As long as a competent analyst running the finite element analysis is given sufficient data and has good knowledge of the system under study, the results of FEA will be reliable, although they should always be validated with experimental or real condition information.

Author details

Ahmad Ivan Karayan and Homero Castaneda
Department of Chemical and Biomolecular Engineering, The University of Akron, USA

Deni Ferdian, Sri Harjanto, Dwi Marta Nurjaya and Ahmad Ashari
Department of Metallurgy and Materials Engineering, University of Indonesia, Indonesia

Deni Ferdian, Sri Harjanto, Dwi Marta Nurjaya and Ahmad Ashari
Center for Materials Processing and Failure Analysis (CMPFA), University of Indonesia, Indonesia

6. References

[1] F. Williamson, *Richard Courant and the finite element method: A further look,* Historia Mathematica, Vol 7(4), 1980, pp. 369-378.

[2] H. C. Martin, Large Deflection and Stability Analysis by Direct Stiffness Method, JPL Technical Report No. 32-931, California Institute of Technology, August 1966.
[3] R. W. Clough, Early History Of The Finite Element Method From The View Point of a Pioneer, International Journal For Numerical Methods In Engineering, 60, 2004, pp. 283–287.
[4] A. Hillerborg, M. Modéer, P.-E. Petersson, *Analysis of crack formation and crack growth in concrete by means of fracture mechanics and finite elements,* Cement and Concrete Research,Vol 6(6), 1976, pp. 773-781.
[5] R. Huiskes , E.Y.S. Chao, *A survey of finite element analysis in orthopedic biomechanics: The first decade,* Journal of Biomechanics, Vol 16(6) 1983, pp. 385-409.
[6] O.C. Zienkiewicz, D.R.J. Owen, D.V. Phillips, G.C. Nayak , Finite element methods in the analysis of reactor vessels, Nuclear Engineering and Design, 20(2) 1972, pp. 507–541.
[7] R.H. Gallagher, Finite element analysis in brittle material design, Journal of the Franklin Institute, 290(6) 1970, pp. 523–537.
[8] R.E. Miller Jr., B.F. Backman, H.B. Hansteen, C.M. Lewis, R.A. Samuel, S.R. Varanasi, Recent advances in computerized aerospace structural analysis and design Computers & Structures, 7(2) 1977, pp. 315–326.
[9] M.Ortiz, Y.Leroy, A.Needleman, *A finite element method for localized failure analysis, Computer Methods in Applied Mechanics and Engineering,* Vol 61(2) 1987, pp. 189-214.
[10] J. Mackerle, *Finite-element analysis and simulation of machining: a bibliography (1976–1996),* Journal of Materials Processing Technology, Vol 86, Issues 1-3, 1998, pp. 17-44.
[11] K.B. Broberg, *The foundations of fracture mechanics,* Engineering Fracture Mechanics,Vol16(4) 1982, pp. 497-515.
[12] Borst, R. De; Sluys, L.J. Muhlhaus, H.-B. Pamin, J. *Fundamental Issues In Finite Element Analyses Of Localization Of Deformation Engineering Computations,* Int J for Computer-Aided Engineering, Vol 10(2), 1993 , pp. 99-121(23).
[13] Griffith: Phil Trans Roy Soc 1921, v221, pp. 163-198.
[14] Y.Prawoto, *Quantitative failure analysis using a simple finite element approach.* Journal of Failure Analysis and Prevention, 10 (1) 2010. pp. 8-10.
[15] http://en.wikipedia.org/wiki/Nastran
[16] http://www.odonnellconsulting.com/forensicfea.html
[17] D. Momcilovic, Z. Odanovic, R.Mitrovic, I.Atanasovska, T.Vuherer, Failure analysis of hydraulic turbine shaft, Engineering Failure Analysis, 20 (2012) pp. 54–66.
[18] A.I. Karayan, A. Hersuni, D. Adisty, A. Yatim, Failure analysis of seawater inlet pipe, Journal of Failure analysis and Prevention, 11(2011) pp. 481-486.
[19] A.I. Karayan, Failure Analysis of Trunkline: An Internal Report, Center for Materials Processing and Failure Analysis (CMPFA), 2011.
[20] American Petroleum Institute (API) 5L Standard.
[21] http://octane.nmt.edu/WaterQuality/corrosion/CO2.aspx
[22] Garverick L, Corrosion in the petrochemical industry, pp. 92, ASM International, 1994.

[23] Sing B, Krishnathasan K. Pragmatic effects of flow on corrosion prediction, NACE corrosion confeence and expo 2009, Paper No. 09275.

Section 4

Applications of FEA in "Machine Elements Analysis and Design"

Chapter 11

Overview in the Application of FEM in Mining and the Study of Case: Stress Analysis in Pulleys of Stacker-Reclaimers: FEM vs. Analytical

Jairo A. Martins and István Kövesdy

Additional information is available at the end of the chapter

1. Introduction

The determination of stresses, deformations and the proper evaluation of calculations outputs are of extreme importance on the mechanical components and to assembly's effectiveness in the Mining Environmental. For instance, components of heavy duty plant machinery like Car Dumpers, Apron Feeders, Stacker-Reclaimers, etc..., and important components of Stacker-Reclaimers like Pulleys; which are under high responsibility must be calculated and designed properly and carefully. The unknown or ignorance of the complete environmental or data inputs (loads and constraints) where the component is applied, can bring tremendous damages to society and jeopardize entire businesses, mainly whether lives are involved. However, new technological tools, like Finite Element Method (FEM), have brought an even higher level to a better understanding of the complex products, those which have several parts in its conception. Like Klauss [2] describes, the Finite element methods are now widely used to solve structural, fluid, and multiphysics problems numerically. The methods are used extensively because engineers and scientists can mathematically model and numerically solve very complex problems. FEM is considered though a step further on the path on designing products, saving weight, consequently costs of design and manufacturing by the better understanding the pieces behaviors and performance prediction.

To evaluate the stresses in mechanical parts and/or components there are basically two manners; by the analytical approach and/or finite element method. This last, considered the most recent and complete tool to evaluate stresses and strains [2].

An example of the FEM simulation is shown briefly in this chapter when designing pulleys to Stacker-Reclaimers. We selected a standard pulley and generated it by analytical model (Redundant Structure Model) as well as by Finite Element Method (FEM) under linear analysis. The analytical formulas presented in the text are those belonged to the classical mechanical engineering background. In fact, the analytical calculation has presented success along the time once most of the products in the field have performed properly. The model is considered robust enough to deliver products under high quality of project and which considers the material and manufacturing data in order to determine the allowable stress by safety factors.

This paper describes the limited resources when calculating pulleys (for Stacker-Reclaimers or belt conveyors) by analytical methods in comparison with the advantages of the Finite Element Method and its comparatively minor imprecision. The analytical calculation, particularly, presents an issue considered a constraint to overcome, which is related to the energy contribution by linear elastic deformation, of each component to the final sum of the stress x strain in the assembly.

To FEM simulation, the software Inventor 2010 [3] was used to develop the model, meanwhile the calculation by FEM made by Autodesk Simulation [4].

2. Overview of products modeled in 3D and simulated by FEM

In the mining business the usage of software's in modeling components and machines 3D and afterwards simulated by FEM has increased potentially within the last decades. Machines like, Car Dumpers (Figure 1), Apron Feeders (Figure 2) and Stackers-Reclaimers (Figure 3) and their main components are modeled 3D and simulated by FEM software's. Several software's, specialized in modeling are available in the market, for example, Solidworks, Ideas, Inventor etc..., and in terms of FEM simulation, Nastran, Ansys and Autodesk Simulator are at the edge of this technology.

2.1. Advantages and drawbacks of modeling 3D and simulating by FEM in mining

Due to the upgrade in the way of designing products along the last decades, the technologists and engineer's had to change their minds when studying products in their initial phase of product design. Till the last decade the drawings were done basically in 2D environment, manually by clip boards and later on by CAD in computers. These required high imagination and capacity to evaluate technically and precisely the components in spatial views and in free space. However, the possibility of error by interferences between parts was relatively high and when happened, high costs were involved due to the required interven in the manufacturing or in the field. With the event of modeling by software, the need of another way of thinking about components and/or machines was strictly required. Despite of apparently complex, at first sight, due to the change of the way of designing, the FEM brings some advantages which worth, as follow;

a. Mitigate or eliminate interferences, by visual analysis and software testing, decreasing substantially the re-work in manufacturing or in field;
b. the possibility of designing lean shapes to particular application and loads;

c. having lower weights and consequently more effectiveness in energy savings to rotary/dynamical parts;
d. providing refined visual presentation (3D) of stresses and displacements suffered by the parts. Therefore bringing excellent power of analysis for engineers;

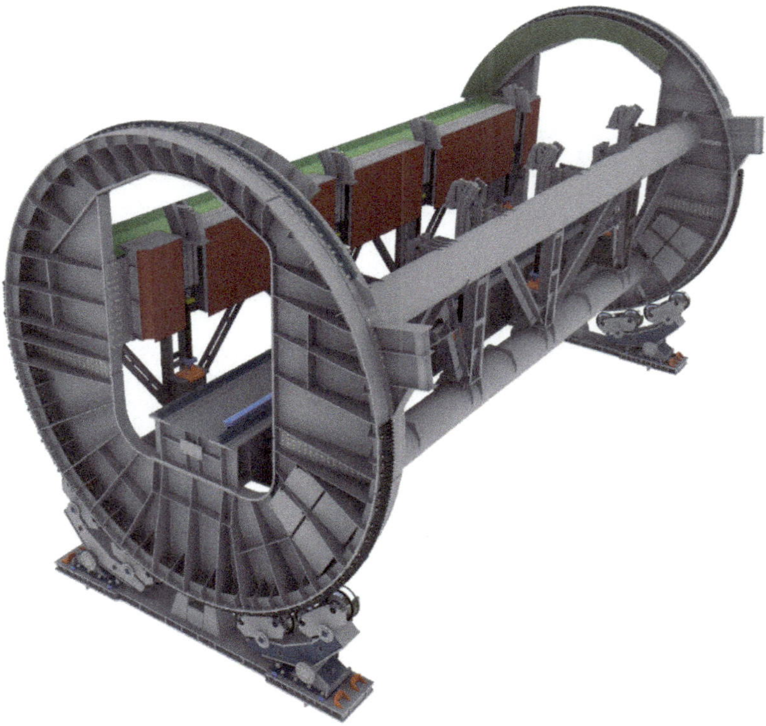

Figure 1. Car Dumper – overview, modeled by Inventor

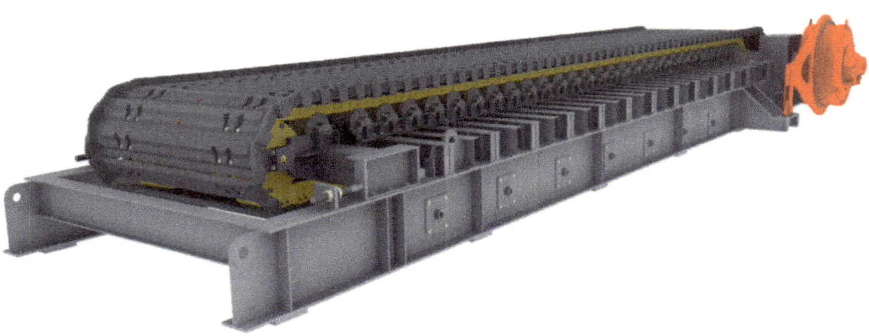

Figure 2. Apron Feeder – overview, modeled by Inventor

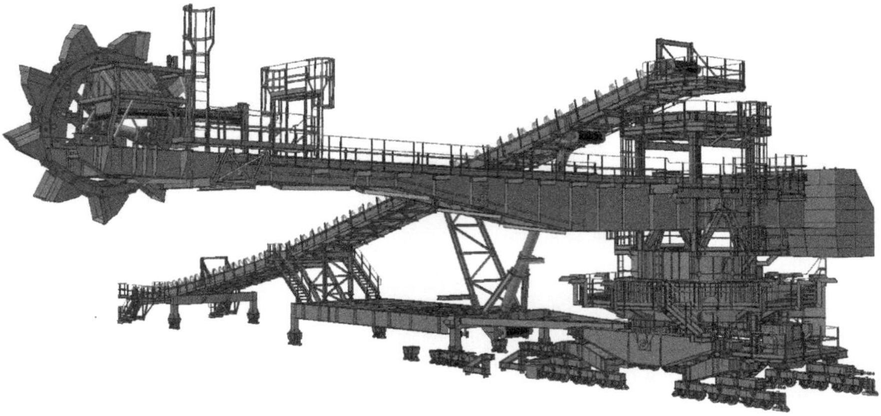

Figure 3. Stacker-reclaimer machine, overview – modeled by Inventor

Note – the 3D model in FEM however requires more deep knowledge of technologists and engineers in Stress x Strain analysis, stress tensors/matrixes, material properties, isotropy and anisotropy, stress states, residual stresses, von-Mises, Mohr circle and basic mechanics evaluations criteria. Even with the software advantages, the output in the FEM models still remains as the engineering duty;

a. beauty pictures to present products in commercial and marketing scenarios (high sensation of reality);
b. in order to produce manuals to operation and/or maintenance.

There also drawbacks in the FEM simulations, like;

a. limits when interferences between parts are present in the model, which require non-linear analysis;
b. usually residual stresses are present in the real component but are neglected in the model;
c. small details in the big picture sometimes need to be handled or suppressed in order to have allow enough capacity to run the model, even when powerful computational machines are used.

In order to overcome such deficiencies in the FEM calculations, safety stresses are applied and fatigue coefficients used within fatigue models like; Goodman, modified Goodman, Gerber and/or Soderberg (15-16) .

3. Description of an analytical method

Most of the formulas of strength of materials express the relations among the form and dimensions of a member, the loads applied thereto, and the resulting stress or deformation. Any such formula is valid only within certain limitations and is applicable only to certain problems. An understanding of these limitations and of the way in which formulas may be combined and extended for the solution of problems to which they do not immediately apply, requires knowledge of certain principles and methods that are stated briefly in the pulleys calculations ahead. In determining stress by mathematical analysis, being analytical or FEM, for example, it is customary to assume the material as elastic, isotropic, homogeneous, and infinitely divisible without change in properties and in conforming to Hooke´s law, which states that strain, is proportional to stress. On the other hand, these assumptions despite of imposing certain limitations upon the conventional methods of stress analysis must be used in the form of safety factors. This precaution has given satisfactory results for nearly all problems in engineering, being in analytical or FEM models.

The pulley is basically composed by; expansion ring (when applicable), hub, shaft, disc and cylinder, as seen in the Figure 5 below. The calculation of individual components is still not an issue nowadays and classical formulas may be applied without main difficulties. But when there is an increase of components quantity and the interaction among them takes place, the analytical method cannot predict the real and accurate interaction, energy shared by each component in the assemble, due the imposed deformations. In other words, the proportion of deformation of each individual into the ensemble is a very complex to determine accurately and manually.

3.1. Division of forces in assemblies and redundant structures, pulley application

The concepts of force´s flow are useful in the visualization of paths taken by the forces lines when crossing machines or structures from the load points till the support points. Whether the structure is simple and statically determined, the equations of equilibrium are enough to determine the reactions. On the other hand if redundant supports exist, it means additional supports to those required to satisfy the static equilibrium conditions, those simple equations are not enough anymore to explain the intensities (magnitude) in any one of the reactions. It happen due to the support works as a separated "spring", deflecting under load, proportionally to its stiffness, in a manner that all reactions are shared by all supports under an unknown way. Whether a stiffer rigid or under a rigid fixed deflection are in parallel with a less stiff spring or under a flexible deflection, the rigid deflection will absorb a higher portion of the loading. But whether a stiffer rigid or under a rigid fixed deflection are in series with a less stiff spring or under a flexible deflection, the loads absorbed are similar. The importance of such simple concept is applicable to all machines and real structures where exist the combinations of parts ("springs"), in series or parallel (7).

As seen in the Figure 4 the springs can be arranged in parallel arrangements as well as in series. If the springs are arranged parallel the deflections are the same but the total force F is divided between the spring 1 and spring 2, as follow;

$$F = F_1 + F_2 \tag{1}$$

once, $y_1 = y_1 + y_2$

$$\frac{F}{y} = \frac{F_1}{y_1} + \frac{F_2}{y_2} \quad (2)$$

from where is obtained;

$$k_p = k_1 + k_2 \quad (3)$$

to k_p as being the spring constant to each spring in parallel

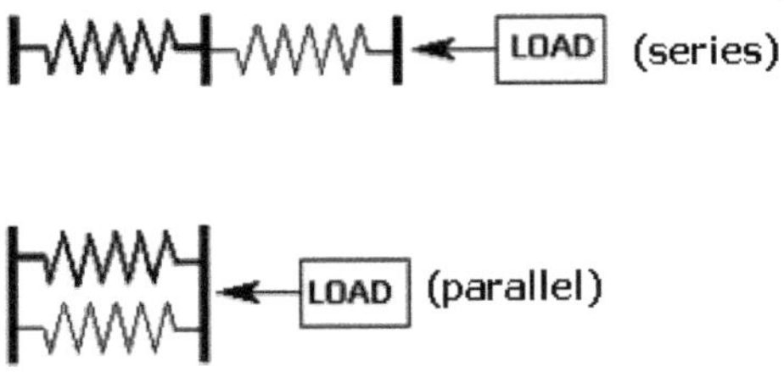

Figure 4. Different springs (stiffness – in parallel and series)

When the springs are arranged in series, the force F is the same to both springs, but the deflection of the spring 1 and spring 2 are associated to compose the total deflection, it means;

$$\frac{y}{F} = \frac{y_1}{F_1} + \frac{y_2}{F_2} \quad (4)$$

from where is obtained;

$$k_s = \frac{1}{\frac{1}{k_1} + \frac{1}{k_2}} \quad (5)$$

to k_s as the combined constant to springs in series

The diagram in the Figure 5 shows the pulley main components with imposed load. This load is transferred to all components and the total energy required to absorb such energy is composed by the sum of individual deformation. This deformation is directly related to the bending imposed in the Cylinder, the Disc and the Shaft. The sum of those deformations can be described as follow;

$$\delta_T = \delta_C + 2.\delta_D + \delta_S \quad (6)$$

δ_T is the total deformation, δ_C the deformation of cylinder, δ_S the deformation of shaft and the δ_D deformation of discs. The deformation of hub is considered zero due to its superior stiffness.

In the case of pulleys the system can be considered the same as explained with springs, what means, the cylinder and the shaft are in series and the discs are parallel each other but in series with the other components. The assembling equation then can be arranged as follow;

$$k_T = \frac{1}{\frac{4.L^3{}_E}{6.E.\pi.R_E{}^4} + \frac{L^3{}_C}{6.E.0{,}4.d^3.\delta} + b.\pi.(R_d + r)} \quad (7)$$

where L_E it the distance between the block bearings in the shaft, L_C it the length of cylinder, d is the cylinder internal diameter, b the disc thickness, R_d the radii of disc, r the internal radii of disc, E the Young Modulus, R_E the shaft radii and δ the thickness of the disc. The final deflection of the ensemble is determined by;

$$y_T = \frac{P}{k_T} \tag{8}$$

being y_T the total deflection in mm, P the resultant load applied in N and k_T the ensemble constant (N.mm).

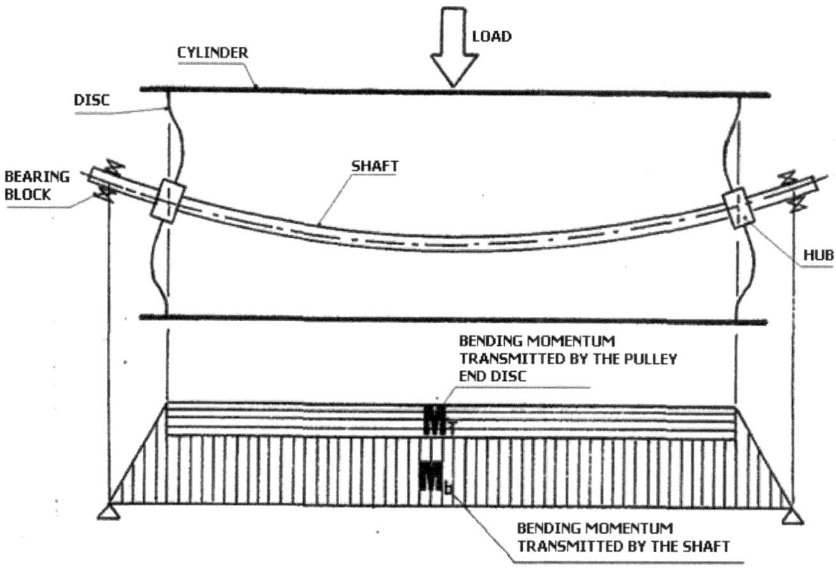

Figure 5. Diagram of pulleys loaded.

It is quite easy to identify the contribution of each element by a simple comparison between the pulleys components in the formula above and in the Figure 5. The load is transmitted by the pulley cylinder toward the discs, which suffer the high deformation due to its low inertia, then to the shaft which is bent due to the reactions at the bearing blocks. This is a normal condition found in driven pulleys, the drive pulleys contain an additional load, torque, transmitted from the shaft to the discs and lately to the cylinder. The drive pulley won´t be covered at this chapter.

4. Description of Finite Element Method (FEM)

4.1. Method

Like described previously, the finite element method (FEM) is a very powerful technique for determining stresses and deflections in complex structures when compared with analytical methods. With this method the structure is divided into a network of small elements connected to each other at node points. Finite element method grew out of matrix methods for the analysis of structures when the widespread availability of the digital computer made it possible to solve system of hundred of simultaneous equations (8). The FEM is then a computerized method for predicting how a real-world object will react to forces, heat, vibrations, etc… in terms of whether it will break, wear out or function according to design. It is called "analysis", but in the product design cycle it is used to predict what will happen when the product is used (5).

4.2. Nodes and elements

A node is a coordinate location in space where the Degrees Of Freedom (DOFs) are defined. The DOFs of a node represent the possible movements of this point due to the loading of the structure. The DOFs also represent which forces and moments are transferred from one element to the next one. Also, deflection and stress results are usually given at the nodes. An element is a mathematical relation that defines how the DOFs of one node relate to the next. Elements can be lines (beams or trusses), 2-D areas, 3-D areas (plates) or solids (bricks and tetrahedra). The mathematical relation also defines how the deflections create strains and stresses. The degrees of freedom at a node characterize the response and represent the relative possible motion of a node. The type of element being used will characterize which DOFs a node will require. Some analysis types have only one DOF at a node. An example of this is temperature in a thermal analysis. A structural beam element, on the other hand, would have all the DOFs shown in Figure 6. "T" represents translational movement and "R" represents rotational movement about X, Y and Z axis direction, resulting in a maximum of six degrees of freedom.

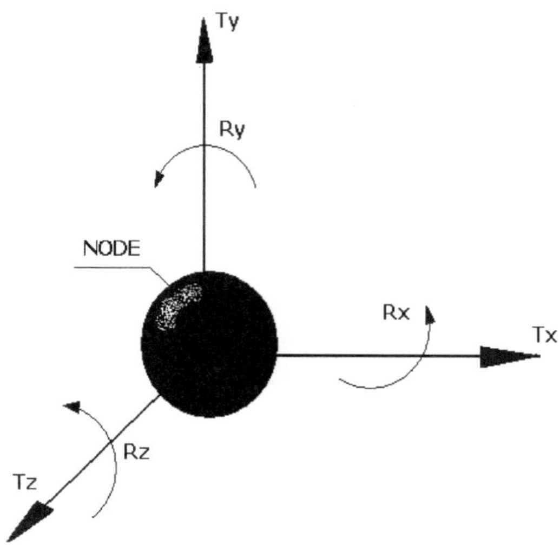

Figure 6. Degrees of freedom of a node (DOFs)

The elements, on the other hand, can only communicate to one another via common nodes. Elements therefore must have common nodes to transfer loads from one to the next, such as in the Figure 7 below.

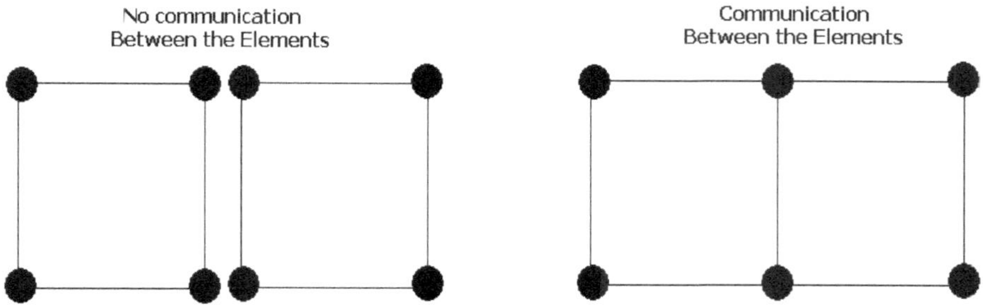

Figure 7. Communication through Common Nodes

Computer programs usually have many options for types of elements to choose, below the most usual elements (9):

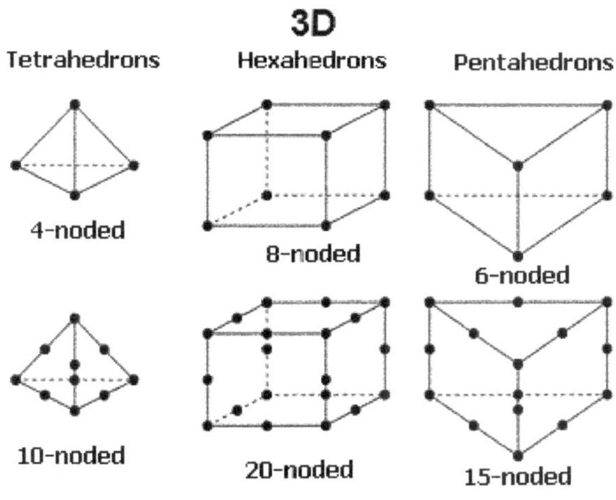

Figure 8. Most usual 3D elements

Since the applied load vector and element stiffnesses are known from the user input, the equation can be solved using matrix algebra by rearranging the equation as follow for the displacement vector:

$$\{x\} = [K]^{-1}.\{f\} \tag{9}$$

where; {f} is the vector that represents all of the applied loads. [K] is the assemblage of all the individuals' element stiffness (AE/L) and {x} is the vector that represents the displacement. A is the area, E is the Modulus of Elasticity and L the length, and {f} = $\int_0^L \varepsilon . dx$.

The strains are computed based on the classical differential equations. Stress can then be obtained from the strain using Hooke's law. These basic equations do not require the use of a computer to solve. However, a computer is needed when complexity is added (4).

4.3. How to build the model

Each individual piece is modeled 3D and then the final assembling built by each part gathering in the final component (product - pulley), see Figure 9 below. The boundary conditions have constraints between the shaft and the hub, the hub and the disc and the disc and the cylinder; all constraints are bonded surfaces. The shaft has at the extremes joint constraints due the presence of block bearings. The bearing blocks usually are composed by spherical roller bearings when pulleys for Stacker-Reclaimers are the case.

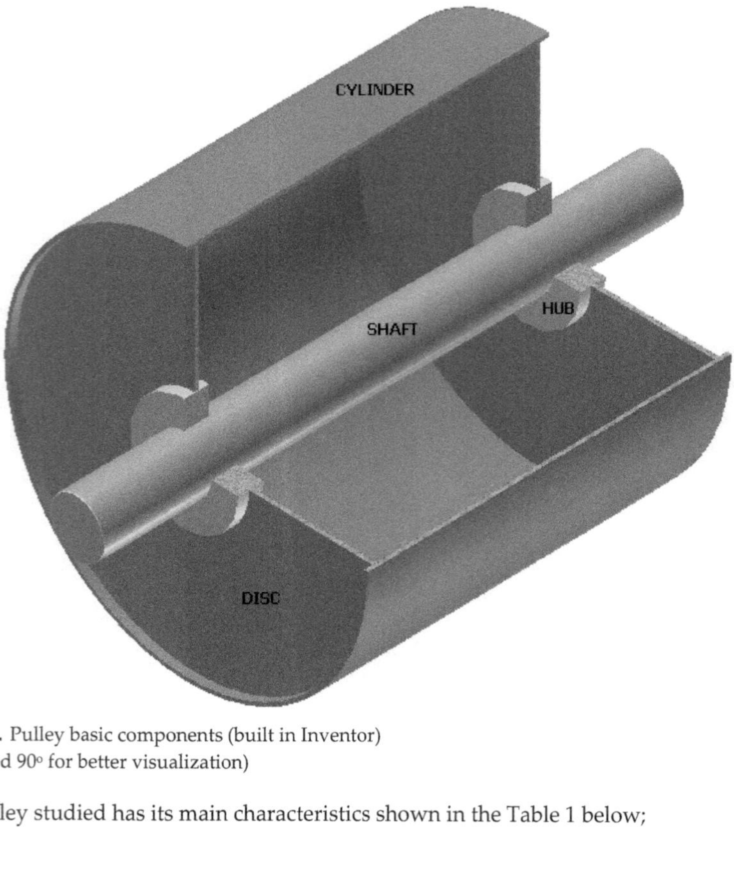

Figure 9. Pulley basic components (built in Inventor)
(sectioned 90º for better visualization)

The pulley studied has its main characteristics shown in the Table 1 below;

Component	Material	Diameter (mm)	Length (mm)	Thickness (mm)
Shaft	SAE-1045	150	1600	diameter 150
Discs	ASTM A-36	300 x 980	-	12
Cylinder	ASTM A-36	1000 (outside)	1000	10
Hub	ASTM A-36	150 x 300	-	70
Bearing block centre distance	-	-	1600	-

Table 1. Pulley main characteristics

When importing solid models that have thin parts, it is often better and simpler to analyze them using plate elements (5-10). Autodesk Simulation can be used to convert thin CAD solid models to plate elements. A plate element is drawn at the midplane of the part. Pulleys are commonly conditioned as described; it has solid elements, like shaft and hubs; and plate elements like discs and the cylinder. As shown ahead the difference is not too substantial but depending on the discrepancy of dimensions, comparatively between parts, the values (stresses outputs) can differ considerably. The DOFs associated with the plate elements are drawn in the Figure 10 and 11 that follows. Note that the out-of-plane rotation (Rz) is not taken into account because of plate theory, thus the plate elements have 5 DOFs.

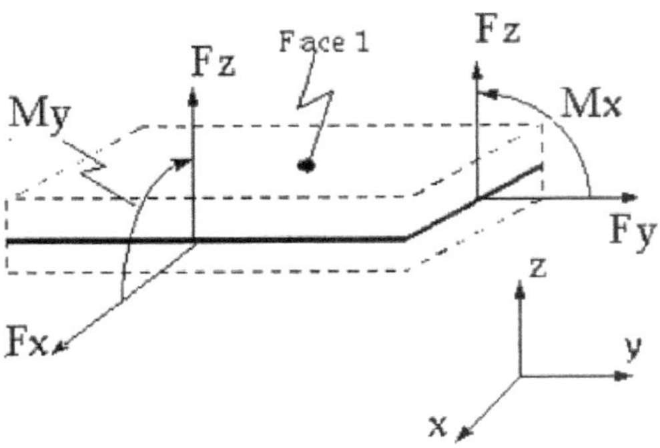

Figure 10. DOFs of midplane elements

The components were calculated by Autodesk Simulation and each component received a particular 3D element type and meshing configuration as follow; the shaft received a brick condition with material AISI 1045 as-rolled, the discs received a midplane condition, material ASTM A36, isotropic, the cylinder simulated midplane condition, material ASTM A36, isotropic. The constraints were determined in the shaft region of block bearings (joint constraint). The command, which simulate block bearings with spherical roller bearing, is universal joint, which constraint the DOFs at Tx, Ty and Tz as well as Rz (longitudinal to the shaft length) in the simulation to the first side, and the DOFs Tx, Ty and Rz to the opposite side.

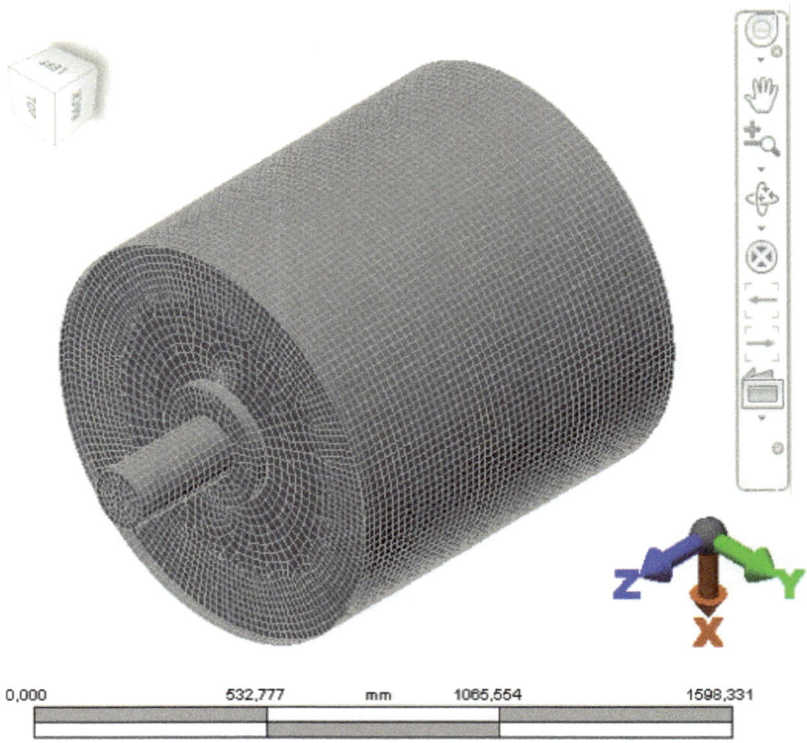

Figure 11. Pulley meshing – Autodesk simulation

The elements in the joints (block bearings) were considered with a very high stiffness's, which guarantee not interference in the stresses results in the model (Figure 12).

In terms of loading, there was a force applied perpendicular to the surface, which resulted in a variable pressure (parabola) around 180º of cylinder, represented by the following equation;

$$P = 0{,}47.R^2 + 0{,}47 \tag{10}$$

where P = pressure (MPa) R = pulley radius (mm), 0,47 = pressure (MPa).

The variable pressure is shown in the figure 13 below. The load applied on the cylinder outside and around 180º was 316kN. The analysis was done based on the previous description in the Autodesk simulation, being the von Mises stresses analyzed for each component, as follow by the Figures 13 to 18.

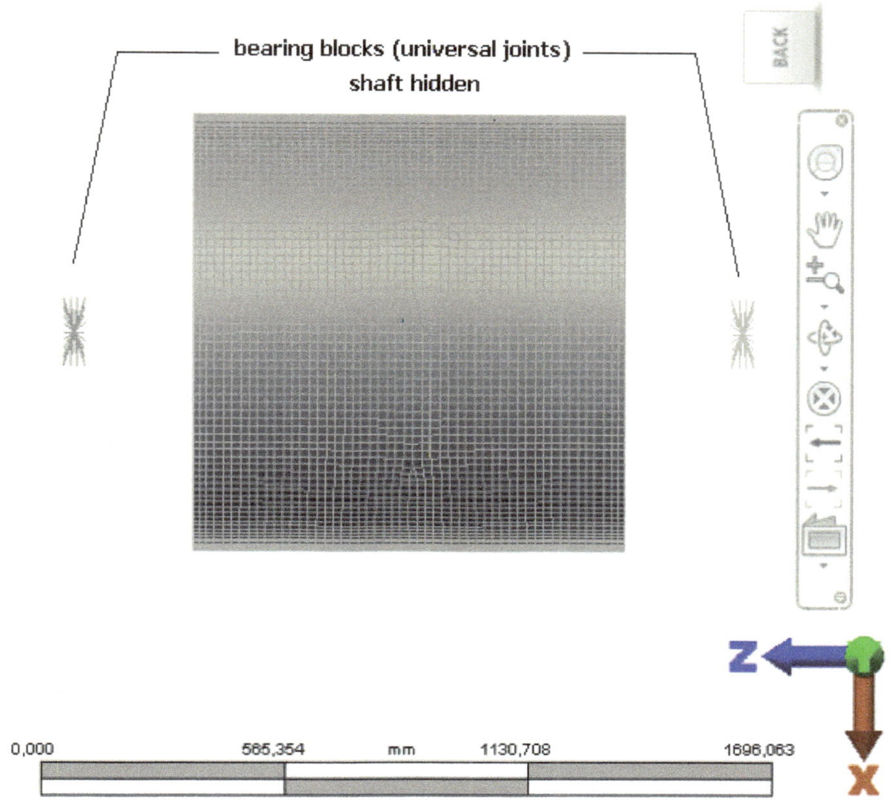

Figure 12. Constraint –Pin (universal joints)

Based on the fact the pulleys applications are dynamic (cyclic loading) the fatigue limit for each material was utilized in comparison with the stress range in reference to the equivalent stress (von Mises) by FEM (11-16). The Table 2 and Figure 19 reveal the main stresses on the pulley components. All the stresses are compared to the fatigue limit once this is the main phenomena the components is submitted. The stress range is calculated toward the von Mises stress (10-11).

Component	Material	von Mises (MPa)	Stress range (MPa)	Fatigue limit (MPa)
Shaft	SAE-1045	90	180	230
Discs	ASTM A-36	128	252	200
Cylinder	ASTM A-36	45	90	200

Table 2. Stresses on the main components (MPa)

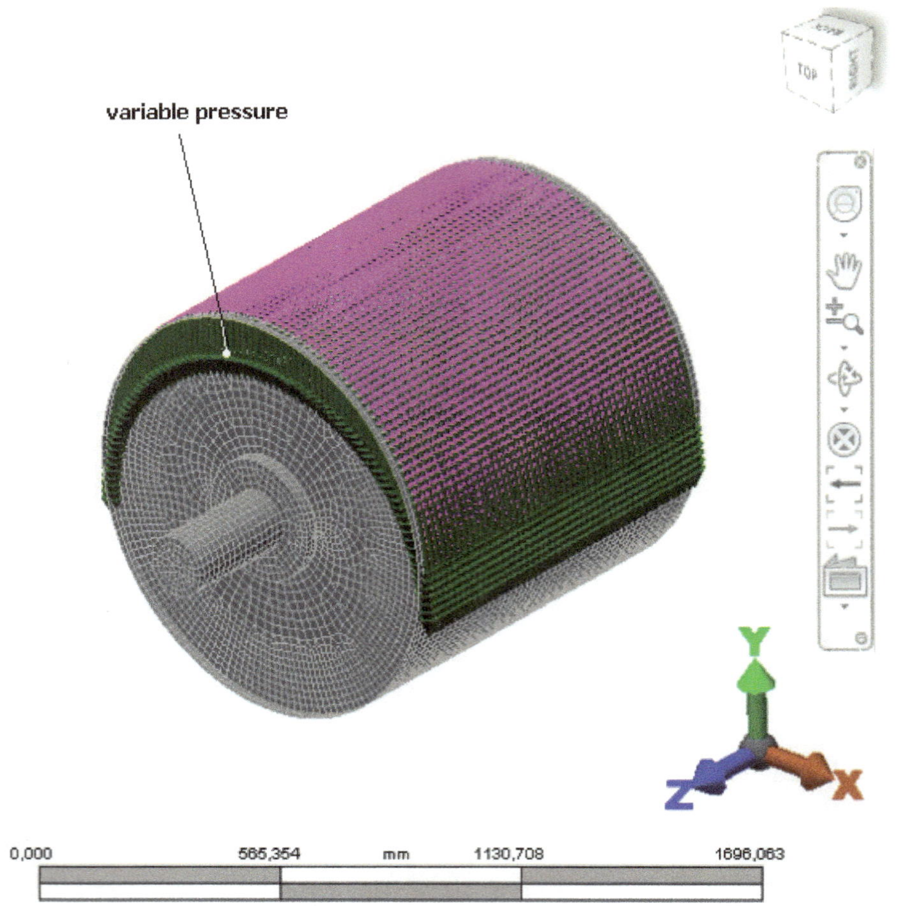

Figure 13. Variable pressure

All the stresses are under the fatigue limit except the discs, which overtake the limit over 52MPa. At this case the re-analysis of the discs thickness should be done and the thickness most of times increased or another type or thickness of disc applied. After the re-calculation, as expected, all assemble components have a new and different stress level, being highly recommendable afterwards the revaluations due to the fatigue limit consideration.

The values found in the analytical model (Table 2) were also compared with the FEM and are described in the Table 3 below.

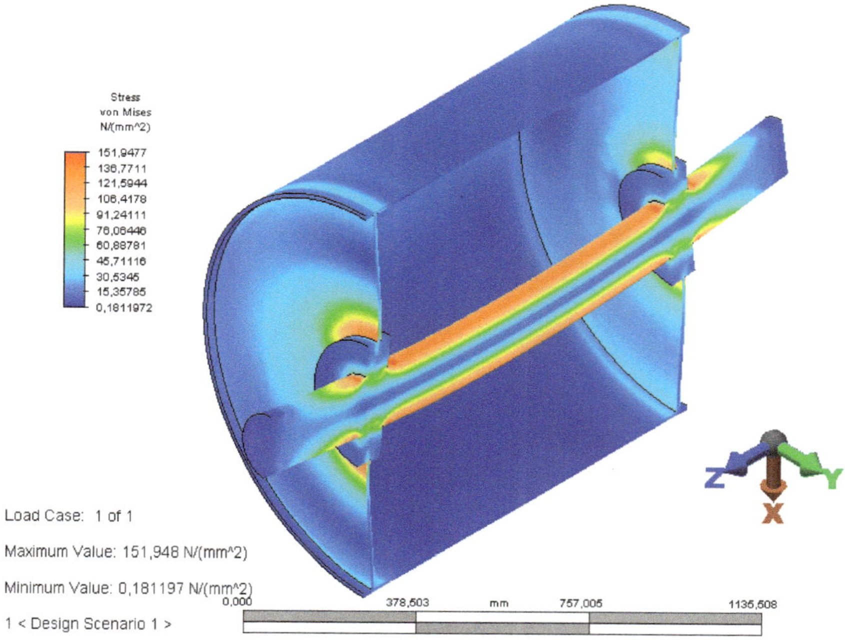

Figure 14. Stresses on the shaft (MPa)

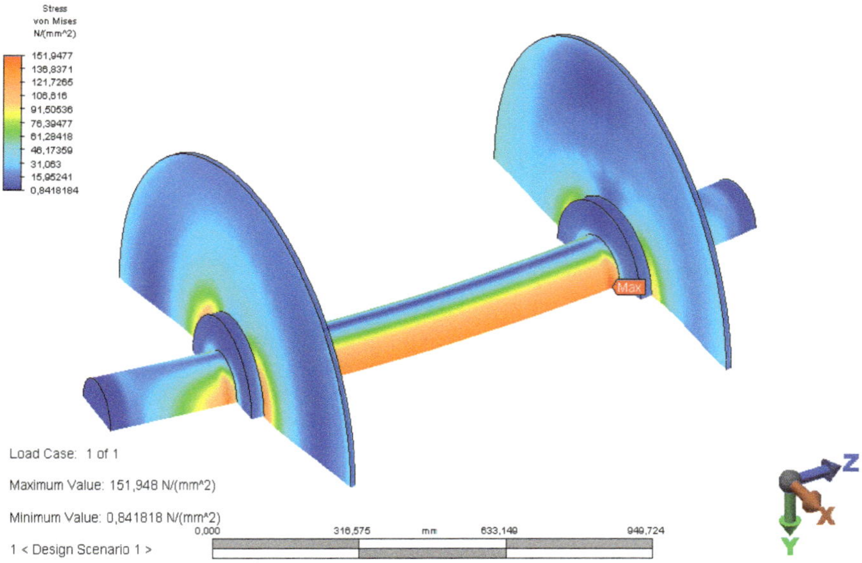

Figure 15. Stresses on the discs (MPa)

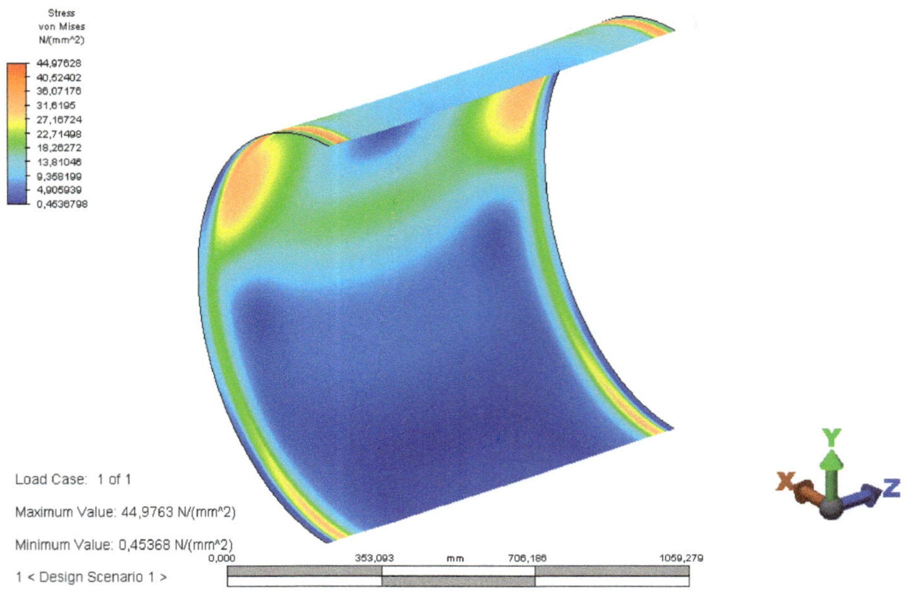

Figure 16. Stresses on the cylinder (MPa)

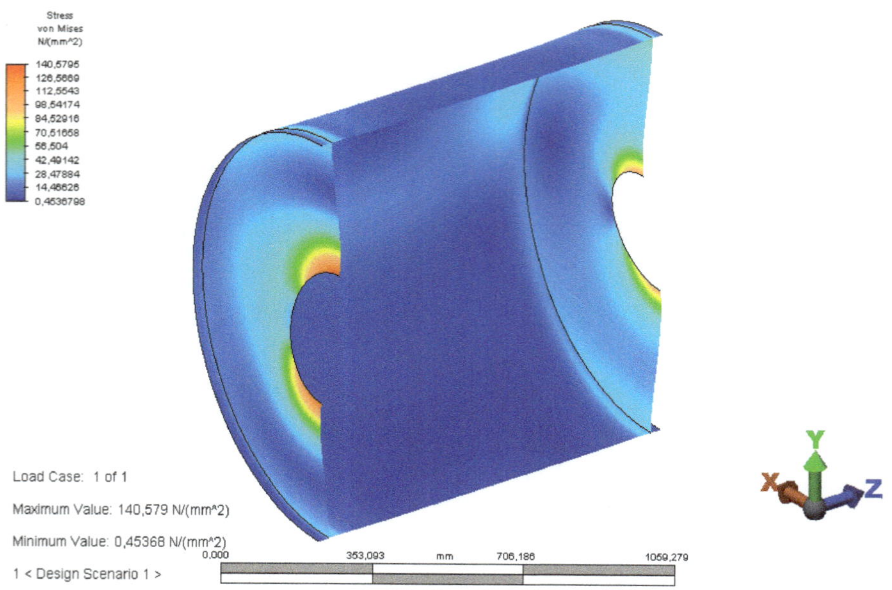

Figure 17. Stresses on the discs (MPa)

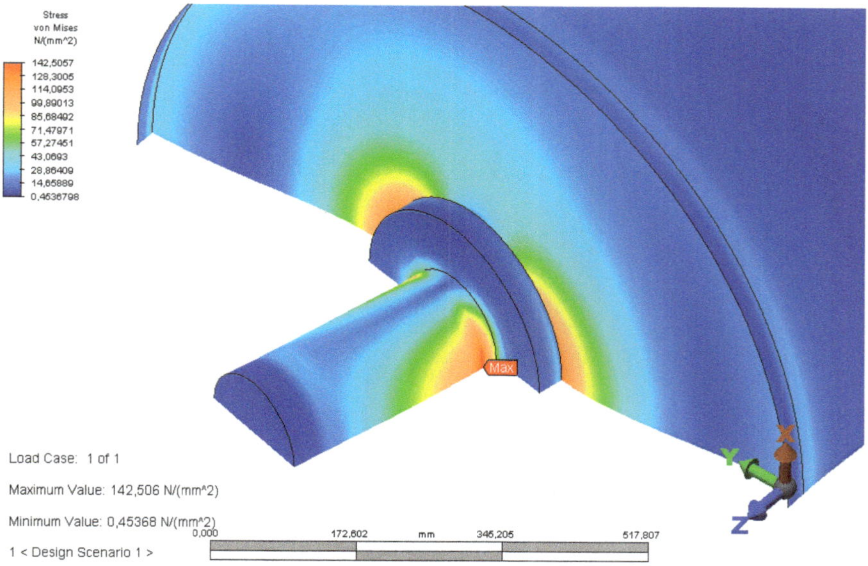

Figure 18. Stresses at the interface hub and shaft (MPa)

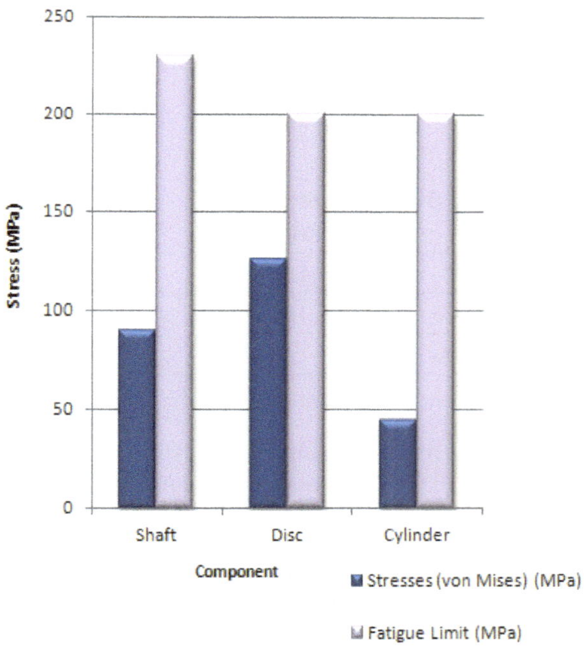

Figure 19. Stresses on the components (MPa)

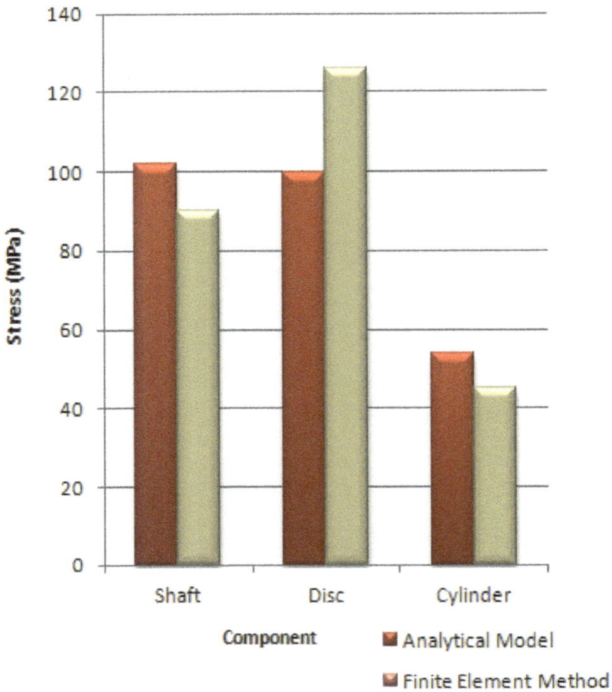

Figure 20. Stresses on the components (MPa)
Analytical versus FEM

Component	Material	Stress (MPa)	MEF (MPa)	Difference (%)
Shaft	SAE-1050	102	90	-10%
Disc	ASTM A-36	100	128	+26%
Cylinder	ASTM A-36	54	45	-33%

Table 3. Difference between analytical and MEF methods

There are differences in the results between the Analytical and FEM models (11). The equivalent stresses on the shaft are closer, around 10% difference, showing the lower value found in FEM, the discs are those which have medium difference and around 26%, being the stresses on the FEM higher than the analytical model and the third is the cylinder which had its lower value found in the FEM and around minus 33%. The differences are not too high but in certain cases should be taken into account when safety factors are in the limit due lean projects purposes. Neither the analytical model nor FEM are described in details once the idea is to bring the basic concepts used in components designing.

5. Conclusions

- The analytical calculation methods are still being used by most of components and machines suppliers;
- the analytical model requires safety factors in order to cover uncertainties in the processes like stresses due plastic deformations and/or complex thermal processes like weldings;
- the analytical model, as known, is not graphical like Finite Element Method (FEM) and sometimes considered obscure (not too complete like FEM) in terms of outputs. The data are not accurate like those presented by FEM either;
- Several impacts are in still in phase in the Mining business since the contemporaneous usage of models 3D and calculations by FEM;
- such changes from drawing 2D to 3D have brought the shift on the way of drawing, requiring less spatial thought than before but on the other hand more accurate drawings and awareness due interferences;
- the finite element methods is a powerful tool to calculate most of components and machines in the Mining area and nowadays being a reality for some companies of high technology;
- there are advantages when using FEM in terms of easy presentation of results due graphics and easy values (stresses and strains) obtained in different directions;
- the usage of FEM provide more sophisticated ways to analyze the calculations in terms of stresses and strains and displacements different directions, states and intensities;
- even with easier results brought by FEM, they are not free of analysis and positioning. The knowledge of mechanics in a deeper way in terms of intensity, state and direction of stresses and strains are more eminent nowadays;
- there are also drawbacks in the FEM like in the analytical, which require the use of safety factors. Stresses due different processes which generate stresses like plastic deformations and residual stresses due welding are not totally overcome in the method
- the case shown present a pulley used in Stacker-Reclaimers in order to analyse the differences found in both models, analytical and by finite element method. Any of them is wrong but they present certain differences in terms of stresses;
- the best way to evaluate the results is measuring the stresses by strain gages, which is not demonstrated in this paper;
- the evaluation and the comparison between the measured values and the calculated ones, being by analytical model or FEM are necessary;
- the FEM does not discharge the analytical model once the last has not presented problems in the field. On the other hand it must be replaced to more sophisticated tools like FEM, which brings several benefits beyond of more precision, productivity, friendly analysis environmental and cost savings due less weight and prior interference analysis.

Author details

Jairo A. Martins* and István Kövesdy

Machines and Components Product Engineering, Metso Minerals, Sorocaba, Brazil

---

* Corresponding Authors

Acknowledgement

The acknowledgements are addressed to CNPq (National Council for Scientific and Technological Development – Brazil), Mr. Rubens Costa, Vice President, Mining Operations, South America at Metso Brazil, Mr. Misael Ramalho and the Bulk-Machines Team and Mr. Vanderson L. Zangerolamo from MAPDATA due their valuable technical support.

6. References

[1] Metso Minerals, Manual de Transportadores Contínuos. Fábrica de Aço Paulista, 4ª edição 1991.
[2] Klaus-Jürgen Bathe. Finite Element Method. Published Online: 13 JUN 2008. Copyright © 2007 by John Wiley & Sons, Inc. DOI: 10.1002/9780470050118.ecse159.
[3] Roark, J.R, Young W.C. Formulas for Stress and Strain. McGraw-Hill Kogakusha, Ltda., 1966.
[4] Inventor 2010, User´s manual, 1st edition 2010.
[5] Autodesk® Simulation Mechanical. Finite element course manual, linear elastic simulations, 2011.
[6] Ivo Babuška. The finite element method with Lagrangian multipliers. Numerische Mathematik . Volume 20, Number 3, 179-192, DOI: 10.1007/BF01436561.
[7] Collins, J. Projeto Mecânico de Elementos de Máquinas. Rio de Janeiro: LTC, 2006.
[8] Dieter, G.E. Mechanical Metallurgy. McGraw-Hill Inc. 1986.
[9] The Finite Element Method – Theory. http://illustrations.marin.ntnu.no/structures/analysis/FEM/theory/index.html
[10] Martins, J. A. Kovesdy, I; Romão, Estaner, Ferreira I. Case study of an evaluation of a Stacker Boom Luffing Pulley by Irwin's model and under BS 7910 guidance. Int. J. of Mining and Mineral Engineering 2011 Vol.3, No.4 pp.267 - 277. DOI: 10.1504/11.45469
[11] J.A. Martins, I. Kövesdy, I. Ferreira. Fracture analysis of collapsed heavy-duty pulley in a long-distance continuous conveyors application. Engineering Failure Analysis. Volume 16, Issue 7, October 2009, Pages 2274–2280.
[12] Bäckström M. (2003). Multiaxial Fatigue Life Assessment of Welds Based on Nominal and Hot-spot Stresses. VTT Publications 502, ISBN 951 38 6234 8.
[13] Det Norske Veritas (2005). Fatigue Assessment of Ship Structures. DNV Classification Notes, N.30-7, July 2005.
[14] Hobbacher A. (2007). Recommendations for Fatigue Design of Welded Joints and Components. International Institute of Welding IIW, Doc. XIII-2151-07/XV-1254-07.
[15] Robert A Adey, John M. W. Baynham, Sharon Mellings, Tom Curtin*Fatigue Life and Crack Growth Prediction Using FEM Data. Computational Mechanics BEASY, Ashurst Lodge, Southampton, Hampshire, SO40 7AA, UK *Computational Mechanics Inc, 25 Bridge Street, Billerica, MA 01821
[16] Aliabadi M.H., Rooke D.R., Numerical Fracture Mechanics, Computational Mechanics Publications, Southampton, U.K. 1991.

Chapter 12

Finite Element Analysis Coupled with Feedback Control for Dynamics of Metal Pushing V-Belt CVT

Toshihiro Saito

Additional information is available at the end of the chapter

1. Introduction

Nowadays automobiles are required to meet environmental requirements, such as lower exhaust emissions and higher fuel economy. One of the key factors for improving the overall efficiency of a vehicle is the efficiency of its transmission.

A CVT has a greater potential for improving fuel economy than a step-type automatic transmission (AT), because of its integrated control with the engine [1]. That is, CVTs are capable of continuously tracing engine operating ranges with high fuel efficiency. Another advantage is that CVTs allow vehicles to drive without lowering the driving torque or the engine rpm while shifting the gear ratio.

However, when the transmission efficiency of a CVT by itself is compared with a step-type AT, CVT is known to have lower efficiency because its driving torque is transferred by means of contact and friction [2]. The transmission efficiency of a CVT is determined by friction loss at its oil pump and metal pushing V-belt. The oil pump must produce enough pulley pressure so that the metal V-belt mounted between two pulleys does not slip. A higher pulley pressure, however, means a greater friction loss at the oil pump [3].

As for the metal V-belt, gradually lowering pulley pressure while maintaining a constant transmission torque increases the transmission efficiency of the belt by itself, as long as it does not slip on the pulleys. However, the transmission efficiency begins to drop under a certain operating condition. This implies the existence of an optimum operating condition that maximizes the transmission efficiency of the belt [4]. To find this condition, it is important to predict friction loss at each portion of the metal V-belt during CVT operation.

A considerable amount of research has so far been made on methods for calculating friction loss that occurs at each part of the V-belt, but many of them use simple equations that are based on assumptions and not linked with dynamic belt behavior [5]. Accordingly, although these methods simulate the tendency of friction loss on each belt part, they are not sufficient for examining the influence on friction loss of metal belt shape and pulley rigidity.

A commercially available multi-body simulation (MBS) code used to be chosen for the dynamic belt behavior [7]. In the MBS model, each V-belt block was treated as a separate rigid body. Contact normal forces were modeled between adjacent blocks, using realistic stiffness and damping properties. Contact normal forces, along with both radial and circumferential friction forces, were also modeled between the block edges and pulley faces. Block geometry and material properties were used to arrive at contact stiffness and damping characteristics. As such, the MBS model contained a relatively small number of degrees of freedom. At the results, the contact forces between the block edges and pulley faces could be only obtained, not the contact pressure distribution.

Therefore, the following describes how CVT-ratio control logic was included in the previously reported technology for simulating the dynamic behavior of a metal pushing V-belt, and presents how the new simulation can closely reproduce actual operation. Thus, this paper reports a technology developed to predict the transmission efficiency of a CVT drive system comprising a metal pushing V-belt and pulleys.

2. Development aims

2.1. Maintaining speed ratio with feedback control

Figure 1 shows the main section of the CVT used in this development. A metal V-belt is mounted in the V groove on two pulley shafts, and a pair of movable pulleys are mounted on the shafts to face each other. The movable pulleys are shifted in the axial direction by line oil pressure supplied from the inside of the shaft. When the CVT is in operation, feedback control is exercised to maintain an arbitrary speed ratio between the two pulleys, varying the oil pressure applied to the large-diameter (in terms of belt mounting position) pulley while maintaining a constant pressure for the small-diameter pulley. When taking an example of the top ratio, the drive pulley speed, driven pulley torque, driven pulley oil pressure, and target ratio are input to the feedback control system. The system then outputs the driven pulley speed, drive pulley torque, and drive pulley oil pressure. The input-output relation differs from the metal V-belt behavior simulation technology developed previously, so it was necessary to incorporate a new feedback control into the simulation in this project.

2.2. Pulley shaft thrust load ratio

Pulley shaft thrust load is obtained as the sum of two values: the product of oil pump line pressure and its acting area, as well as the product of centrifugal oil pressure and its acting area. The ratio between drive pulley thrust load and driven pulley thrust load, referred to as the pulley shaft thrust load ratio, has a positive correlation with the speed ratio; therefore, this ratio is uniquely determined once the speed ratio is set. Because the developed simulation outputs pulley shaft thrust loads, its prediction accuracy can be verified by comparing simulated and measured values.

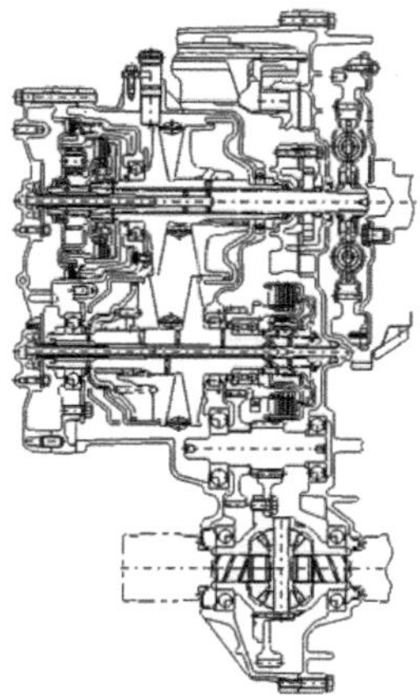

Figure 1. CVT cross section

2.3. Transmission efficiency

Transmission efficiency can be obtained by multiplying the ratio between drive shaft torque and driven shaft torque by the speed ratio, as expressed in Equation (1). Here, each torque value can be obtained as the product of a tangential friction force, generated between the V-face of each pulley and metal V-belt elements, and the effective element V-surface radius. As such, it is necessary to accurately predict the direction of a friction force acting on each element V-surface and its effective radius under that condition. The effective radius of the element V-surface is influenced by its contact pressure distribution. This means that it is also necessary to consider the elastic deformation of the element V-surface.

For this reason, to simulate the transmission efficiency of the metal pushing V-belt, the following were set as development objectives:

1. Implement feedback control equivalent to speed ratio control performed in an actual vehicle
2. Quantify sliding velocities and friction forced at various element contact areas
3. Calculate friction losses at various portions of the metal V-belt

3. Technology of predict transmission efficiency of metal pushing V-belt

3.1. V-belt model consideration of element deformation

The transmission efficiency of a metal pushing V-belt is mainly determined by friction losses that occur between its elements and pulleys and between the elements and rings. Friction loss has a correlation with the product of the friction force and sliding velocity of a friction surface. Accordingly, a key to predicting the transmission efficiency is to simulate the friction forces and sliding velocities of the element V surfaces in contact with the pulley. These friction forces and sliding velocities have distributions on the pulley, and these distributions cannot be simulated accurately unless element deformation is taken into account (Figure 2). Because the effective radius of elements would be influenced by the contact pressure distribution. For this reason, an element stress prediction model [6] previously designed to consider element deformation was modified to make a model of the metal V-belt used in this study.

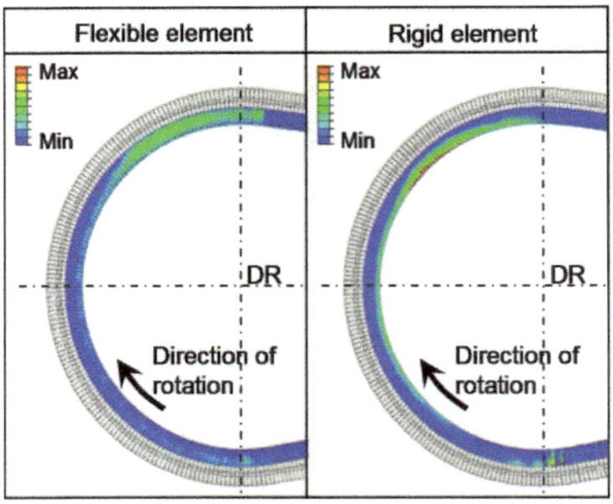

Figure 2. Comparison of friction force distribution of Element V-surfaces on DR pulley

Figure 3 and Figure 4 show the model. Two flexible solid elements are employed in the direction of thickness for individual elements to reproduce bending deformation. Three flexible solid elements are used to divide the R section of the element neck, where stress is thought to concentrate. Reproduction of detailed form of the nose and hole sections has been prioritized, and therefore modeled with rigid elements. Four flexible shell elements have been used in the direction of width to model each layer of rings to enable reproduction of the crowning of the rings. In addition, the ring-element, element-element, ring-ring and element-pulley contact surfaces have been defined, and appropriate friction characteristics have been assigned in each case.

To make the load conditions around the belt pulleys closely resemble the layout in an actual CVT, beam elements were used to express the shafts. In this model, the shafts are supported at the bearing positions. Also, a gear is provided to mesh with the driven (DN) shaft to reflect reaction forces applied by the gear. Because the belt mounting diameter varies depending on speed ratios, deflection rigidity calculated for the mounting positions of each ratio was applied to the pulley V-face. Regarding the relation between each shaft and the movable pulley, the model defines a fitting clearance at their engagement position, as well as a backlash in the rotational direction at the roller position.

Figure 5 shows the flow of analyzing the metal V-belt. In this flow, the belt is initially placed at the perfect-circle position under no stress, and then both pulleys are moved to a specified shaft distance. Next, misalignment is applied to one of the pulleys. Then, the drive pulley is gradually accelerated to reach a target speed, while pulley thrust pressure is being applied. In the meantime, reverse torque is gradually applied to the driven pulley.

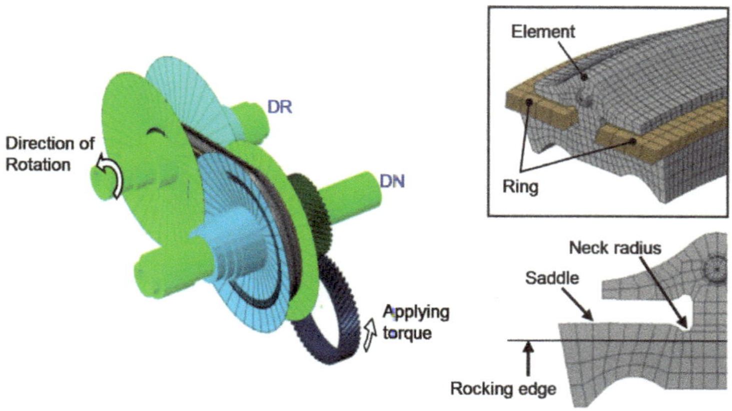

Figure 3. Metal V-belt model for predicting transmission efficiency considering shaft deformation

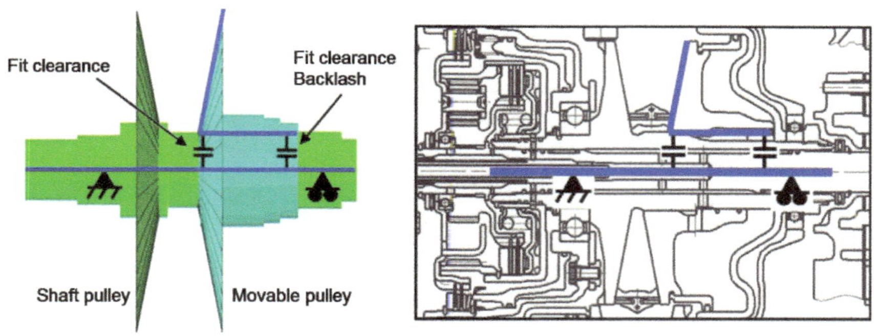

Figure 4. Boundary condition around pulley shaft

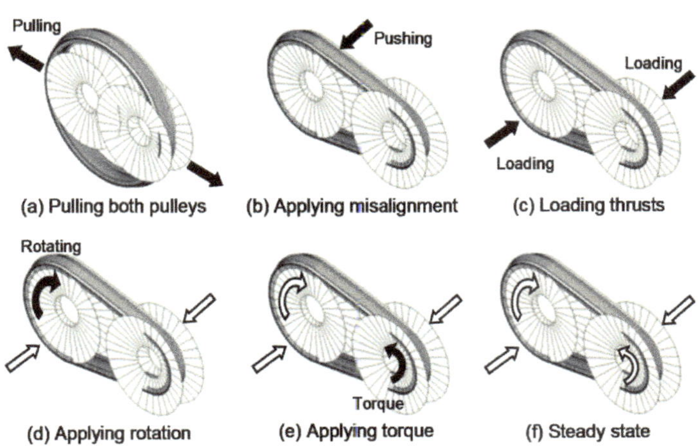

Figure 5. Analytic procedure

3.2. Belt transmission efficiency prediction using pulley thrust pressure control

The transmission efficiency η of the V-belt can be obtained by the following equation:

$$\eta = \frac{\omega dn \times Tdn}{\omega dr \times Tdr} = i\frac{Tdn}{Tdr} \tag{1}$$

ωdr : drive (DR) shaft speed
ωdn : driven (DN) shaft speed
i : ratio
Tdr : DR shaft torque
Tdn : DN shaft torque

Thus, the DR shaft torque, DN shaft torque, and ratio must be obtained to predict efficiency η. The conventional element stress prediction model required simulations to be made based on pulley thrust pressures measured in an actual CVT. Unlike element stress measurement, the transmission efficiency can be easily measured in an actual vehicle, but it would be impractical for a simulation to require actual measurements to predict the efficiency. Accordingly, a pulley thrust controller used in the previous research [7] was implemented in the simulation to apply pulley thrust pressure that depended on each operating condition.

Figure 6 is a block diagram of the thrust controller. Enclosed in the dashed box in this figure is a traditional proportional-integral (PI) controller. This controller calculates the actual ratio from ωdr and ωdn of the belt model, and adjusts pulley thrust pressure by means of a feedback loop until the ratio reaches a target value Itarget. Controller gains were re-defined because the previous metal V-belt model was replaced with the element stress prediction model.

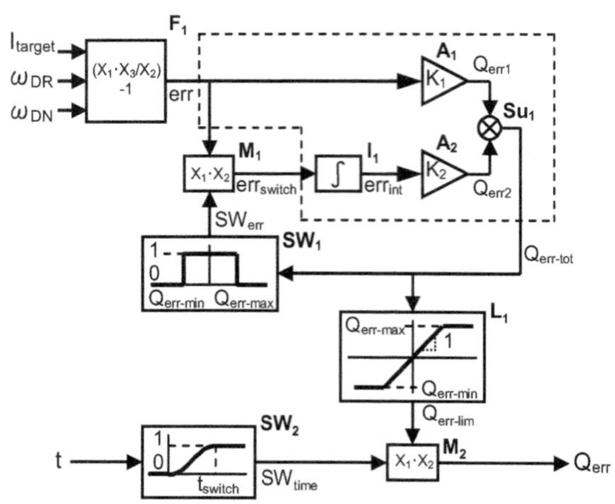

Figure 6. Thrust controller block diagram

Figure 7 shows calculations made with the metal V-belt model incorporating the above pulley thrust pressure control. In this simulation, ωdr, Tdn, and the driven pulley thrust pressure Qdn were given as input, and the drive pulley thrust pressure Qdr was controlled to make the actual ratio reach the target value. Because speed ratio error must be considered in deciding the Qdr control value, pressure was applied before starting the rotation. Then, ratio control was started after making sure that the actual ratio had been read accurately. As demonstrated in Figure 7, the speed ratio is maintained at the target value when the pulley thrust pressure is controlled by the ratio controller.

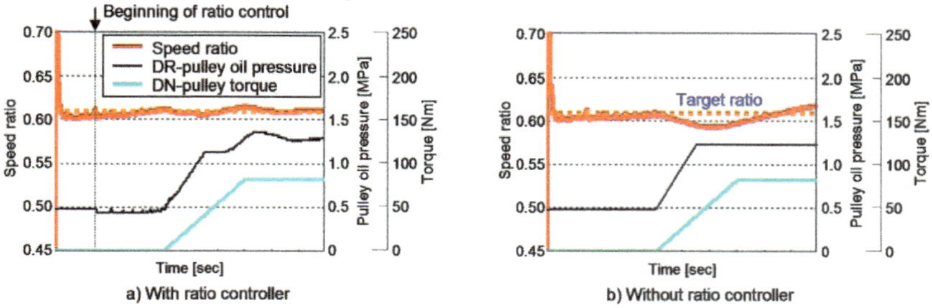

Figure 7. Effect of ratio controller

Figure 8 shows the belt transmission efficiency obtained from the computation result of this model. As shown here, the simulation implementing pulley thrust pressure control enables transmission efficiency prediction at a target speed ratio.

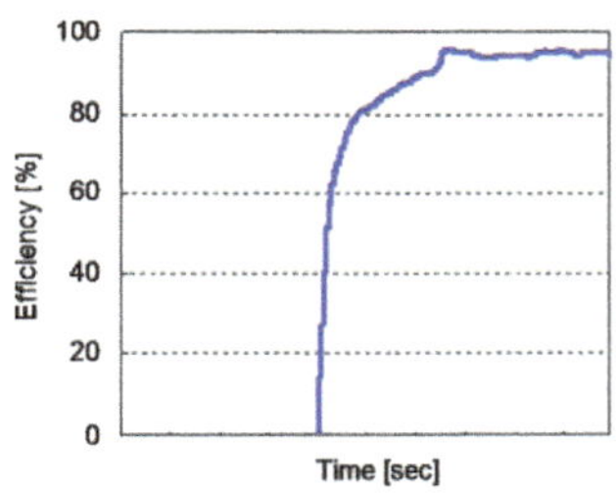

Figure 8. Calculated transmission efficiency of metal V-belt with ratio controller

4. Belt transmission efficiency at different ratio

In a CVT equipped with a metal pushing V-belt, the transmission efficiency is known to peak at the speed ratio of 1.0 (MID). The efficiency lowers gradually while a vehicle is decelerating (shifting to LOW) or accelerating (shifting to OD). The developed simulation was used to calculate friction loss at each portion of the metal V-belt at different speed ratios.

4.1. Accuracy of belt efficiency prediction

Figure 9 shows belt transmission efficiency calculated for and measured at the MID, LOW, and OD ratios under a certain operating condition. Figure 10 shows the calculated and measured pulley shaft thrust load ratios. Both graphs indicate the same tendency, which verifies the validity of simulations at all ratios.

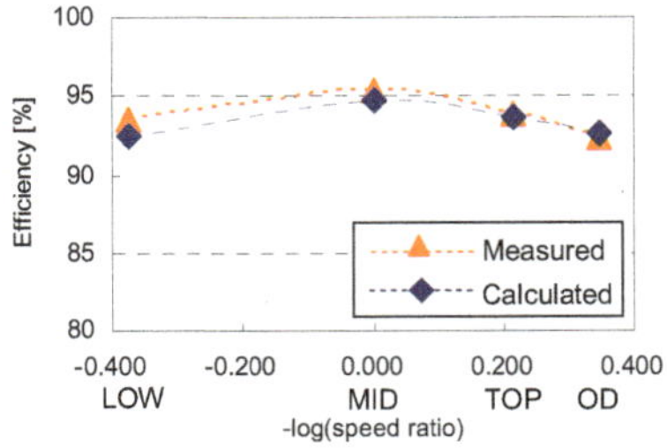

Figure 9. Belt transmission efficiency

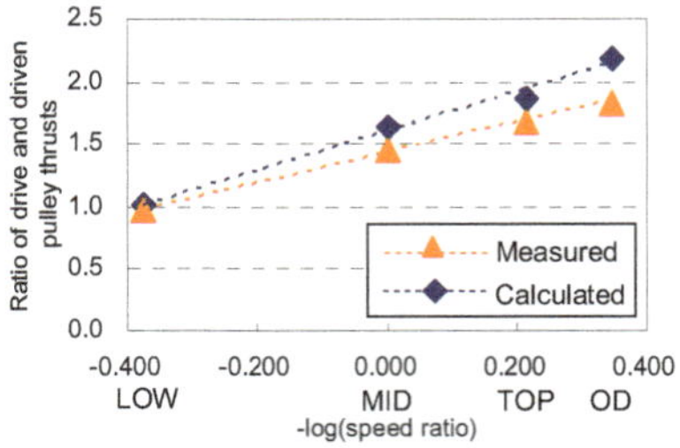

Figure 10. Ratio of drive and driven pulley thrusts

4.2. Friction loss of metal V-belt at each ratio

As mentioned before, the transmission efficiency of a metal pushing V-belt is mostly determined by friction loss that occurs between its elements and pulleys and between the elements and rings. Calculations were therefore made to compare friction losses of these areas at different ratios.

Figure 11 shows friction forces acting on the element V-surface and the element-pulley speed difference during one rotation of the belt. Similarly, Figure 12 shows friction forces acting on the element saddle surface and the element-ring speed difference. Both friction forces and speed differences are in the torque transmission direction. The speed difference is calculated for the same belt mounting position at each ratio. Figure 11 shows the speed difference between the element V-surface speed and the pulley speed, and Figure 12 the difference between the element saddle speed and the speed of the innermost ring.

When a speed difference occurs between two parts to which a friction force is applied, this speed difference may be assumed to be sliding velocity. Thus, friction loss can be obtained by multiplying the sliding velocity and the friction force. Figure 13 shows friction losses calculated in this way for the element V-surface and the saddle surface at each ratio. When attention is paid to friction loss on the element V-surface, loss is large near the drive pulley exit at the LOW ratio. In contrast, at the MID and OD ratios, friction loss is greater at the pulley entrance. On the driven pulley, loss is greater near the pulley exit, regardless of the ratio. When studied in relation to sliding velocity, friction loss becomes greater at locations where the sliding velocity is higher at all three ratios.

As for friction loss on the saddle surface, loss is smaller at the MID ratio than the other ratios. This is because, at the MID ratio, the belt mounts on a medium-diameter position on the pulley, where the sliding velocity between the pulley and the element is lower. At the other ratios, friction loss on the saddle surface is greater when it is on the smaller-diameter pulley.

A similar comparison was made with friction loss that occurs between adjacent elements and between adjacent of rings (Figure 14, Figure 15, and Figure 16).

As shown in each figure, friction loss is smaller than the loss calculated for the two aforementioned areas. This is because the sliding velocity is smaller between adjacent elements and adjacent rings where a friction force is present.

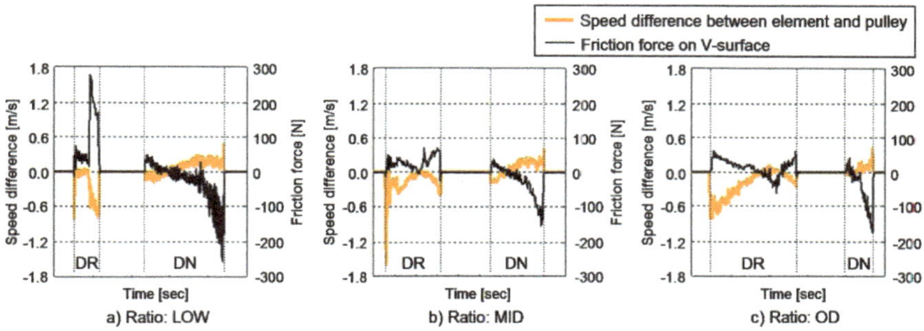

Figure 11. Element V-surface friction force and element-pulley speed difference

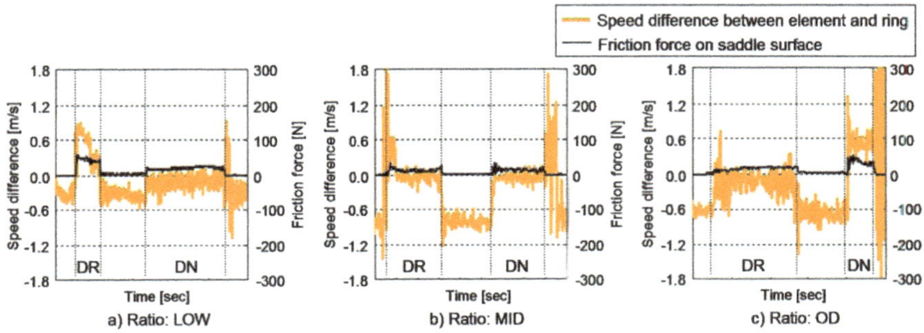

Figure 12. Element saddle-surface friction force and element-ring speed difference

Figure 17 shows the comparison of friction losses per unit time at the three speed ratios. When this figure is studied with Figure 9, it is clear that friction loss is greater at ratios with lower transmission efficiency. The comparison also reveals that the transmission efficiency is affected by friction losses on the element V surface and the saddle surface. When attention is paid to the proportion of friction losses at different ratios, friction losses on the element V-surface and the saddle surface are almost equal at the LOW ratio. At the OD ratio, friction loss on the saddle surface accounts for a dominant proportion.

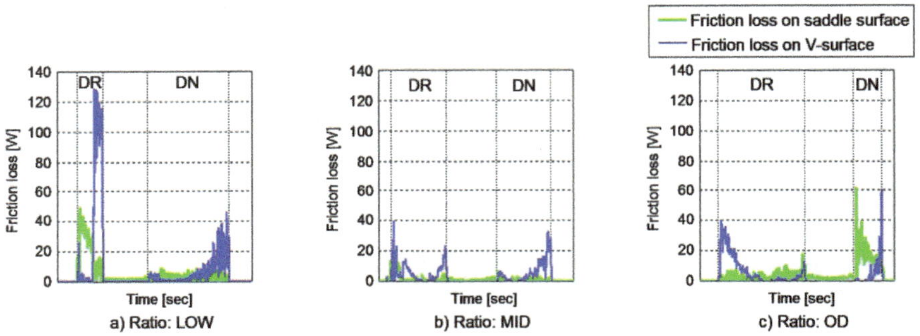

Figure 13. Friction loss on element friction surface

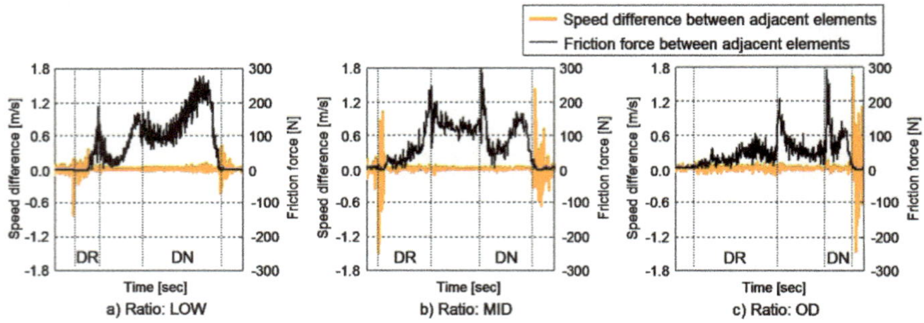

Figure 14. Friction force and speed difference between adjacent elements

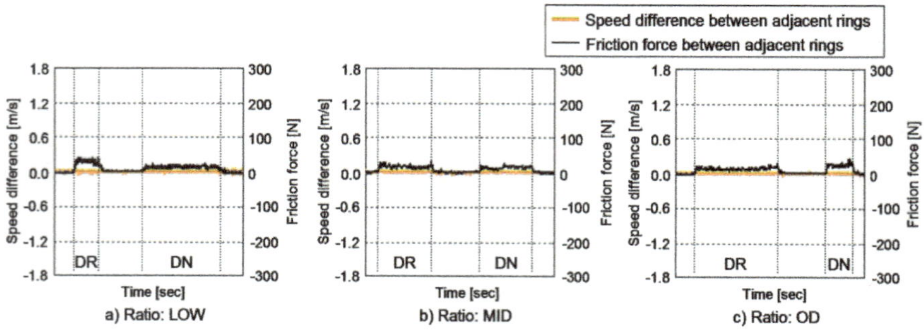

Figure 15. Friction force and speed difference between adjacent rings

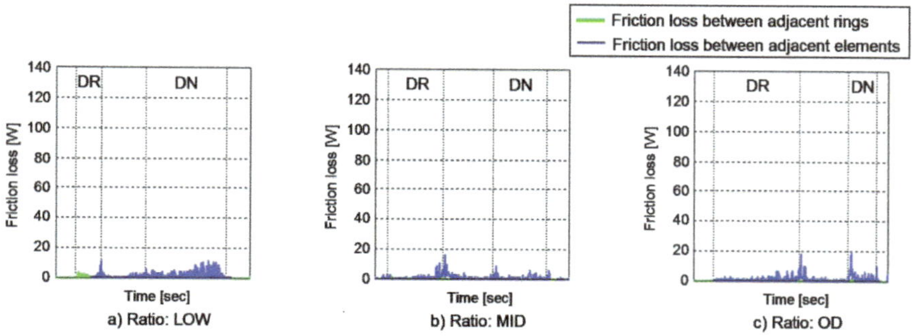

Figure 16. Friction loss between adjacent elements and between adjacent rings

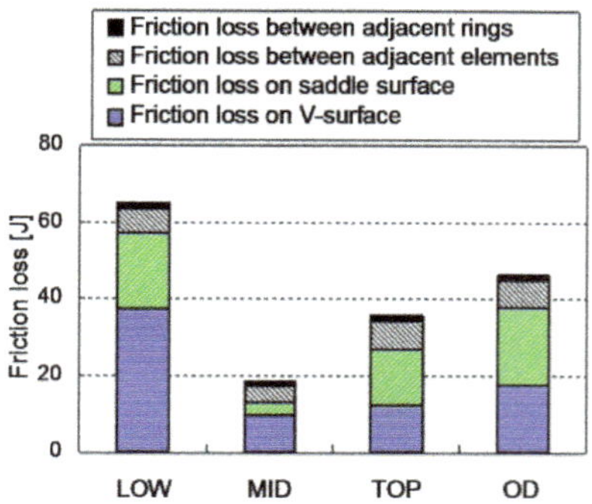

Figure 17. Belt friction loss per unit time

5. Application for actual CVT configuration

Figure 18 shows an actual CVT configuration to be calculated the transmission efficiency. Replacement of the former single-piston configuration with double pistons aligned in the axial direction has resulted in approximately 1.8 times more thrust at identical oil pressures. This has reduced line pressure when the CVT is in its frequently used overdrive ratio, thus reducing the pump workload. In this mode, the elements and the pulleys are modeled as elastic bodies to help enable their deformation to be considered. The number of divisions in the circumferential direction was increased and a coefficient of friction was set to help enable study of the effect of the fit clearance of the fitted parts of the pulley shaft and the movable pulley and of the roller spline backlash. Figure 19 shows Von Mises stress distribution except the gears. Most of the parts are now modeled as flexible bodies.

The transmission efficiency is now predicted using pulley thrust pressure control with appropriate control gains. (Figure 20) The speed ratio is kept at a target speed ratio during this simulation.

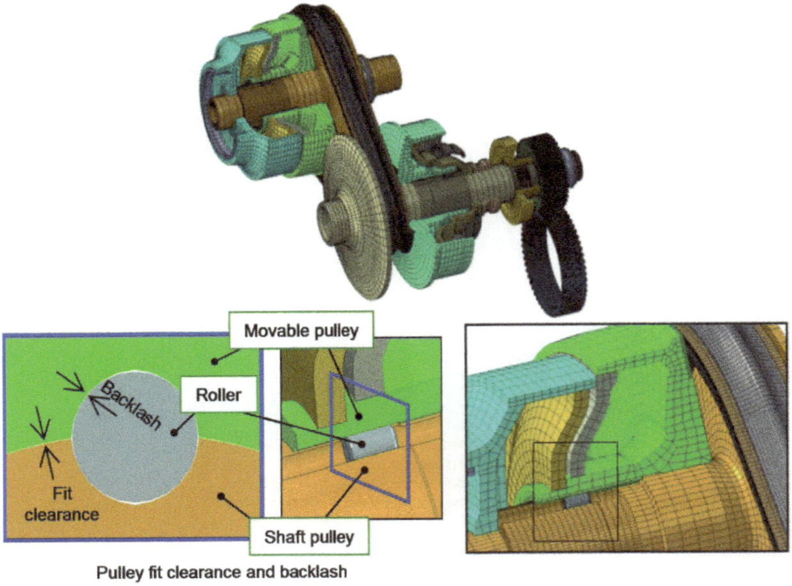

Figure 18. Metal V-belt model considering pulley V-surface deformation

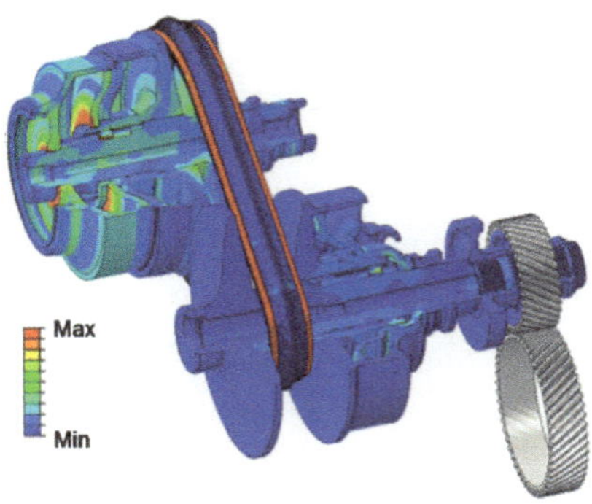

Figure 19. Stress distribution of Metal V-belt and pulleys

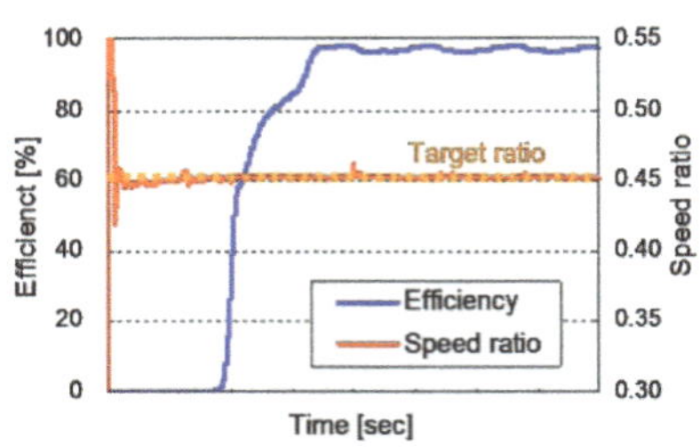

Figure 20. Calculated transmission efficiency of metal V-belt with FEM pulleys

5.1. Effect of fit clearance on transmission efficiency

In CVTs equipped with metal pushing V-belts, transmission efficiency is known to peak at the shift ratio of 1.0 (MID). Transmission efficiency declines gradually while the vehicle is decelerating (shifting to LOW) or accelerating (shifting to OD). However, the effect of the clearance of the section in which the external shape of the pulley shaft and the internal diameter of the movable pulley are in contact and of the roller on the transmission efficiency and the strength of the metal pushing V-belt is not necessarily clear. The method discussed here was therefore employed, using the OD ratio, in order to calculate friction loss for each part of the belt when the clearance of the section in which the external shape of the pulley shaft and the internal diameter of the movable pulley are in contact ("large-diameter clearance" below) and the clearance in the direction of rotation determined by the roller ("backlash" below) were varied.

Figure 21 and 22 show that when the large-diameter clearance of the fitted sections becomes narrower, friction loss on the element V-surfaces is reduced, and the transmission efficiency of the belt increases. Fig. 23 shows the changes in the winding diameter of the belt at this time. When the large-diameter clearance increases, the changes in the belt winding diameter also increase in magnitude. In other words, the belt slips in the radial direction, thus reducing the amount of friction force available for transmission, with the result that transmission efficiency declines. In addition, a comparison of the surface pressure distribution on the element V-surfaces shows that when the large-diameter clearance becomes greater, the surface pressure distribution tends towards the inside of the radius (Fig. 24). This is believed to be an effect of the fact that when the belt is transmitting torque, its effective radius is reduced, and torque transmission efficiency declines.

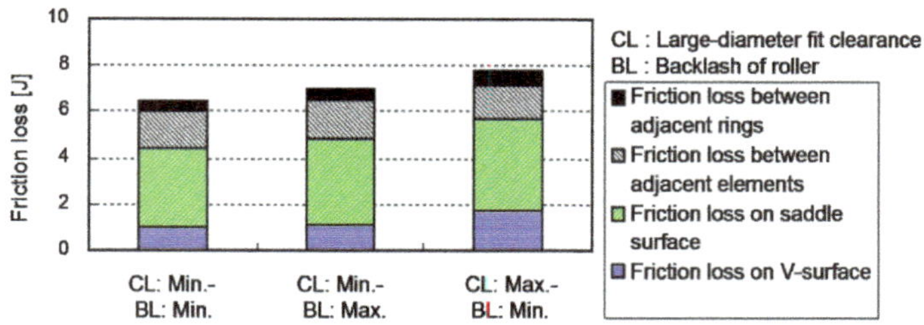

Figure 21. Fit clearance and friction loss on belt surface

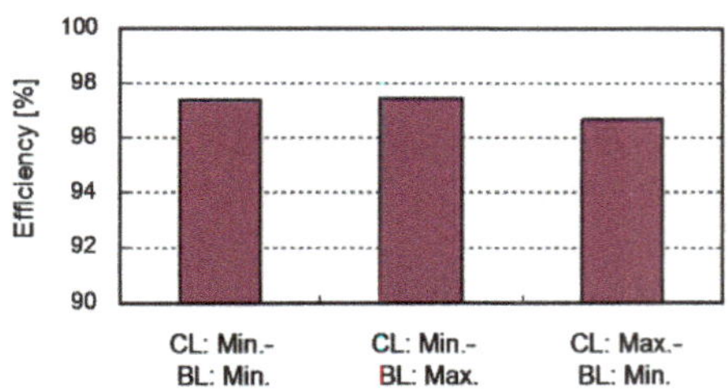

Figure 22. Fit clearance and belt transmission efficiency

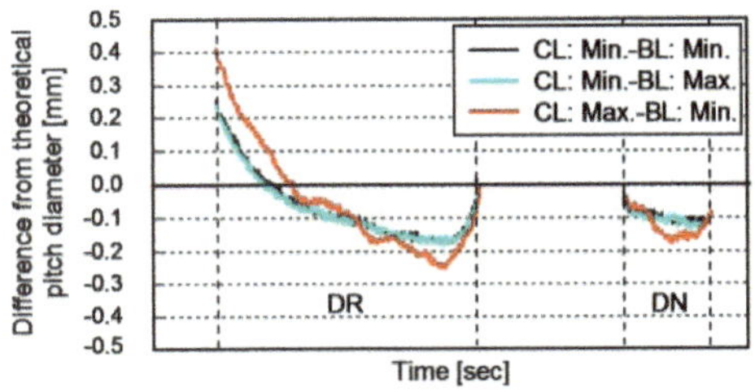

Figure 23. Fit clearance and belt pitch diameter

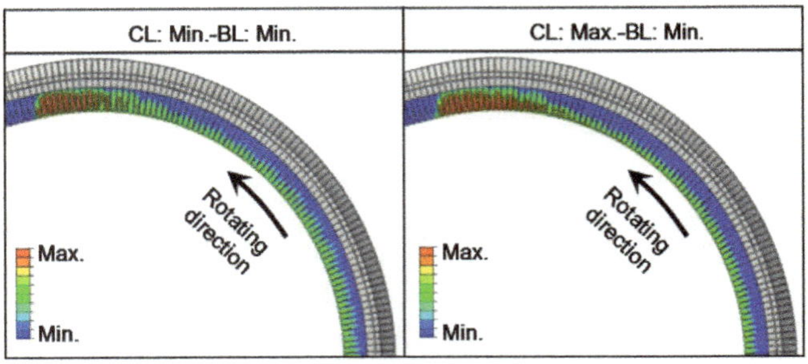

Figure 24. Fit clearance and distribution of element V-surface pressure

5.2. Effect of pulley stiffness on belt transmission efficiency

Of the parts that make up a CVT, the size and weight of the pulleys is particularly high, creating the need for the development of lighter-weight pulleys. Effects on the strength of metal pushing V-belts due to changes in pulley stiffness resulting from reduction in weight have also been reported [5,7]. As in the analyses conducted in the previous chapters, the effect on the transmission efficiency and strength of the belt when the pulley stiffness was varied was therefore considered. Altering the form of the pulleys in order to quantitatively vary stiffness would represent a challenge, and Young's modulus for the pulleys in the model discussed above was therefore altered in order to vary the stiffness. The Young's modulus of the pulleys was varied by ±20 against Young's modulus for iron, and the fit clearance was minimized.

As Figure 25 and 26 show, when the pulley stiffness is reduced, friction loss on the element V-surfaces increases, and the transmission efficiency of the belt declines. Figure 27 shows changes in the winding diameter of the belt at this time. A comparison of these results with the results of former chapters shows that the effect produced by reducing the stiffness of the pulleys displays an identical tendency to that produced by increasing the fit clearance, but the magnitude of the change is smaller.

6. Discussion

These analyses focus on the steady state response under some operating conditions. However, some results show the vibration, such as speed difference between element and ring in Figure

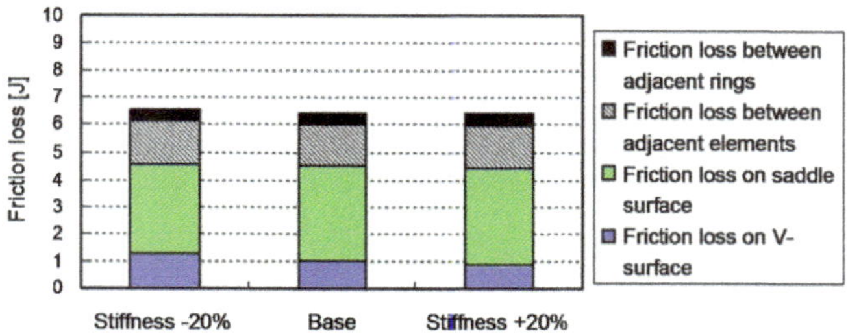

Figure 25. Pulley stiffness and friction loss on belt surface

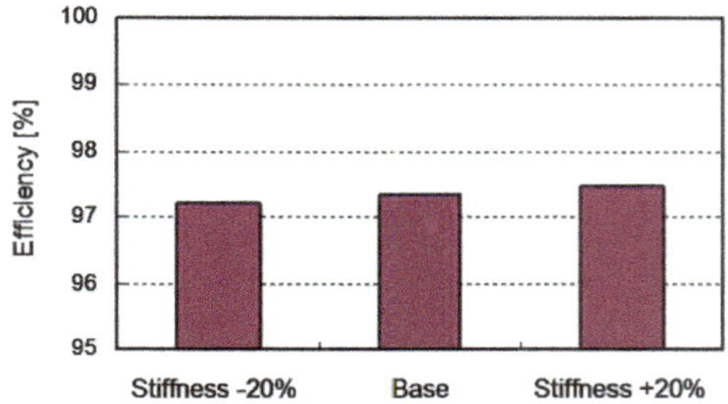

Figure 26. Pulley stiffness and belt transmission efficiency

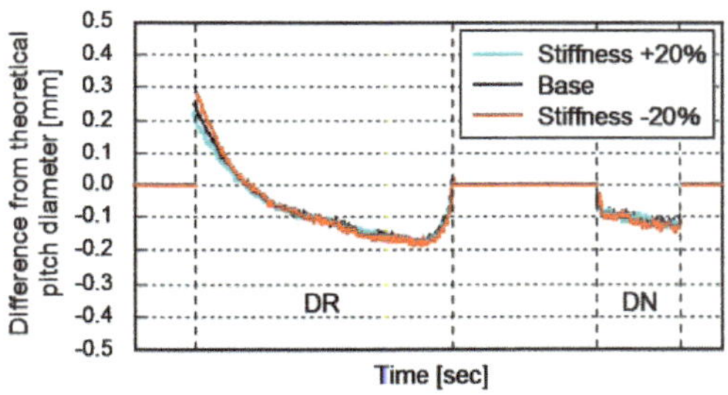

Figure 27. Pulley stiffness and belt pitch diameter

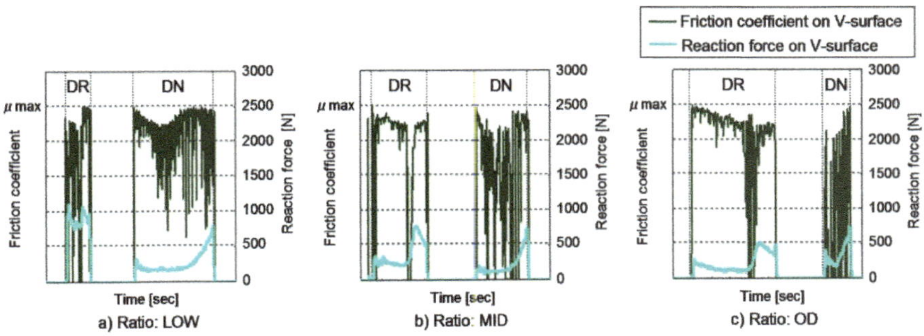

Figure 28. Friction coefficient currently used on element V-surface

12. These are two reasons for the vibration. One is the impact forces are generated when the element gets into and out the pulleys. Another is the each element behavior itself. Figure 28 shows the element reaction force on V-surface and the friction coefficient on V-surface. The friction coefficient in these analyses shows the vibration in the pulleys. This presents that each element could move individually to keep the contact and friction for adjacent parts.

7. Conclusion

1. Feedback control to maintain speed ratios in an actual CVT has been implemented in a metal V-belt behavior simulation, making it possible to predict the CVT transmission efficiency under an arbitrary operating condition.
2. This simulation has successfully quantified sliding velocities and friction forces of element contact areas along the entire length of the V-belt.
3. The simulation was used to calculate friction losses from sliding velocities and friction forces at different ratios—loss on the V-surface, loss on the saddle surface, loss between elements, loss between rings, and their proportions.
4. The simulation technique was also available for actual CVT configuration to predict the transmission efficiency using flexible pulley components.

Author details

Toshihiro Saito
Honda R&D Co., Ltd. Automobile R&D Center, Japan

8. References

[1] Yamamoto, K., Sakaguchi, S., Kishida, M., Kimura, E. and Abe, H. "Development of Integrated Engine-CVT Control System," Honda R&D Technical review, Vol. 11, No. 1
[2] Akehurst, S., Vaughan, N. D., Parker, D. A. and Simner, D. "Modelling of loss mechanisms in a pushing metal V-belt continuously variable transmission. Part 1: torque losses due to band friction," Proc. Instn Mech. Engrs vol. 218 Part D: J. Automobile Engineering
[3] Veenhuizen, P. A., Bonsen, B., Klaassen, T.W.G.L., Albers, P.H.W.M., Changenet, C. and Poncy, S. "Pushbelt CVT efficiency improvement potential of servo-electromechanical actuation slip control," 2004 CVT Congress
[4] Van der Noll, E., Van der Sluis, F., Van Dongen, T. and Van der Velde. A. "Innovative Self-optimising Clamping Force Strategy for the Pushbelt CVT," SAE Paper, 2009-01-1537
[5] Narita, K. and Priest, M. "Metal-metal friction characteristics and the transmission efficiency of a metal V-belt-type continuously variable transmission," Proc. IMechE vol.221 Part J: J. Engineering Tribology
[6] Saito, T. "Application of Stress Simulation Under Transient Condition for Metal Pushing V-belt of CVT," SAE Paper, 2008-01-0415
[7] Saito, T. and Lewis, D. A. "Development of a Simulation Technique for CVT Metal Pushing V-belt with Feedback Control," SAE Paper, 2004-01-1326

Chapter 13

Development of Graphical Solution to Determine Optimum Hollowness of Hollow Cylindrical Roller Bearing Using Elastic Finite Element Analysis

P.H. Darji and D.P. Vakharia

Additional information is available at the end of the chapter

1. Introduction

Technological progress creates increasingly arduous conditions for rolling mechanisms. Advances in many fields including gas turbine design, aeronautics, space and atomic power, involve extreme operating speeds, load, temperatures, environments which increases power and load on machinery and demand high strength to weight ratio of the rolling element bearings. Also bearing stiffness is an important parameter in the designing. Bearing design calculations require a good understanding of the Hertzian contact stress due to which high stress concentration is produced which greatly influence the fatigue life and dominate the upper speed limits as in the case of solid rolling elements. Since being originally introduced, cylindrical rolling element bearings have been significantly improved, in terms of their performance and working life. A major objective has been to decrease the Hertz contact stresses at the roller–raceway interfaces, because these are the most heavily stressed areas in a bearing. It has been shown that bearing life is inversely proportional to the stress raised to the ninth power (even higher). Whereas making the rollers hollow which are flexible enough reduces stress concentration and finally increase the fatigue life of bearing.

Investigators have proposed that under large normal loads a hollow element with a sufficiently thin wall thickness will deflect appreciably more than a solid element of the same size. An improvement in load distribution and thus load capacity may be realized, as well as contact stress is also reduced considerably by using a bearing with hollow rollers. Since for hollow roller bearing no method is available for the calculation of hollowness, contact stresses and deformation. The contact stresses in hollow members are often calculated by using the same equations and procedures as for solid specimens. This approach seems to be incorrect.

Initially in the present work author has carried out sufficient literature review (Somasundar & Krishnamurthy, 1984; Harris & Aaronson, 1967; Bamberger, Parker & Dietrich, 1976; Bhateja & Hahn, 1980; Murthy & Rao, 1983; Hong & Jianjun, 1998; Zhao, 1998; Yangang, Raj & Qingyu, 2004; Darji & Vakharia, 2008) and market survey to understand the practical application of hollow cylindrical roller bearing and its advantages in comparison with solid roller bearing. It is concluded that bearing manufacturer are production these type of bearing as per the requirement, but no standard formula or catalogue is available through which user can directly select hollow cylindrical roller bearing. Thus no standard formula (method) is available to find the optimum hollowness for the given loading condition and dimensions of bearing. Hollowness of the roller bearing is mainly dependent of applied load, dimensions of roller and endurance limit of the material used. Calculation of exact contact pressure for the hollow roller requires a finite element approach, and this has not been carried out yet. Present work is aimed to identify optimum hollowness irrespective of the geometry of the bearing and applied load.

To meet the requirement, in the first part of the present work contact analysis has been carried out for contact between roller and flat. Dimensions of the rollers are calculated using equation of equivalent diameter corresponding to the five different cylindrical roller bearing i. e. 2206, 2210, 2215, 2220 and 2224 to get the large data range. Value of applied load is taken from minimum to maximum. Finite element analysis is carried out for the same roller-flat contact and results are compared with analytical solution given by Hertz. This step is required to check the feasibility of FE procedure. In the second part of the work FE analysis has been carried out for the same applied load and material for all five cases, only the change is rollers are taken hollow. Roller hollowness is ranging from 10% to the hollowness for which bending stress at the inner bore should not exceed endurance limit of the material is taken for the consideration. Flexural fatigue failures occurred in hollow roller when the maximum bending stress at the bore cross the limit of endurance limit of the material. The fatigue cracks always began in the bore of the hollow roller. Those that propagated to the roller surface resulted in surface cracks and spallig and finally it fails the bearing. Around seventy FE analysis are done to generate the large data range. Finally graphical solution has been proposed to identify optimum hollowness irrespective of geometry of bearing and material properties.

2. Analytical study of solid cylindrical roller bearings

In the present work five different cylindrical roller bearing of NU 22 series are selected. First of all load distribution (Fig. 1) is calculated by applying load equal to static load carrying capacity of the bearing and the values of contact pressure, deflection, contact width and von Mises stress induced in the roller-race interface are determined (SKF General Catalogue, 1989; Design Data, 1994; Harris, 2001; Shigley, 1983; Nortron, 2010; I. S. 9202, 2001; Harris & Kotzalas, 2007; Horng, Ju & Cha, 2000; Demirhan & Kanber, 2008; Kania, 2006). Taking modulus of elasticity E = 201330 N/mm^2 and Poission ratio v = 0.277 for the bearing material AISI 52100 steel (Guo & Liu, 2002). These five cylindrical roller bearings are selected in such a way that the size of roller should be in step of 5 mm approximately. So we can get the wide data range of load distribution for further analysis. Considering this point in the present work 2206, 2210, 2215, 2220 and 2224 bearings are selected for analytical analysis. Contact behavior of all these bearing is studied using Hertz theory.

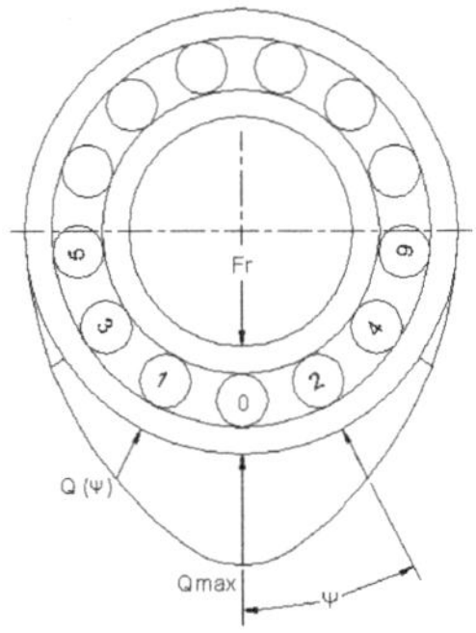

Figure 1. Load distribution in roller bearing

Before executing the FE analysis for cylindrical roller bearing to understand the contact behavior, it is possible to execute the same contact behavior for roller-flat interaction in place of roller-race as shown in Fig. 2. It is very easy to check the contact behavior of roller and flat. To check the contact behavior of roller and flat, if we will take the same diameter of roller which is used in the corresponding roller bearing then contact width will be changed. So comparison of contact behavior of roller and flat with roller-race as in case of bearing is not possible. Using the equation for equivalent diameter in the present work roller diameter is identified in such a way that for the same loading condition contact width of roller and flat interaction will remain same as the interaction of roller with race in bearing. These equivalent diameter of roller is designated by 'Roller 1' for NU 2206 bearing, 'Roller 2' for NU 2210 bearing, 'Roller 3' for NU 2215 bearing, 'Roller 4' for NU 2220 bearing and 'Roller 5' for NU 2224 bearing. Contact interaction between roller and flat plate is shown in Fig. 1, which is a part of the contact between inner-race and roller in the cylindrical roller bearing as described. This contact interaction is studied in detail by analytically using Hertz theory in the present work.

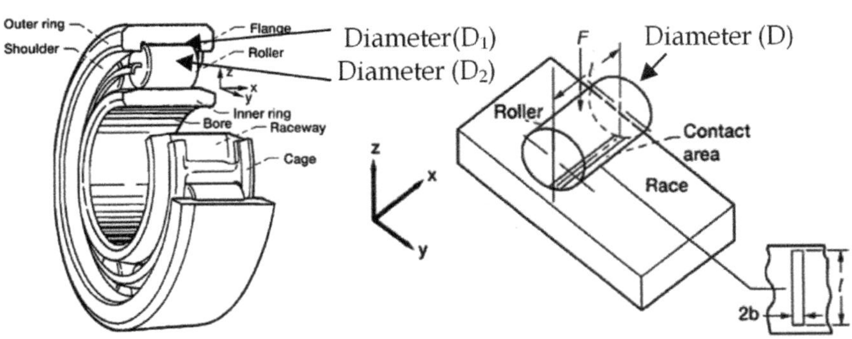

Figure 2. Schematic of contact profile of roller on flat race

Table 1 shows the value of all these analytical results corresponding to roller-flat contact.

Roller position	Load (Q) N	Contact width (b)mm	Contact pressure (p_{max}) N/mm^2	von Mises stress (σ_{VM}) N/mm^2	Deformation (δ) mm
0	5414.28	0.1321	2175.58	970.31	0.01206
1 & 2	4730.5	0.1235	2033.57	906.97	0.01068
3 & 4	2890.5	0.0965	1589.61	708.97	0.00685
5 & 6	517.68	0.0408	672.72	300.03	0.00146

Table 1. Analytical results for 2206 bearing : Equivalent diameter – 6.62 mm (Roller 1)

Roller Position	Load (Q) N	Contact width (b) mm	Contact pressure (p_{max}) N/mm^2	von Mises stress (σ_{VM}) N/mm^2	Deformation (δ) mm
0	10397.88	0.1929	2451.83	1093.51	0.01917
1 & 2	9404.92	0.1835	2331.82	1039.99	0.01752
3 & 4	6656.75	0.1544	1961.77	874.95	0.01283
5 & 6	2823.74	0.1005	1277.7	569.86	0.00593

Table 2. Analytical results for 2210 bearing : Equivalent diameter – 8.58 mm (Roller 2)

Roller position	Load (Q) N	Contact width (b) mm	Contact pressure (p_{max}) N/mm^2	von Mises stress (σ_{VM}) N/mm^2	Deformation (δ) mm
0	21620.59	0.2713	2307.3	1029.05	0.02581
1 & 2	19801.61	0.2596	2208.11	984.81	0.02385
3 & 4	14716.21	0.2238	1903.56	848.99	0.01826
5 & 6	7444.23	0.1592	1353.88	603.83	0.00989

Table 3. Analytical results for 2215 bearing : Equivalent diameter – 12.82 mm (Roller 3)

Roller position	Load (Q) N	Contact width (b) mm	Contact pressure (p_{max}) N/mm^2	von Mises stress (σ_{VM}) N/mm^2	Deformation (δ) mm
0	50656.34	0.4112	2615.63	1166.57	0.04333
1 & 2	45818.86	0.3911	2487.6	1109.47	0.03959
3 & 4	32430.3	0.329	2092.83	933.4	0.02901
5 & 6	13756.7	0.2143	1363.06	607.93	0.01341

Table 4. Analytical results for 2220 bearing : Equivalent diameter – 17.14 mm (Roller 4)

Roller position	Load (Q) N	Contact width (b) mm	Contact pressure (p_{max}) N/mm^2	von Mises stress (σ_{VM}) N/mm^2	Deformation (δ) mm
0	73318.39	0.4946	2622.82	1169.78	0.05224
1 & 2	66316.76	0.4704	2494.44	1112.52	0.04773
3 & 4	46938.59	0.3957	2098.59	935.97	0.03497
5 & 6	19911.01	0.2577	1366.81	609.6	0.01616

Table 5. Analytical results for 2224 bearing : Equivalent diameter – 20.56 mm (Roller 5)

The induced von Mises stress in the cylinder/roller is less then the yield strength 1410.17 N/mm^2 of roller material AISI 52100 steel (Guo and Liu, 2002).

3. FE analysis of solid cylinder and flat contact

3.1. Existing FE models

Since the first mathematical treatment of the contact problem of ideally smooth elastic solids, presented by Hertz in 1882, significant progress has been made in the field of contact mechanics. In particular, the deformation characteristics of semi-infinite elastic media subjected to concentrated and distributed surface traction have been elucidated, and analytical solutions for the contact pressure distributions and subsurface stress fields have been obtained for elastic bodies of different shapes and various interfacial friction conditions (Timoshenko & Godier, 1970). The results of these studies have been invaluable in the design of durable mechanical components, such as rolling element bearings (Komvopoulos & Choi, 1992). Existing FE Models like GW Model (Greenwood & Williamson, 1966), KE Model (Kogut & Etsion, 2002) and JG Model (Jackson & Green, 2005) are studied and finite element analysis for the present case is carried out.

3.2. Finite element analysis details

3.2.1. Model description

In order to validate the relationship of load vs deflection, load vs contact width etc., an FE model of an un-profiled roller contacting a flat plate was set up. A sketch of the problem is presented in Fig. 3.

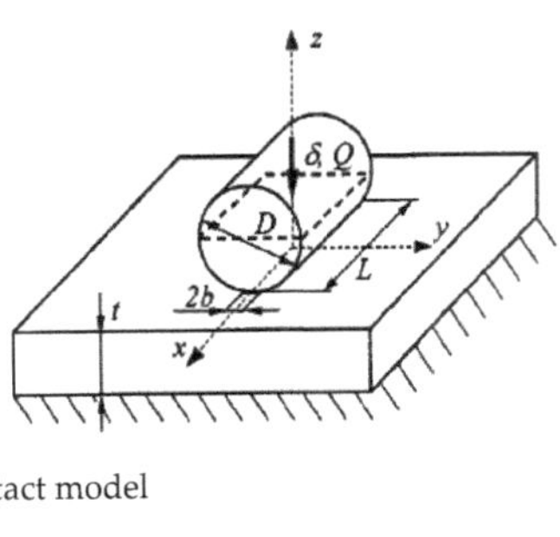

Figure 3. Sketch of roller-plat contact model

A commercial package ANSYS 9.0 was used to solve the non linear contact problem. Initially for Roller 1, first of all three dimensional axis symmetric model was developed to form the single asperity contact between half cylinder and flat plate as shown in Fig. 6. For bearing 2206 taking equivalent diameter of roller as 6.62 mm and length 12mm. Dimensions of flat plat are taken as 12 x 8 x 2 mm^3. The circular surface of cylinder and contact flat surface of plate was discretized by SOLID 185 elements. SOLID185 is used for the 3-D modeling of solid structures. It is defined by eight nodes having three degrees of freedom at each node: translations in the nodal x, y, and z directions see Fig. 4. The element has plasticity, hyperelasticity, stress stiffening, creep, large deflection, and large strain capabilities. It also has mixed formulation capability for simulating deformations of nearly incompressible elastoplastic materials, and fully incompressible hyperelastic materials.

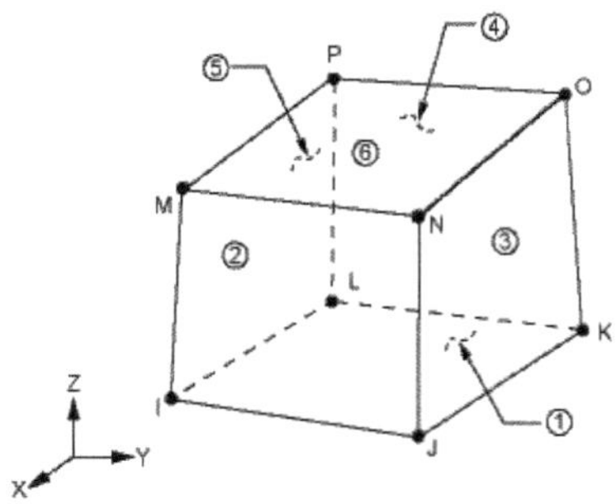

Figure 4. SOLID 185 geometry

3.2.2. *Mesh convergence*

A converged solution is one that is nearly independent of meshing errors. An extremely coarse mesh would give a very approximate solution, which is far from reality. As the mesh is refined by reducing the size of the elements, the solution slowly approaches an exact solution. It should be noted that, in theory, the solution will not be exact until the mesh size is zero, which is obviously impossible. However, it is possible to fix a tolerance to the solution error and this can be achieved by solving the problem on several meshes. In order to ensure that the solution obtained is as close as possible to reality, solutions should be obtained from several meshes starting with a very coarse mesh and finishing with a very fine mesh. Once these solutions are available, many key quantities can be compared and plotted against mesh densities (or number of points) as shown in Fig. 5.

In order to investigate the convergence of the solutions, all models have been solved with increasing numbers of elements. The elements around the roller contact region are subdivided into number of elements as shown in Fig. 7. Although the stresses and displacements at different regions are investigated in this work the convergence check has been made for only point A as shown in Fig. 7 It is a common point of contact of roller and flat plate where induced von Mises stress should be investigated. The von Mises stress (σ_{VM})

converged with coarse to fine meshes as shown in Fig. 5. The number of elements and nodes of models will increase as size of model will increase.

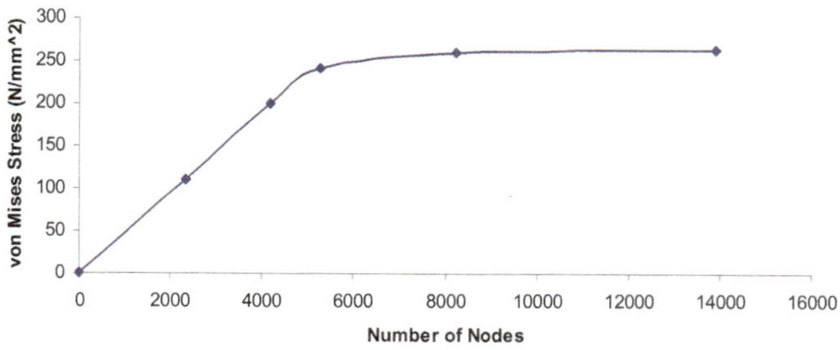

Figure 5. Von Mises stress vs number of nodes

It is very clear from the Fig. 5 that for the last three points value of von Mises stress is approximately same and its value are 242.09 N/mm^2, 259.94 N/mm^2 and 263.52 N/mm^2 for the corresponding values of 0.08, 0.05 and 0.03 element edge length.

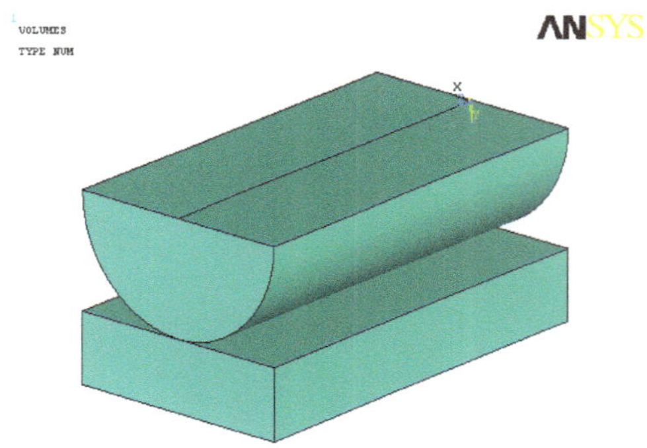

Figure 6. Finite Element Model

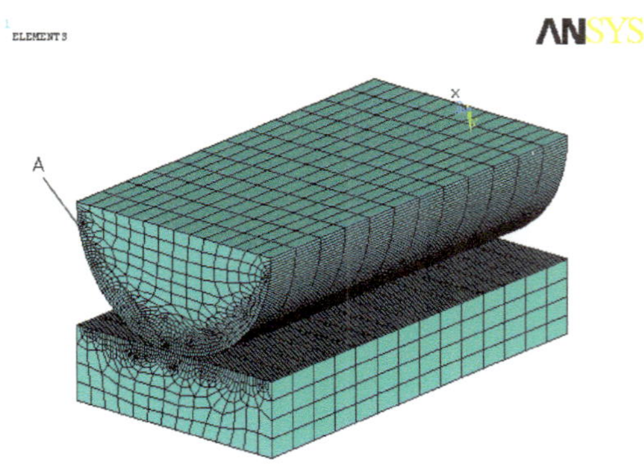

Figure 7. Densely meshed regions of the contact model

The region of most interest is adjacent to the contact interface and has the greatest concentration of elements for lower interferences. Away from the contact region, the mesh becomes coarser to minimize the computational efforts. In the present all FE models in contact region element edge length is taken as 0.08 and in other area it is taken as 0.5, the details of which is shown in Fig 8. The total numbers of elements generated are 14120 and nodes generated are 16977 for this model.

3.2.3. Contact model

In order to create contact models in ANSYS, a contact pair of elements must be created a contact element and a target element. ANSYS has general guidelines as to what line, surface, or volume these elements should be applied. Perhaps the most critical feature is the mesh size. For example, a large target element size and very fine contact element will not work. The sizes of the contact and target elements should be fairly close to one another. It is possible to get a solution to converge, but the results will most likely be incorrect. That is why there is a densely meshed region in both the bottom part of the half-cylinder and on the surface of the block shown in Fig. 8.

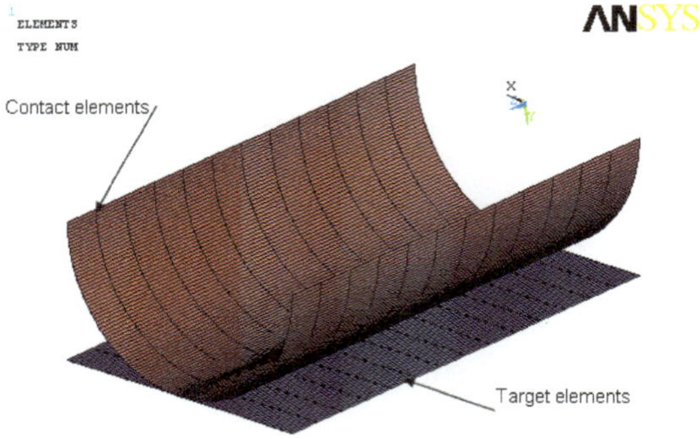

Figure 8. Contact and target elements

3.2.4. Boundary condition and application of pressure

The boundary conditions are presented in Fig. 9. The nodes on the bottom surface of the flat plat all degree of freedoms are restricted and rigidly constrained from translating in the x, y and z direction. Where as on top surface of half cylinder uniform pressure of 68.16 N/mm^2 is applied which is related with Qmax (5414.28 N) for bearing no. 2206

3.2.5. Solutions

The solutions have been carried out by means of a PC. The hardware configuration consists or an Intel Pentium IV 2.53 GHz CPU with 1 GB of RAM. The models were solved in round 25 minutes to 7 hours. All models have been solved as 3D static with Newton Raphson option.

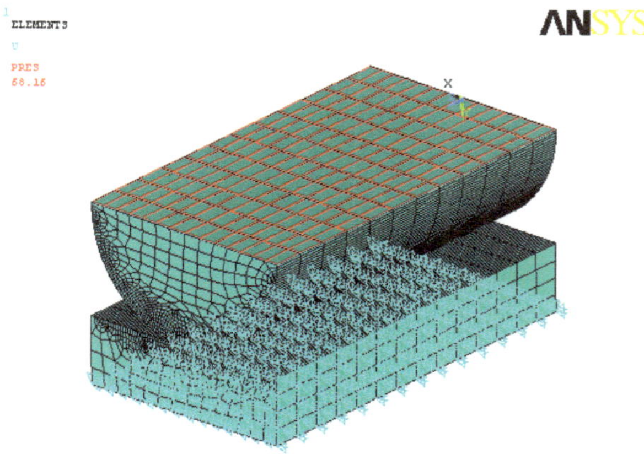

Figure 9. Applied pressure and boundary condition

3.2.6. Evaluation of the finite element model

To examine the appropriateness of the finite element mesh and modeling assumptions, such as the dimensions and fineness of the mesh and the imposed boundary conditions, finite element results for an elastic half-space indented by a rigid cylindrical asperity were compared with analytical results for line contacts. The FE model was first verified by comparing its output with the analytical results of the Hertz solution in the elastic regime. The verification included the contact pressure, contact stresses, deformation and contact width. For the evaluation of Finite Element Model initially von Mises stress criteria is taken for the consideration because it is the final output of analytical study as discussed in section 2. Also it is an important stress which should remain within limit with respect to yield stress of the material. Figure 10 shows the contour plot of von Mises stress for the applied load of 5414.28 N. As it is clear from the figure that at contact zone induced stress is higher which is marked by red colour. Figure 11 shows the detail view of contact zone with node numbers. Value of von Mises stress is to be identifying for the node no 5258 which is on contact surfaces and it is 1031.1 N/mm^2. Whereas analytical result gives 970.31 N/mm^2 (Table 1). Thus the von Mises stress of FE model differs from the Hertz solution by 5.8% which is acceptable for the present analysis. The small differences between analytical and FEA solutions near the contact edge may be attributed to the fineness of the mesh. The favorable comparison of the results illustrates the suitability of the finite element model for the present analysis involving only global variables, such as von Mises stress, contact pressure, contact width and deformation.

Figure 12 shows the contour plot for shear stress distribution. It is clear from the FE analysis that value of induced shear stress is 457.88 N/mm^2 and analytical result gives 485.16 N/mm^2 with the percentage error of 5.6 % only.

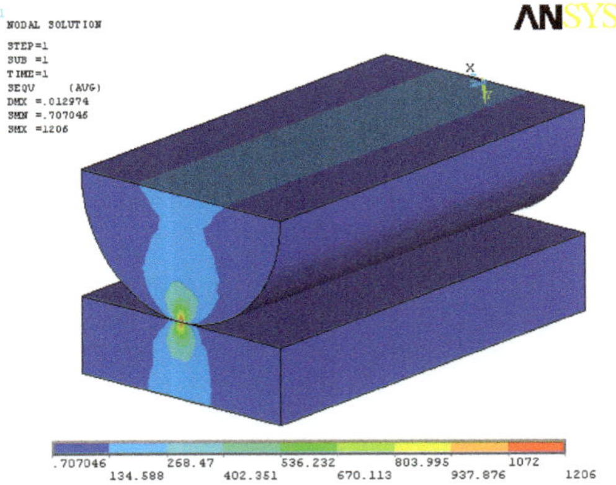

Figure 10. von Mises stress distribution over solid cylinder-flat

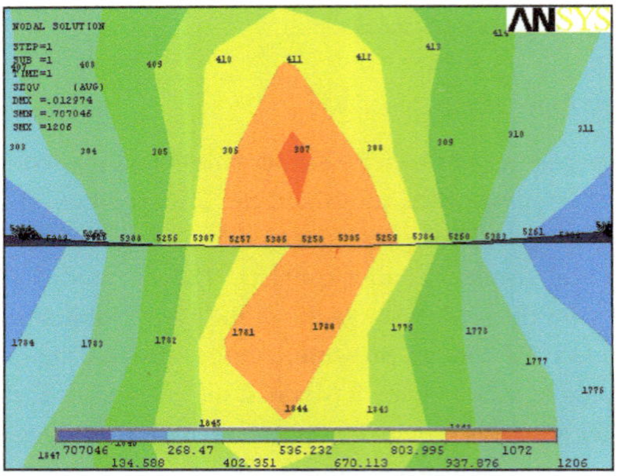

Figure 11. Detail of contact zone for von Mises stress distribution

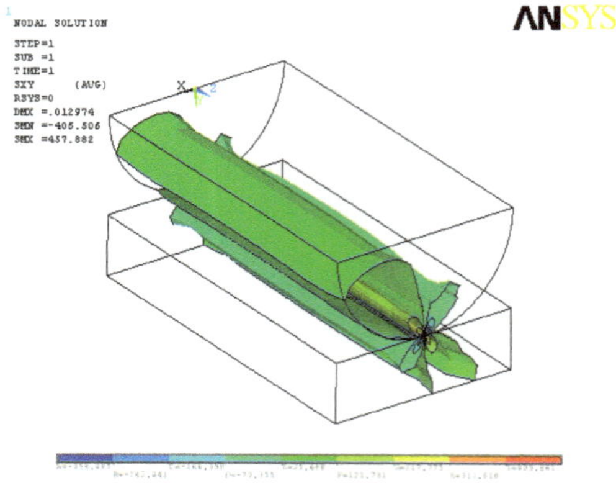

Figure 12. Contour plots for shear stress

Now using similar contact model and boundary conditions FE analysis has been carried out for all five rollers. Applied loads are taken as per the calculated load distribution among the rollers. Table 6 shows the validation of meshing scheme employed by comparing the analytical results with FEA.

Figure 13 to 16 shows the graphical comparison of Table 6. Four important parameters von Mises stress, Contact Pressure, Deformation and Contact Width are plotted for different applied load.

Load (N)	Von Mises stress, σ_{VM} (N/mm^2)		Contact width, b (mm)		Contact pressure, p (N/mm^2)		Deformation, δ (mm)	
	Analytical	FEA	Analytical	FEA	Analytical	FEA	Analytical	FEA
Roller 1								
5414.28	970.31	1031.1	0.1321	0.1412	2175.58	2034.5	0.01206	0.01288
4730.5	906.97	958.92	0.1235	0.1322	2033.57	1898.2	0.01068	0.01148
2890.5	708.97	719.39	0.0965	0.1066	1589.61	1437.9	0.00685	0.00742
517.68	300.03	242.09	0.0408	0.0467	672.72	587.2	0.00146	0.00184
Roller 2								
10397.88	1093.51	1083.7	0.1929	0.2097	2451.83	2255.4	0.01917	0.02028
9404.92	1039.99	1018.9	0.1835	0.2003	2331.82	2136	0.01752	0.0186
6656.75	874.95	840.21	0.1544	0.1726	1961.77	1754.5	0.01283	0.01376
2823.74	569.86	519.65	0.1005	0.1218	1277.7	1053.9	0.00593	0.006469
Roller 3								
21620.59	1029.05	1110	0.2713	0.2991	2307.3	2092.5	0.02581	0.02643
19801.61	984.81	1050.2	0.2596	0.2877	2208.11	1992.4	0.02385	0.02451
14716.21	848.99	914.4	0.2238	0.2529	1903.56	1684.6	0.01826	0.01899
7444.23	603.83	616.54	0.1592	0.1879	1353.88	1146.5	0.00989	0.01057
Roller 4								
50656.34	1166.57	1245.6	0.4112	0.4659	2615.63	2308.2	0.04333	0.04357
45818.86	1109.47	1176.8	0.3911	0.4466	2487.6	2177.8	0.03959	0.04
32430.3	933.4	957.3	0.329	0.3860	2092.83	1783.4	0.02901	0.02989
13756.7	607.93	592.31	0.2143	0.2723	1363.06	1072.4	0.01341	0.01442
Roller 5								
73318.39	1169.78	1237.3	0.4946	0.5677	2622.82	2285	0.05224	0.05235
66316.76	1112.52	1174.7	0.4704	0.5442	2494.44	2155.9	0.04773	0.04811
46938.59	935.97	963.4	0.3957	0.4706	2098.59	1764.6	0.03497	0.03604
19911.01	609.6	582.93	0.2577	0.3306	1366.81	1065.4	0.01616	0.01758

Table 6. Validation of meshing scheme employed

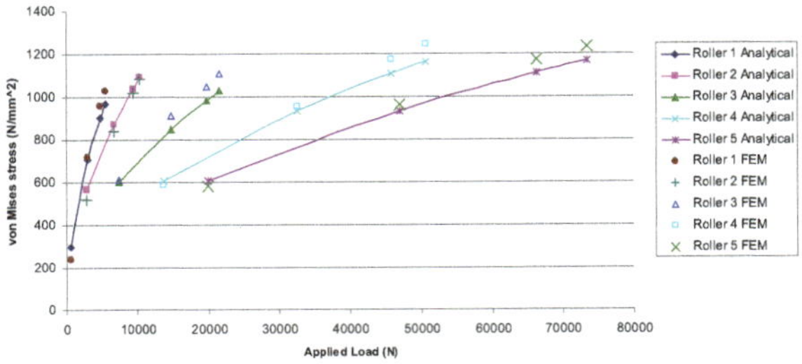

Figure 13. von Mises stress vs Applied load

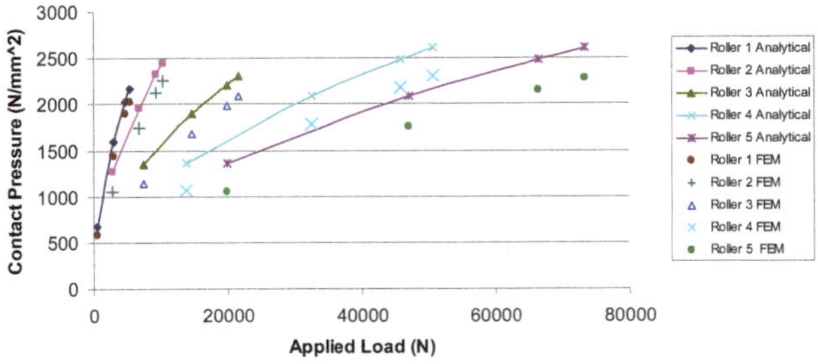

Figure 14. Contact Pressure vs Applied load

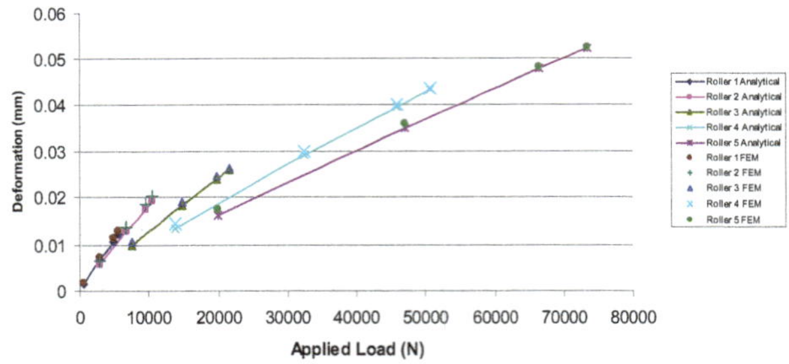

Figure 15. Deformation vs Applied load

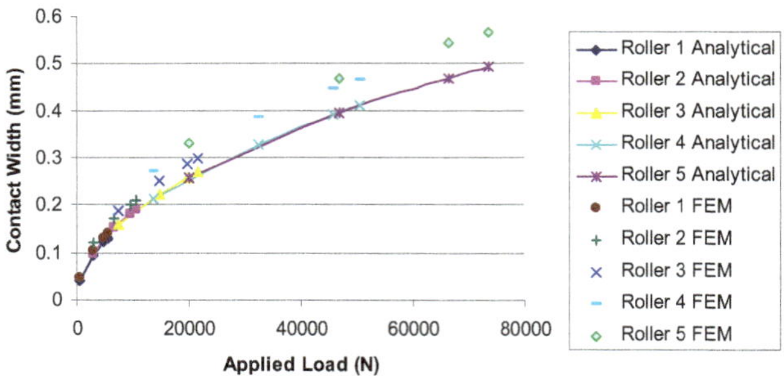

Figure 16. Contact Width vs Applied load

3.2.7. Summary of FE analysis

The finite element method was used to analyze the contact mechanics aspects of nominally flat single-asperity surfaces and to identify the effect of important parameters like von Mises stress, contact pressure, contact width and deformation for the given load. On the basis of the presented results, the following major conclusions can be drawn.

The smaller error in the FE model is attributed to overall balance (static equilibrium) enforced by the FEM package. The smaller differences between analytical and FEA solutions near the contact edge may be attributed to the fineness of the mesh.

On the basis of the results discussed, it may be concluded that the finite element configuration shown in Fig. 10 and the invoked modeling approximations are acceptable for the purpose of the present analysis.

Figure 14 to 17 shows that the agreement between analytical and finite element results from different rollers and various load is appreciably good. The maximum disagreement between the FEA value and analytical values occurs at the lowest applied load. The accord between the FEA and analytical results gets progressively better as higher applied load. Thus smaller the interference the smaller number of contact elements are in effect, leading to a large error and visa versa.

4. Finite element analysis for hollow roller and flat contact

4.1. Finite element model

Different hollowness percentage ranging from 10% to 80% (in step of 10%) has been investigated for Roller 1 of diameter 6.62 mm which is equivalent of 8 mm diameter roller co-relate with bearing 2206. Figure 17 shows the finite element model for 40 % hollowness. Same surfaces as taken in the contact model of solid roller and flat i.e. outer surface of roller and top surface of flat plate are selected for contact element and target element respectively.

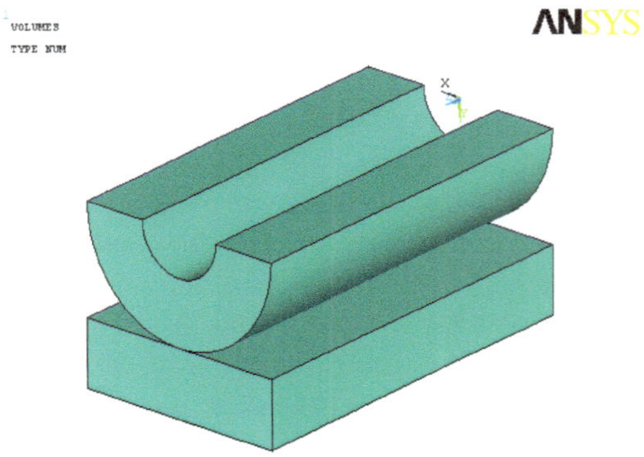

Figure 17. Finite element model for 40% Hollowness

4.2. Meshing

Now in this case also for the contact surface element edge length is taken as 0.08. Also as discussed by Murthy & Rao (1983) that, in addition to the contact stresses at the outer contact zone, the hollow specimens are subjected to tangential stresses (bending stress) at inner surface. Thus for the inner surface element edge length should be high and is taken as 0.08 as shown in Fig. 18. For other remaining area it is taken as 0.5.

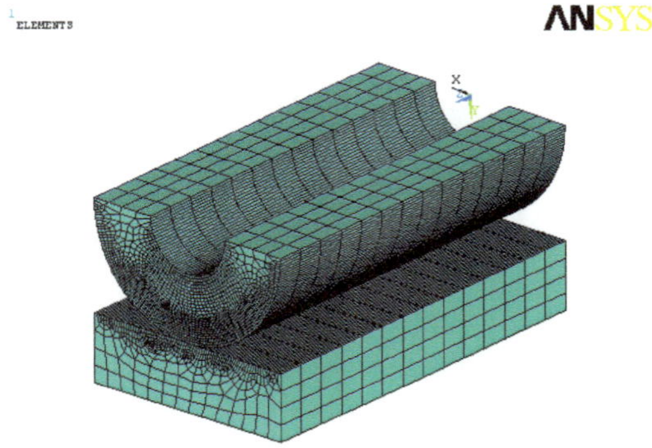

Figure 18. Densely meshed regions of the contact model of hollow roller and flat

Same boundary condition and pressure is applied as discussed in section 3.2.4. For each hollowness 10% to 80%, FE model is developed as shown in Fig. 19, maximum applied load 5414.28 N is taken and results are observed.

5. Results and discussion

Due to thin section very less material is available to resist the force so von Mises stress is increase after 60% hollowness. Also at this stage plastic deformation will take place and failure will occur due to permanent deformation, which is not desirable and should be avoided.

An added criterion for evaluation in a bearing with hollow rollers is the roller bending stress. To evaluate the life integrals, the value of the fatigue limit stress must be known for the bearing component material. This can be determined by endurance testing of bearings or selected components. Performance analyses were conducted, using the von Mises stress as the fatigue failure-initiating criterion. Based on this subsequent study fatigue limit stress for bearing material AISI 52100 is 684 N/mm^2 (Harris and Kotzalas, 2007). From Table 7 it is very clear that bending stress is continuously increase from 377.07 N/mm^2 to 1721.4 N/mm^2 as hollowness increase from 10% to 80% respectively. But the practical limit of this stress is 684 N/mm^2. So the hollowness should be restricted upto 52% which is clear from Fig. 20.

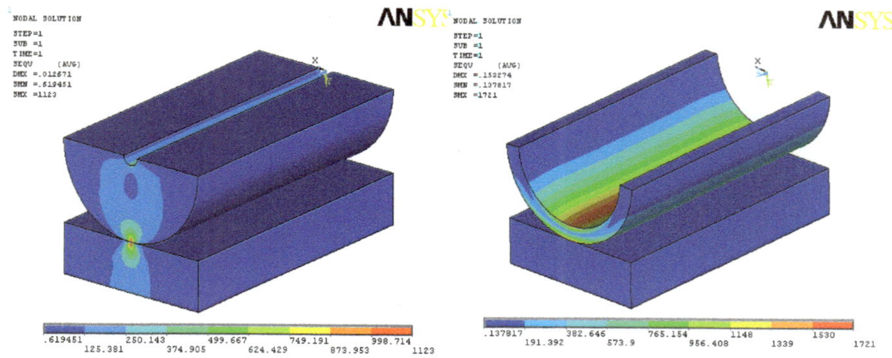

(a) 10% hollowness (b) 80% hollowness

Figure 19. von Mises stress plots for hollow roller

Hollowness %	Contact pressure (N/mm²)	von Mises stress (N/mm²)	Bending stress (N/mm²)	Deformation (mm)
10	1901	952.92	377.07	0.01253
20	1812.3	895.99	422.55	0.01264
30	1679.6	843.84	463.1	0.01345
40	1536.3	785.46	539.91	0.01517
50	1391.9	741.49	650.03	0.01885
60	1247.1	704.71	822.71	0.02663
70	1093.3	832.79	1123.8	0.04516
80	897.31	1379.1	1721.4	0.10102

Table 7. Values of parameters for different hollowness for Roller 1

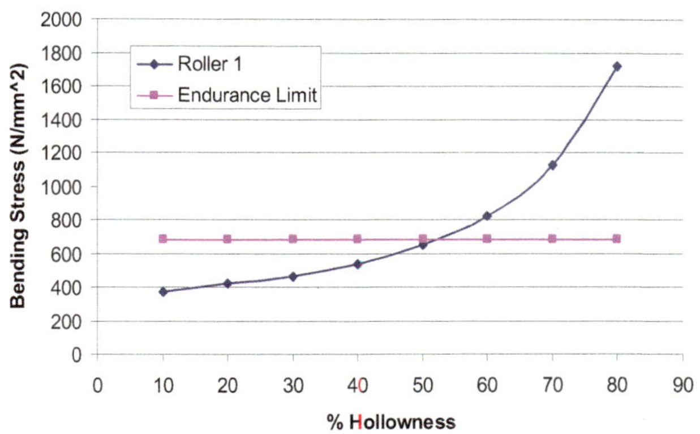

Figure 20. Bending stress vs Hollowness for Roller 1

Thus for the present case of Roller 1, for the applied load of 5414.28 N the % hollowness of the hollow roller should not exceed 52%, otherwise induced stress at the bore of the roller will increase beyond the endurance limit and cause fatigue failure of roller.

Great care must be given to the smooth finishing of the inside surface of a hollow roller during manufacturing as the stress raisers that offer due to poorly finished inside surface will reduce the allowable roller hollowness ratios still further than indicated by Fig. 20.

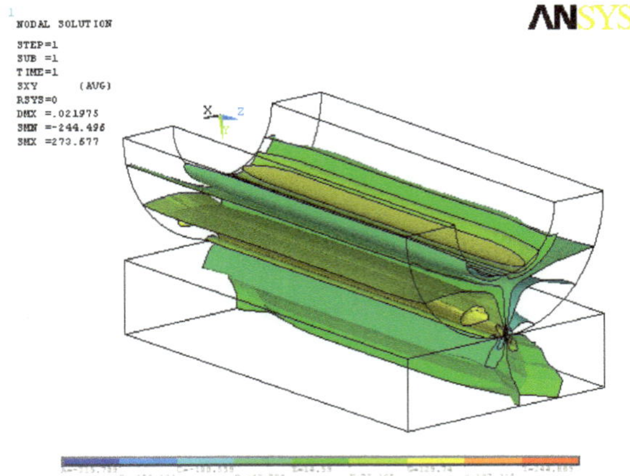

Figure 21. Contour plot for maximum shear stress for hollow roller

Figure 21 shows the contour plot of maximum shear stress for 52% hollowness. The induced shear stress is 273.68 N/mm^2 which is approximately half than the shear stress induced in solid roller for the same load of 5414.28 N. Thus reduction in shear stress gives improvement in fatigue life of bearing.

Figure 22 to 25 shows the effect of hollowness on different parameters.

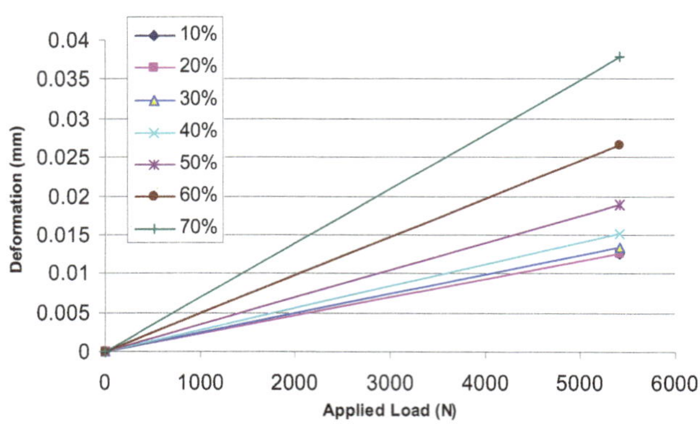

Figure 22. Effect of hollowness on the deformation for same applied load

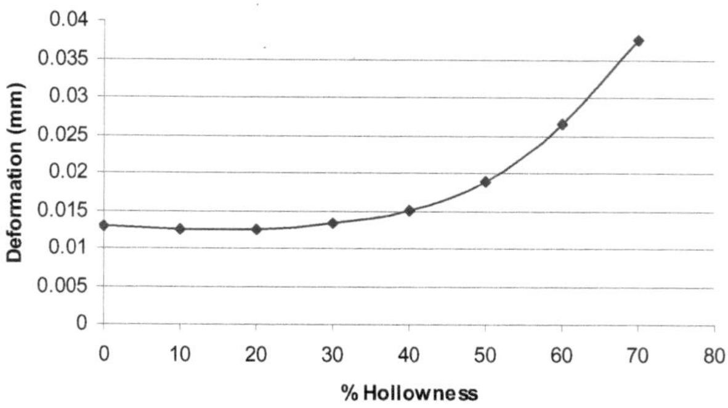

Figure 23. Deformation vs Hollowness

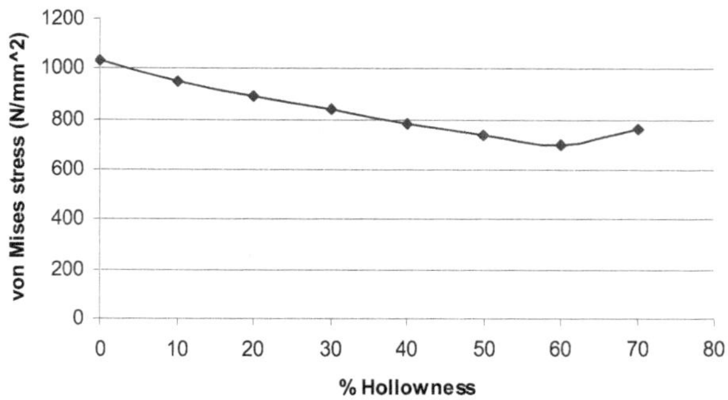

Figure 24. Mises stress vs Hollowness

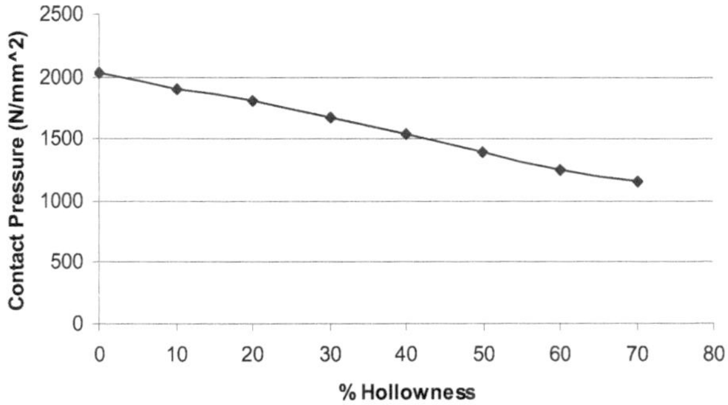

Figure 25. Contact pressure vs Hollowness

The term "hollowness" as referred to here for the rollers in the ratio of the inner diameter to the outer diameter expressed as a percentage and is an important parameter in the design of the hollow roller bearing. It determines not only the overall bearing stiffness but also the amount of preload most desirable, the bearing's load capacity and its life. In fact, hollowness is a control parameter used to optimize the bearing design. In the present case of Roller 1, load is applied in such a way that induced bending stress at inner bore should cross endurance limit of the material i.e. 684 N/mm² for each hollowness. Result of FE analysis is shown in Fig. 26. The roller hollowness values from 10% to 80% have been analyze by Finite Element as discuss above for Roller 1 and the roller load, deflection and stress curves of Fig. 27 have been developed. The dotted line across these curves show the points of constant maximum roller bore stress for values of 684 N/mm².

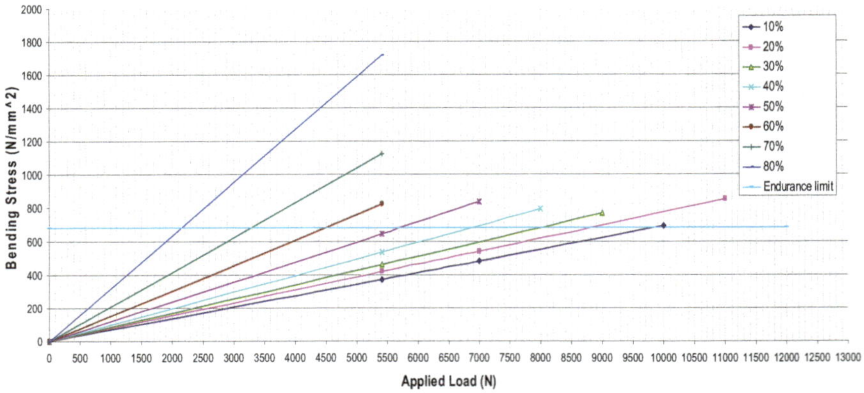

Figure 26. Bending stress crosses the endurance limit for different hollowness

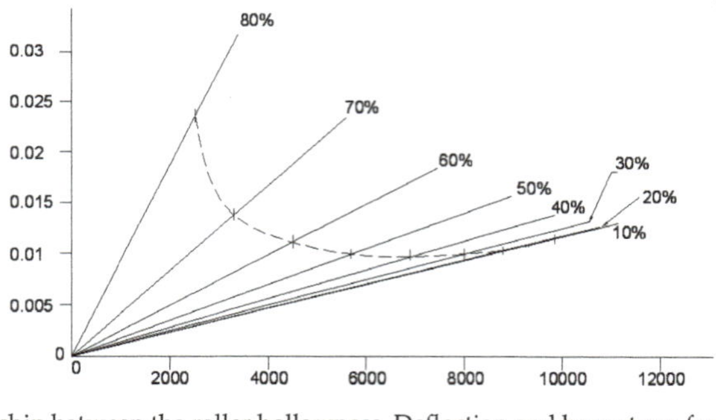

Figure 27. Relationship between the roller hollowness, Deflection and bore stress for 6.62 mm diameter roller

Corresponding to each roller hollowness value, there is a specific optimum load for each bearing design, which is indicated in Table 8.

Figure 27 gives the best solution to find the optimum hollowness for verities of load. But this is not the final solution, because solution given in Fig. 27 is only applicable for Roller 1 i.e. equivalent roller of bearing 2206. As bearing/roller geometry will change again same procedure of FE analysis should be carried out and again optimum hollowness is to be identified. Thus similar analysis can also be carried out for remaining four rollers to find the optimum hollowness and results are given in Table 9. For each roller these analyses have been carried out upto the hollowness where induced bending stress should just cross the endurance limit.

% Hollowness	Max. Applied Load (N)
10	9850
20	8800
30	8000
40	6900
50	5700
60	4500
70	3300
80	2520

Table 8. Maximum applied load for different hollowness for Roller 1

Roller no	Maximum Load (Q_{max}) N	% Hollow-ness	Contact pressure (N/mm^2)	von Mises stress (N/mm^2)	Bending stress (N/mm^2)	Deformation (mm)
Roller 2	10397.88	10	2153.5	1067.8	500.59	0.0195
		20	1984.2	1045.7	548.37	0.02
		30	1812.5	969.11	599.86	0.02149
		40	1652.5	900.87	693.55	0.02461
Roller 3	21620.59	10	1917.9	975.09	468.32	0.026
		20	1769.9	909.63	491.53	0.02671
		30	1629.6	855.49	538.21	0.02871
		40	1484	818.1	619.64	0.03284
		50	1324.6	804.27	746.49	0.04103
Roller 4	50656.34	10	2126.9	1094.9	620.35	0.0428
		20	1951.7	1021.8	634.68	0.04448
		30	1802.2	984.04	689.2	0.04793
Roller 5	73318.39	10	2105.2	1084.6	633.57	0.05157
		20	1922	1009.9	640.66	0.05374
		30	1820.2	993.23	692.28	0.05704

Table 9. Values of parameters for different hollowness for Roller 2, 3, 4 & 5

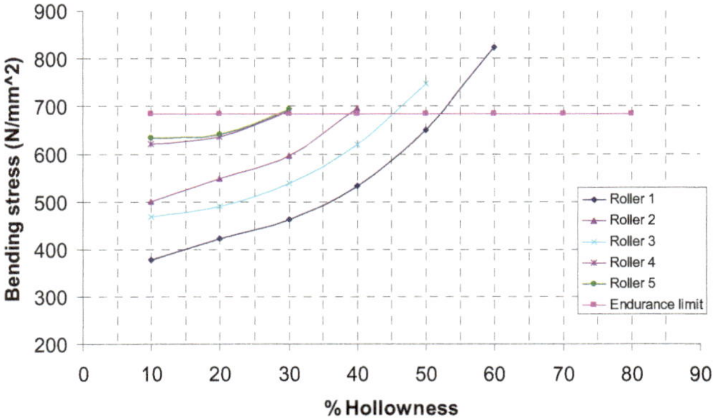

Figure 28. Comparison of hollowness for different rollers

From Fig. 28 it is clear that for

Roller 1 optimum hollowness should be 52% for the applied load of 5414.28 N,
Roller 2 optimum hollowness should be 39% for the applied load of 10397.88 N,
Roller 3 optimum hollowness should be 45% for the applied load of 21620.59 N,
Roller 4 optimum hollowness should be 29% for the applied load of 50656.34 N,
Roller 5 optimum hollowness should be 28% for the applied load of 73318.39 N.

It is very clear from the results and discussion of all five rollers that optimum value of hollowness is dependent on magnitude of applied load, bearing geometry i.e. diameter and length of roller and mechanical properties of material used. If the value of applied load will increase than hollowness should be reduced to maintain the bending stress within endurance limit of the material. Change in bearing geometry will change the applied pressure and resulted into change in hollowness. Thus the solution given in Fig. 26 and 27 is not a generalized solution and it can not be applicable to any bearing geometry for any load. It is applicable to specific type of bearing and for specific load only respectively. If applied load will change one can't use the results shown in Fig. 28 and lengthy FE procedure should be again carried out to get the result in the form of hollowness.

5.1. Generalized graphical solution

To find the optimum hollowness for any material and for any applied load irrespective of bearing geometry, in the present work large data are generated by FE analysis. To get the generalized solution FE analysis for the hollowness percentage ranging from 1% to 95% (after 95% ANSYS solution was not supported) is carried out and following Table 10 has been developed. This table shows the values of bending stress corresponding to the applied pressure.

Table 10 is presented in graphical form in Fig. 29. This diagram shows the value of bending stress for different applied pressure with respect to hollowness ranging from 1% to 95%.

Applied pressure (N/mm^2)	Bending Stress (N/mm^2)											
	1%	5%	10%	20%	30%	40%	50%	60%	70%	80%	90%	95%
99.06	515.11	526	633.57	640.66	692.28							
98.52	497.6	512.3	620.35	634.68	689.2							
86.56	441.46	450.112	500.59	548.37	599.86	693.55						
78.3	419.68	418.9	452.84	496.14	542.76	627.65						
76.66	371.8	401.3	468.32	491.53	538.21	619.64	746.49					
70.21	344.029	359.13	429	450.31	493.08	567.54	683.78					
68.16	330.68	347.21	377.07	422.55	463.1	539.91	650.03	822.71	1123.8	1721.4	3365	4653.6
59.55	303.71	309.66	329.92	369.25	404.64	471.74	567.96	718.81				
55.42	277.1	282.64	321.12	351.66	384.46	444.33						
52.18	263.5	273.95	319.16	334.9	366.59	422.04	508.42					
36.39	189.23	192.867	201.72	225.7	247.51	288.52	347.05	439.33				
26.39	129.311	138.475	161.84	169.76	185.62	213.61	257.35					
23.51	112.848	124.6	136.5	149.44	163.28	188.71						
6.52	33.9	34.88	36.36	40.636	44.509	51.777	62.158	78.697				

Table 10. Value of bending stress corresponding to the applied pressure

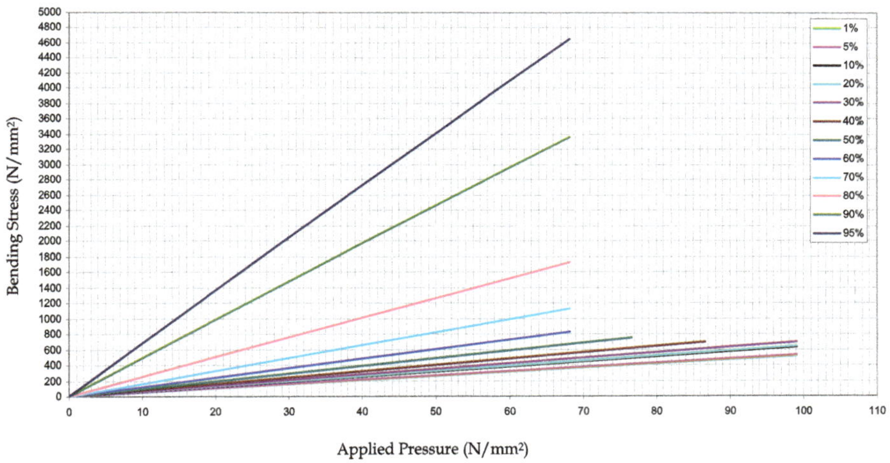

Figure 29. Bending stress vs applied pressure for different hollowness

6. Conclusion

In case of solid roller bearing induced sub-surface stresses are the limiting criteria for the fatigue life of bearing whereas for hollow roller bearing bending stress is the limiting criteria. The bending stresses on the internal diameter of the roller in the plane of the loading forces are the most critical for destructions. In the present work graphical solution was developed to determine optimum hollowness of cylindrical roller bearing for which induced bending stress should be within the endurance limit of the material. Figure 29 shows the generalized diagram for bending stress vs applied pressure. Following are the major outcomes from this diagram.

For the same value of applied pressure, Fig 29 shows that there is very small variation in the value of bending stress by increase the hollowness from 10% to 30%.

If the hollowness increases from 1% to 95% the slop of line will also increase accordingly.

The durability of the bearings with hollow rollers operating on cycles not exceeding the maximum permitted level of bending stresses can be substantially greater than the durability of similar bearings with solid rollers.

For the applied load on equivalent size of roller initially applied pressure is to be calculated. As per the endurance limit of the material used and calculated applied pressure optimum hollowness can be identified from the diagram.

For the particular hollowness diagram gives the maximum limit of applied pressure and hence applied load. The developed graphical solution can be applicable for any material of bearing.

Author details

P.H. Darji
Department of Mechanical Engineering, C. U. Shah College of Engineering & Technology, Surendranagar, India

D.P. Vakharia
Department of Mechanical Engineering, S.V. National Institute of Technology, Surat, India

7. References

Bamberger, E. N., Parker, R. J. and Dietrich, M. W., 1976, "Flexural Fatigue of Hollow Rolling Elements," NASA TN D – 8313, Washington.

Bhateja, C. P. and Hahn, R. S., 1980, "A Hollow Roller Bearing for Use in Precision Machine Tools," Annals of the CIRP, 29, 1, pp. 303 – 307.

Darji, P. H. and Vakharia, D. P., (2008), "Determination of Optimum hollowness for hollow cylindrical rolling element bearing", Proceedings of ASME 2008 International Mechanical Engineering congress and exposition, Boston, Massachusetts, USA, Oct 31-Nov 6, Paper No. IMECE2008-67294. ISBN: 978-0-7918-4873-9

Demirhan, N. and Kanber, B. (2008) Stress and Displacement Distributions of Cylindrical Roller Bearing Rings Using FEM. Taylor & Francis, Mechanics Based Design of Structures and Machines, 36, 86 – 102. ISSN : 1539 – 7734.

Design Data (1994). PSG College of Technology, Coimbatore.

Greenwood, J. A. and Williamson, J. B. P. (1966) Contact of Nominally Flat Surfaces. Proc. R. Soc. London, Ser A. 295, 300 – 319. ISSN 1471-2946.

Guo, Y. B. and Liu, C. R. (2002) Mechanical Properties of Hardened AISI 52100 Steel in Hard Machining Processes. Tran. ASME, J. Manf. Sci. & Engg., 124, 1 – 9. ISSN 1087-1357.

Harris, T. A. (2001) Rolling Bearing Analysis. , 4th ed, John Wiley & Sons, New York. ISBN 0-471-35457-0

Harris, T. A. and Aaronson, S. F., 1967, "An analytical investigation of cylindrical roller bearings having annular rollers," Tribology Transactions, 10, pp. 235 – 242. ISSN 1040-2004.

Harris, T. A. and Kotzalas, M. N. (2007) Rolling Bearing Analysis - Advanced Concepts of Bearing Technology. 5th ed., Taylor & Francis, Boca Raton. ISBN 10: 0-8493-7182-1.

Hong, L. and Jianjun L., 1998, "Analysis of Contact Problems on Hollow Cylindrical Rollers", Tribology, 120, pp. 134 – 139. ISSN 1004-0595.

Horng, T. L, Ju, S. H. and Cha, K. C. (2000) A Deformation Formula for Circular Crowned Roller Compressed Between Two Flat Plates. Tran. ASME, J. Tribol, 122, 405 – 411. ISSN 0742-4787.

I. S. 9202 (2001) Specification for Cylindrical Rollers. Edition 1.1.

Jackson, R. L. and Green, I. (2005) A Finite Element Study of Elasto-Plastic Hemispherical Contact Against a Rigid Flat. Tran. ASME, J. Tribol, 127, 343 – 354. ISSN 0742-4787.

Kania, L. (2006) Modeling of Rollers in Calculation of Slewing Bearing with the use of Finite Elements. Int. J. Mechanism and Machine Theory, 41, 1359 – 1376. ISSN 0094-114X.

Kogut, L. and Etsion, I. (2002) Elastic-Plastic Contact Analysis of a Sphere and a Rigid Flat. Tran. ASME, J. Tribol, 69, 657 – 662. ISSN 0742-4787.

Komvopoulos, K. and Choi, D. –H. (1992) Elastic Finite Element Analysis of Multi-Asperity Contacts. Tran. ASME, J. Tribol, 114, 823 – 831. ISSN 0742-4787.

Murthy, C.S.C. and Rao, A. R. (1983) Mechanics and Behaviour of Hollow Cylindrical Members in Rolling Contact. Wear, 87, pp. 287 – 296. ISSN 0043-1648.

Nortron, R. L. (2010) Machine Design An Integrated Approach. 2nd ed., Pearson Education, Singapore. ISBN 978-81-317-0533-9.

SKF General Catalogue 4000/ IIE (1989) W. Germany. Reg : 47.23000.

Shigley, J. E. (1983) Mechanical Engineering Design. 1st ed, McGraw-Hall Singapore. ISBN 0-07-100292-9.

Somasundar, H. V. and Krishnamurthy, R., 1984, "Surface Durability of Tufftrided Rolling Elements," Wear, 97, pp. 117 – 127. ISSN 0043-1648.

Timoshenko S. P. and Godier J. N. (1970) Theory of Elasticity. 3rd ed., McGraw-Hall Singapore. ISBN : 0-07-085805-5.

Yangang, W., Yi, Q., Raj, B. and Qingyu, J., 2004, "FE analysis of a novel roller form : A deep end cavity roller for roller type bearings," Elsevier, Journal of Materials Processing Technology, 145, pp. 233 – 241. ISSN 0924-0136.

Zhao, H., 1998, "Analysis of Load Distributions within Solid and Hollow Roller Bearings", ASME J. Tribol., 120, pp. 134 – 139. ISSN 0742-4787.

Chapter 14

Optimization and Improvement of Throwing Performance in Baseball Pitching Machine Using Finite Element Analysis

Shinobu Sakai and Hitoshi Nakayama

Additional information is available at the end of the chapter

1. Introduction

Pitching machines for baseball are widely used in venues ranging from professional baseball stadiums to amusement facilities (Adair, 1994). The most important purpose of the pitching machine is to reproduce the throws of an adversary pitcher, which will be useful for the improvement batting technique. The most common commercial pitching machines for baseball are the "arm" type and the "two-roller" type. Some pitching machines can pitch a high-speed ball (fastball) and a breaking ball, but these machines have certain limitations. In particular, it is very difficult to change ball speed and direction simultaneously (Mish et al., 2001). Therefore, the throwing performance of conventional pitching machines used for batting practice is not very high. Balls pitched to change in instant at various speeds and with different pitch types (ex. fastball, curveball, screwball and forkball) are easily achieved by a new pitching machine equipped with three rollers which has been developed by the authors. It is called a "three-roller" type pitching machine (Sakai et al., 2007). With the structure of three rollers, comes the production of a new pitching machine that can pitch balls repeatedly in the way the batter desires, controlling both ball speed and pitch type. However, as observed during our study, the seam of a baseball coming in contact with the rollers, the spin rate and projection angle of the ball delicately change. From the results, it became clear experimentally that from the throw accuracy deteriorates.

In this chapter, the throw simulation of the three roller-type pitching machine is analyzed using a commercial dynamic finite element analysis code (ANSYS/ LS-DYNA). The moving behavior (such as velocity and spin rate) and contact stress state of the ball pitched by the pitching machine are clearly observed. The effect of the throw accuracy by the seam posture of the ball in the machine is examined. In addition, the shapes and materials of the rollers do not negatively affect the throw accuracy based on the seam posture examined. In other words, a robust roller (optimum roller) has been found. Furthermore, the optimum roller was produced on the specifications provided in the analytical results. Based upon throws from the machine using the rollers experimented on, the propriety of the analysis results is inspected.

2. Overview of commercial pitching machine

One important function of a pitching machine for baseball has traditionally been to throw a ball at a very high speed. The most common commercial pitching machines for baseball are the "arm" type and the "two-roller" type (Mish et al., 2001). The principal mechanism of the arm type machine is a spring and crank lever that imitates a human arm when pitching a ball (see Fig. 1(a)). The two-roller type is shown in Fig. 1(b). This machine uses two-roller to project a ball by taking advantage of the frictional force between the two rollers and the ball. Generally, the ball speed can be controlled with this type of machine, but direction changes are very difficult or impossible to achieve. With the two-roller machine type, the spin direction given to a ball is controllable only on the plane in which the roller is turning, because a ball flies only on that plane. This can change the spin rate and speed of the ball by changing the number of the ball's turns on its plane. However, the plane must be spun manually in order to produce arbitrary breaking balls. Additionally, the batter can be expected pitch type of a ball thrown by the two-roller type machine, because when a braking ball is pitched, the projection equipment (two rollers) in the machine must be slanted.

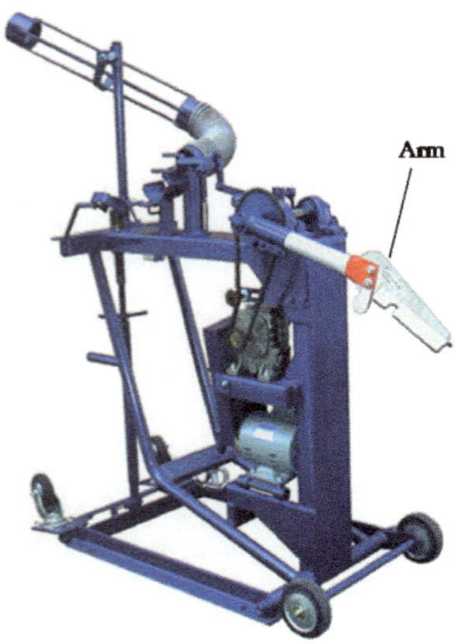

(a)

(b)

Figure 1. Commercial pitching machine. a) Arm type. b) Two-roller type.

Additionally, the throw performance of both types of pitching machine that have so far been developed for use during batting practice is not very high. The control precision of the latest commercial arm and two-roller types pitching machines is only 300 mm in height and 150 mm in width. One of the major reasons for this lack of precision is that baseball balls have a peculiar seam. Pitchers use this seam to throw various types of pitches, however this very seam also decreases the control precision of pitching machines (Mizota et al., 1995, Himeno et al., 1999, Sakai, et al., 2007).

If we wish to develop a high fidelity pitching machine for baseball capable of throwing a wide range of pitches with full freedom and control over both the speed and angular velocity of the ball, it is important that both of these quantities be independent of one another (Oda et al., 2003). Thus, the authors decided to develop a machine with the important functions (throwing freely and producing the breaking balls, speeds and direction a batter desires) that are currently missing from commercial pitching machines.

3. Overview of new type pitching machine

A new type of pitching consists of three rollers arranged around the circumference of a ball in its discharge position, and the rotary direction of the ball can be controlled over a full 360 degrees as the three rollers create three planes on a three-dimensional axis by variously changing the turn frequency of each roller. With this structure, various kinds of throws with variable pitch types and speeds, becomes possible. A schematic of the pitching machine that was developed by the authors is shown in Fig. 2(a), and a photograph of the machine is shown in Fig. 2(b).

This machine employs three rollers, which includes one more roller than the pitching machines typically found today. A ball for baseball is thrown with frictional force by the rubber tire, and this roller is installed around the circumference of a ball in its discharge position at 120° intervals. Three motors are installed, one in each respective roller, in which the number of revolutions can be adjusted from 0 to 3000 min-1, and these motors can be controlled independently. Additionally, this machine has a mechanism that can change the vertical angle Θ from -5° to 5° and the horizontal angle Φ from -6° to 6°, as shown in Fig. 2.

With the new type of pitching machine, which adopts these new mechanisms, a wide range of speeds (from 19.4 up to 44.4 m/s), pitch types (fastball, curveball or screwball) and variable directions, can be pitched as desired. Moreover, each motor is connected to a personal computer (PC) through a controller, and the number of revolutions of each of the motors can be controlled by the PC. In addition, this machine is equipped with various sensors that measure the number of revolutions N_1, N_2 and N_3 of three rollers, the vertical angle Θ the horizontal angle Φ and the initial velocity V of the pitched ball. The pitching machine is capable of throwing a ball with higher accuracy (vertically 200mm and horizontally 100mm) in a wide area at a variety of speeds, employing different pitch types compared to current pitching machines.

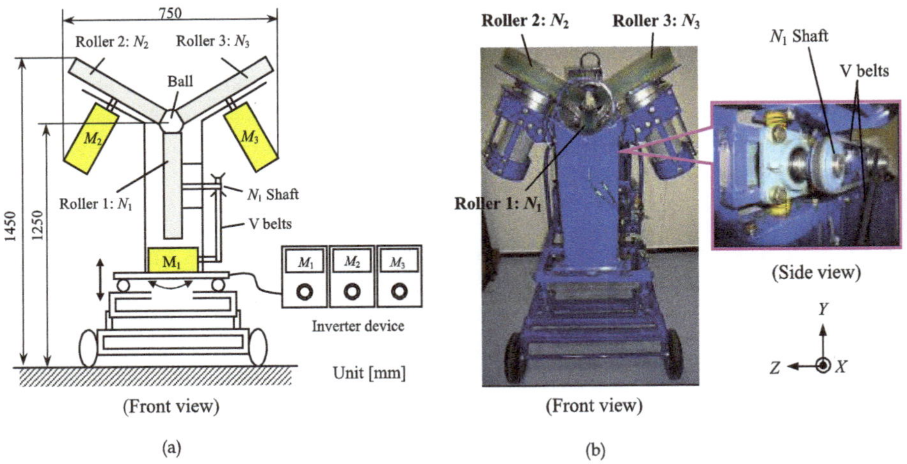

Figure 2. Three-roller type pitching machine. a) Schematic. b) Photograph.

4. Throw analysis of the three-roller type pitching machine

4.1. Finite element models and analysis conditions

In this study, the analysis models of only the roller parts and ball for baseball were made, because they were thought most important in the pitching machine for a thrown ball. Finite element models of a ball for traditional baseball and the roller part of the pitching machine with three rollers are shown in Fig. 3. Each of these measurements is shown in the figure. In this analysis, the aluminum alloy flange was very stiff compared to the ball and the rubber tire, and was treated as a rigid body. The ball was modeled with a viscoelastic agent in consideration of the dynamic characteristics of the ball. The characteristics of the ball were determined using a viscoelastic model of three elements, as shown in Fig. 4. This model shows the relationship between shear modulus $G(t)$ and time t, which are given in the following equation (Hendee et al. 1998, Nicholls et al., 2004, Sakai et al., 2008):

$$G(t) = G_\infty + (G_0 - G_\infty)e^{-\beta t} \tag{1}$$

where G is the relaxed shear modulus, G_0 is the instantaneous modulus, and β is the decay constant. Each material property in Eq. (1) was decided upon according to static and dynamic experiments and a finite element analysis (FEA) conducted on a ball. The material properties of these agents are shown in Table 1 (Nicholls et al., 2006). From the results of the analysis carried out beforehand, it was confirmed that the shape of the seam of the ball influences the throw more so than the materials of the ball. The seam of the ball has the same material properties as the ball's main body. The analysis for throwing the ball was calculated using dynamic finite element analysis code (ANSYS/ LS-DYNA, version 9.0, Theoretical Manual, 2002).

As a condition of the analysis, the initial velocity V_0=1.0 m/s and the initial angular velocity ω_0 =28.56 rad/s of the ball are given for all cases. The termination time was set to 0.1s from the moment of impact of the rubber tire with the ball to the ball's ejection. These analysis conditions were calculated from the image with the pitching experiments filmed using a high-speed video camera. Here, the friction coefficient μ =0.5 was determined from the surfaces in contact with the ball and rubber tire. The value of μ will be described in the next section.

In general, there are two possible spin directions for each ball: two-seam (the seam appearing two times per ball rotation) and four-seam (the seam appearing four times). Additionally, a new model of a ball without a seam (a spherical ball) was created in order to examine the effect of the seam. In this analysis, there were three kinds of pitch type locations, which are simulated as shown in Table 2 (a no-spin ball, a fastball and a curveball). N_1, N_2 and N_3 were revolutions per minute for each roller, and the three roller numerical grand total was fixed at 4500 min-1. The analysis was intended to start from the time the ball was thrown to just after release. Also, the flight trajectory of the ball after release was not considered (Watts et al., 1975, Himeno, 2001, Mizota, 1995).

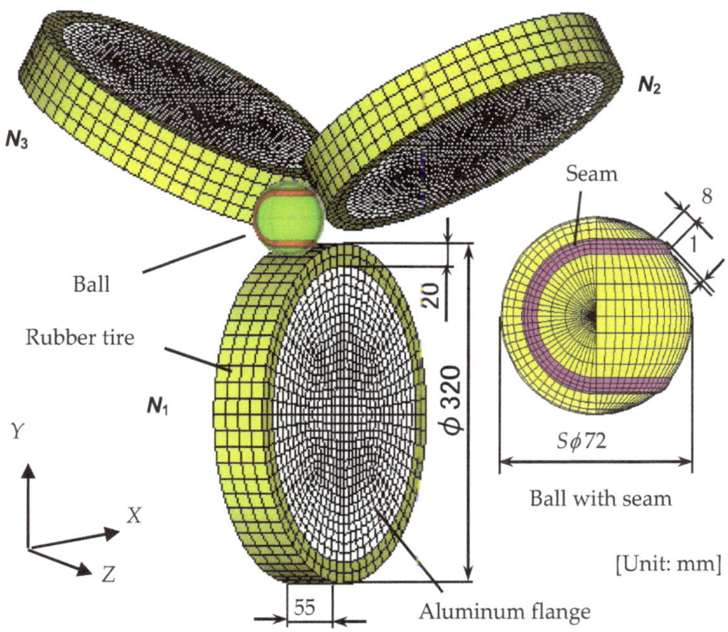

Figure 3. Finite element models of a ball with seams for baseball and three rubber rollers (15 104 elements).

In the analysis, a local coordinate axis was used as the starting point at the center position of a ball in release, as shown in Fig. 5. The projection vertical (θ) and horizontal angles (ϕ) were calculated by each ingredient of velocity V (V_x, V_y, V_z) after the throwing of the ball.

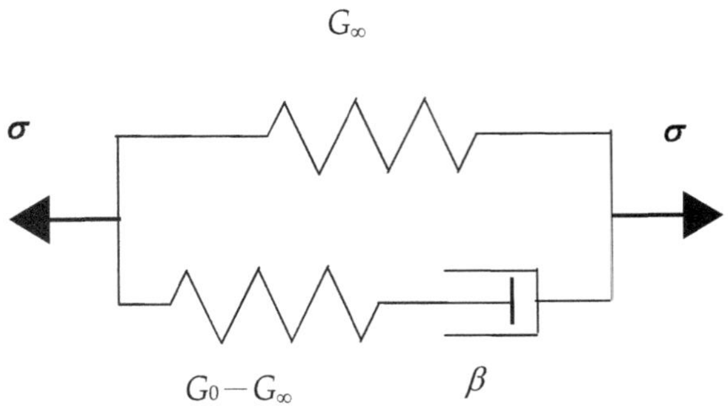

Figure 4. Viscoelastic model using three elements of the ball for baseball.

Property	Rubber tire	Baseball
Density, ρ (kg/m^3)	1 000	835
Young's modulus, E (MPa)	100	—
Poisson's ratio, ν	0.45	—
Istantaneous modulus, G_0 (MPa)	—	46.15
Relaxed shear modulus, G_∞ (MPa)	—	8.85
Bulk modulus, K (MPa)	—	100
Decay constant, β (s^{-1})	—	7 000

Table 1. Material properties of rubber tire and baseball (ball).

unit (min^{-1})

	N_1	N_2	N_3	N_1+N_2+N_3
Case 1 (No-spin ball)	1 500	1 500	1 500	4 500
Case 2 (Fast ball)	1 700	1 400	1 400	4 500
Case 3 (Curve ball)	1 325	1 750	1 425	4 500

Table 2. Analytical conditions: three pitch types (no-spin ball, fastball and curveball).

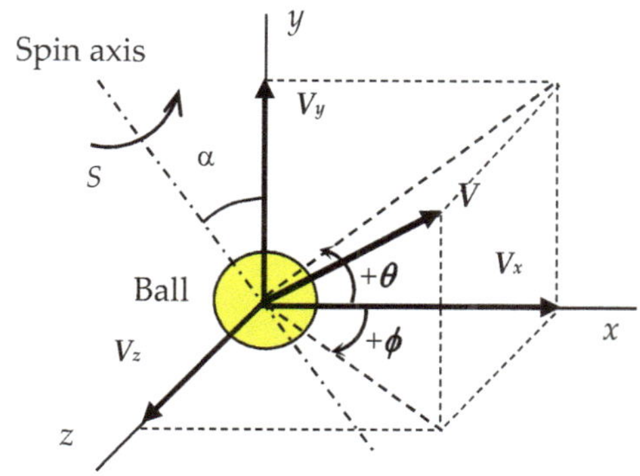

Figure 5. Coordinate system after pitching ball.

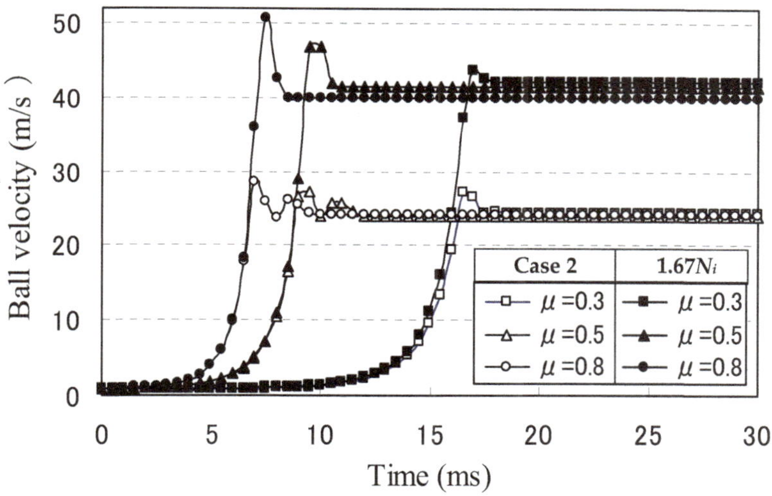

Figure 6. Time series of ball velocity V_x for four-seam fastball (case 2) after pitching.

4.2. Coefficient of friction μ between ball and three rollers

The value of the coefficient of friction μ range of 0.3-0.8 was analyzed. The results from letting μ for the three-roller type pitching machine vary between 0.3, 0.5 and 0.8 are shown in Fig. 6. This figure shows a time series for the x-direction velocity of a ball (V_x) in case 2 (four-seam fastball), and three rollers exhibited 1.67 revolutions. As a result, when the ball came into contact with the rollers, the velocity of the ball suddenly increased.

Conversely, it is understood that a pitched ball maintains a contact speed after its release without regard for value of μ. For this reason, the ball is in a slippery state during the early stages of its contact with the rollers, and contact time changes with changes in the value of μ.

In the moment that the rollers pitches to hold the ball, the contact state of the ball and rollers almost becomes an adherence state in the whole area. Therefore, the ball's course does not vary with the value of μ. From the results, the coefficient of friction μ was determined to be 0.5 (Sakai et al., 2007, Nicholls et al.,2004).

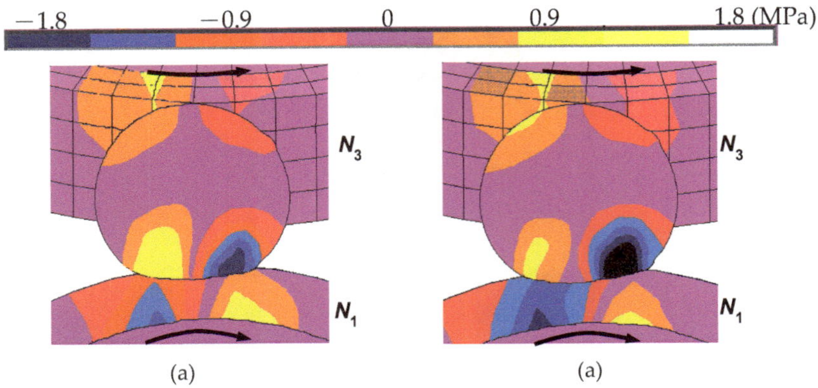

Figure 7. Shear stress distribution in x-y section of a pitched ball. a) No-spin ball. b) Fastball.

4.3. Analysis results and discussion

One example of the analysis results, the shear stress distribution in an x-y section of a pitched two-seam ball is shown in Fig. 7. In the figure, (a) and (b) show the pitch of a no-spin ball and fastball, respectively. It is understood that the absolute value for the fastball is more than the no-spin ball. For this reason, the number of turns of the N_1 roller is faster than other rollers in order to add back spin to a ball for the fastball. The shear stress values in the occurred contact surface neighborhood with the ball and the N_1 roller therefore becomes high.

A time series of the ball's x direction velocity V_x, when the two-seam ball was thrown by the three pitch types (no-spin ball, fastball and curveball) is shown in Fig. 8. The velocity-time curve was almost the same for all the pitch types. The ball's velocity suddenly rises when the ball begins to come into contact with the roller, and is pitched at an almost constant speed after release. However, while not shown here, this is almost the same as the other pitch types and different seam postures. The speed of a pitched ball was understood not to be influenced by the pitch type. Additionally, when the number of a roller is n, the outside radius of a roller is R (m), the relationship between the number of revolutions of each roller N_i (min-1) and the ball's speed V (m/s) can be given by the following equation:

$$V = \frac{\sum_{i=1}^{n} VR_i}{n} = \frac{2\pi R}{60n} \sum_{i=1}^{n} N_i \tag{2}$$

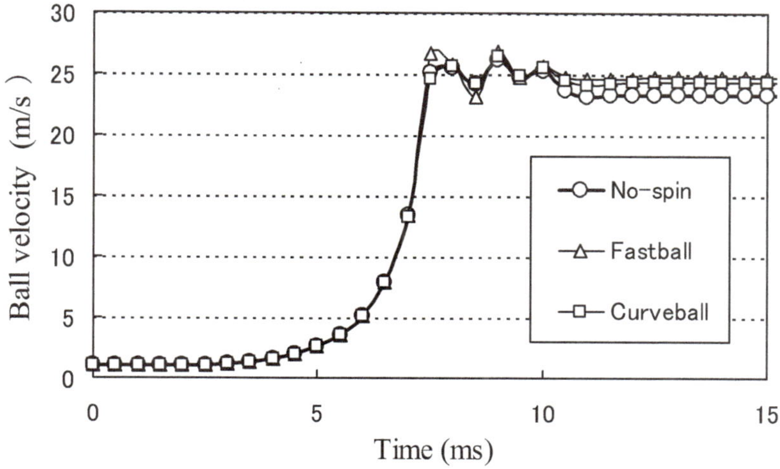

Figure 8. Time series of ball velocity V_x in two-seam ball during release

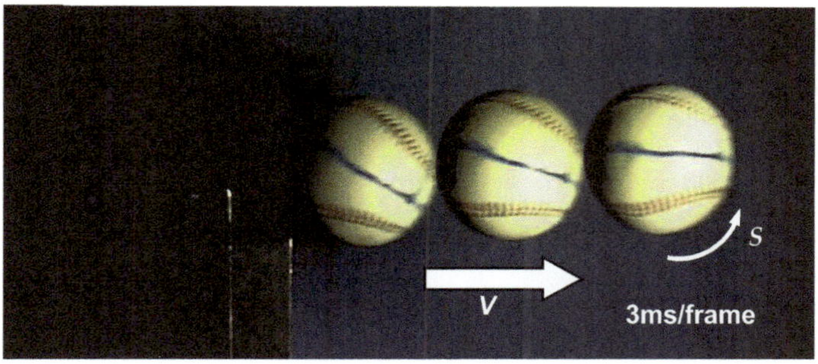

Figure 9. Flight behavior of ball after release using high-speed video camera in Case 2 (Fastball, V = 25.1 m/s.

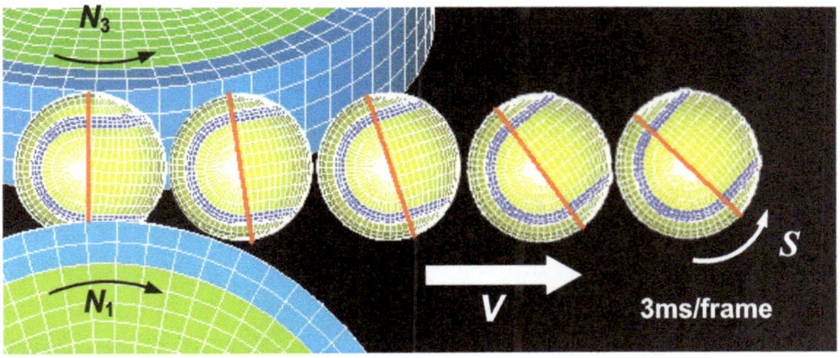

Figure 10. Flight behavior of ball after release by FEA result in Case 2 (Fastball, V=24.7 m/s).

4.4. Comparison of results of throw experiment and analysis

The three-roller type pitching machine shown in Fig. 2 was used, and it was tested to throw a ball in the throw condition shown in Table 2. The state of the throw experiment filmed the behavior of the ball just after release at intervals of 2000 fps (frames per second) using a high-speed video camera (MEMRECAM fx-K3, made by nak Image Tech. Inc.). One example of the throw experiment results, a flight image of the pitch of a four-seam fastball, is shown in Fig. 9 (Nicholls et al., 2003, Chu et al., 2006, Sakai et al., 2007, Takahashi et al., 2008). As well, a state of the flight of the FEA that pitches the ball in the same condition in Fig. 10 is shown.

From both figures, the spin axis of the ball is the axis which is inclined to wards the Z-axis (side of a paper plane) from the vertical direction. It is understood that the ball is pitched spinning around its axis. The speed of the pitched ball V and spin rate S were calculated from both images. The experimental and analytical values were almost the same. Additionally, similar results were acquired in the experiments and the simulation (FEA) of other pitch types and seam postures. These results showed good agreement between the experiments and the FEA.

Figure 11 shows the comparison of the spin rate S of each pitch type when we threw a ball with two- or four-seams in the experiments, as well as the analysis results. The spin rates of the experiment and analysis values are understood to be almost the same in both pitch types. Additionally, a pitch type is decided by the number of turns of the three rollers. It is understood that there are few differences associated with seam posture (Jinji, 2006).

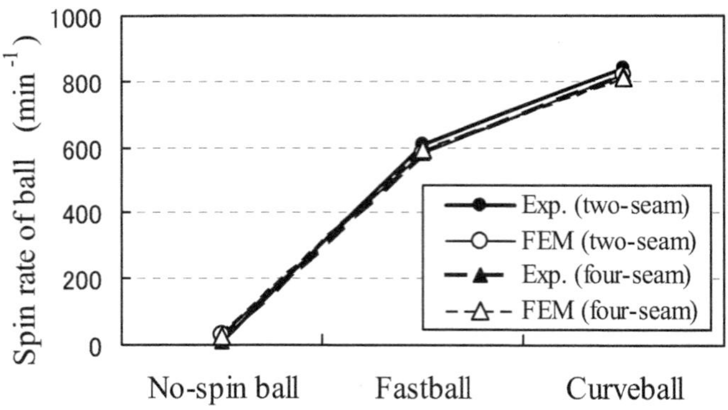

Figure 11. Spin rate of ball after throw for three pitch types.

Figure 12 shows the comparison projection horizontal angle ϕ after a throw of each pitch type. From the results, it is understood that the experiment and analysis values are different by pitch type and seam posture. In this case, the analysis and the experiment values were each compared with each seam posture (two- and four-seam), and the absolute experimental value was slightly greater than the analysis one within the three pitch types. Also, it is understood that both values become almost equal with minor differences.

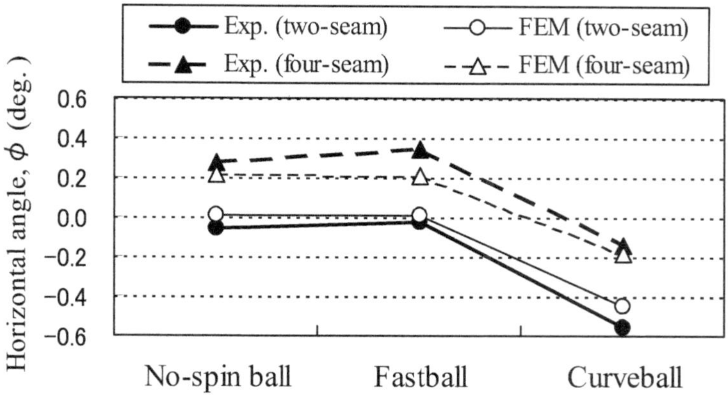

Figure 12. Horizontal angle ϕ of ball after throw for three pitch types.

Conversely, the difference in value between the two- and four-seam balls in the experiment was from 0.3 to 0.4 degrees for each pitch type. Here, in this throw condition (the initial ball-velocity 25m/s), when a difference of ϕ was 0.2 degrees, the trajectory of the pitched ball to a home plate is about 70mm (the size of one ball) slip off in the side direction in theory. This difference is thought is a significant factor contributing to the fall in throw accuracy by the seam of a ball for baseball (Frohlich, 1984, Mehta, 1985, Watts et al., 1975, Mizota et al., 1995, Alaways et al., 2001).

From the results of the throw experiments and simulations, the value of the speed, spin rate and projection angle in the pitched ball are almost the same. It is thought that the analysis model and its results are proper.

5. Analysis of models that changed roller outer diameter, *D*

5.1. Analysis models and conditions

The roller outer diameter ϕ 320mm (standard model; D320) for the present experiment was changed, the analysis models of ϕ 220mm (D220) and ϕ 420mm (D420) were made the roller of the small and large diameter. Both the analysis models are shown in Figs. 13 and 14, respectively. Here, the roller materials and roller distance are the same as the analysis model in the previous section.

In the analysis conditions, the speeds and pitch types of the pitched balls are the same as in the standard model D320. The circumference velocity VR_i of each roller was calculated from the number of each roller turns as shown in Table 2.

Next, the number of each roller turn N_i (min-1) was calculated so that VR_i in the case of D220 and D420 became equal. Both the models were set to the number of those turns. Additionally, the initial velocity of the ball was the same as stated in the previous section, and the throw analysis was then carried out.

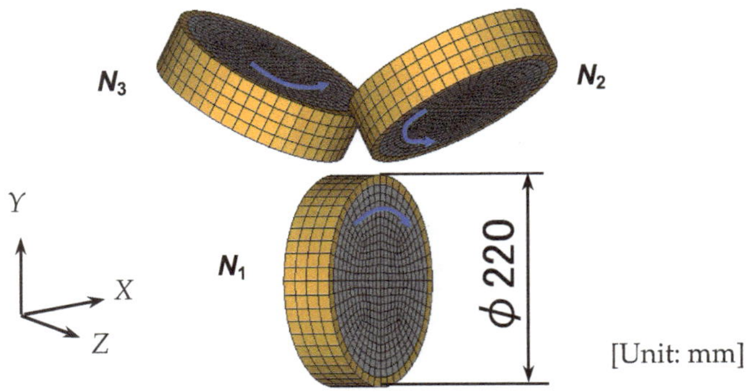

Figure 13. Finite element models of small rollers (D220).

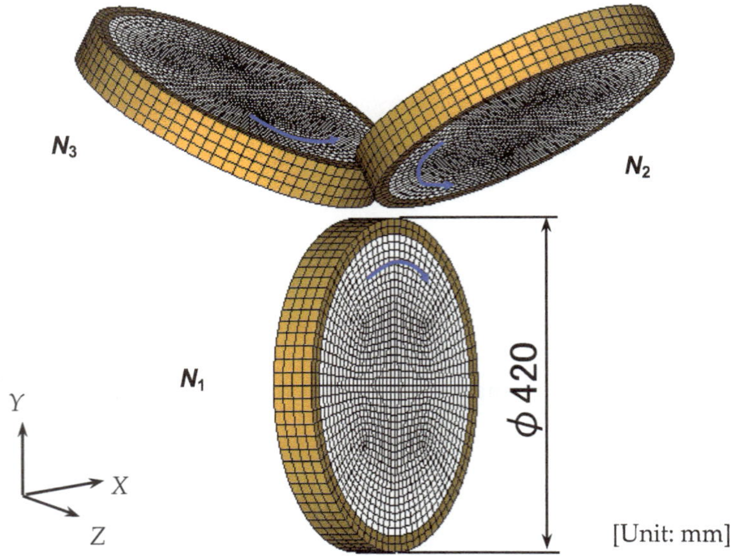

Figure 14. Finite element models of large rollers (D420).

5.2. Results and discussion

One example of the analysis results, the time-history of a two-seam fastball's velocity for each model, is shown in Fig. 15. The number of three rollers turn of each model increased to 1.67 times (ex. D220H), is shown in the same figure. As shown in the figure, there is a small difference in the time at which the ball comes into contact because each roller's the diameter was different. A similar velocity-time curve is drawn for all the models.

Otherwise, the speed of a pitched ball after the throw was almost constant in all models. This is not shown here. This result was identical to when a four-seam ball was thrown. The ball speed was decided almost entirely by the circumference velocity of the roller, and clearly does not depend on the roller's outer diameter.

Figure 16 shows the comparison of several rollers on the spin rate in two-seam balls. The spin rate of the pitched curveball in D220 is compared with other models that are slightly lower. It is understood that the spin rate in D320 and D420 is almost equal for all pitch types. The spin rate is an important factor in deciding a pitch types (breaking balls), because the throw performance of the D220 is lower in comparison with other rollers.

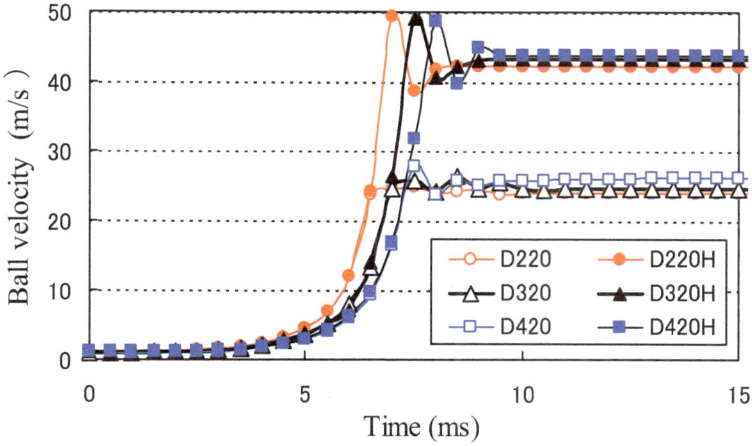

Figure 15. Time series of ball velocity for two-seam fastball during release on several rollers *D*.

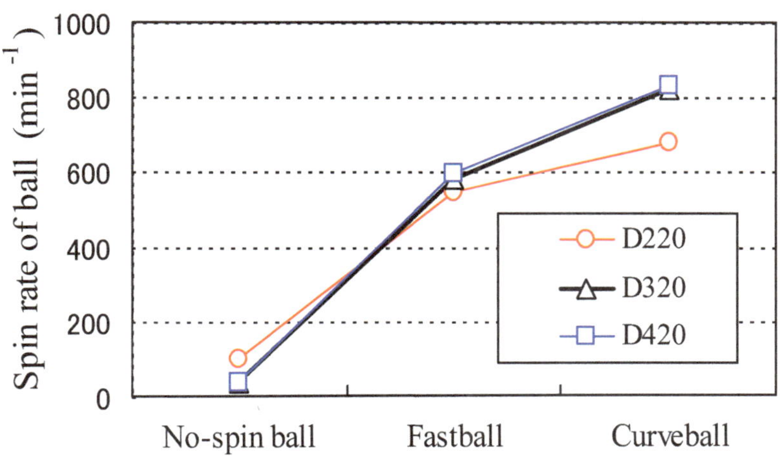

Figure 16. Comparison of several rollers on spin rate of two-seam ball.

Next, throw accuracy by seam posture was evaluated quantitatively. A difference of a vertical and horizontal projection angles θ and ϕ after a ball is thrown is defined as $\Delta\theta$ and $\Delta\phi$ in the next two equations, respectively:

$$\Delta\theta = \left| \theta_2 - \theta_4 \right| \tag{3}$$

$$\Delta\phi = \left| \phi_2 - \phi_4 \right| \tag{4}$$

where the suffix 2 or 4 expresses two- or four-seam balls, respectively. If the values of $\Delta\theta$ and $\Delta\phi$ are small, the throw accuracy of the pitching machine is higher with respect to the seam position and posture.

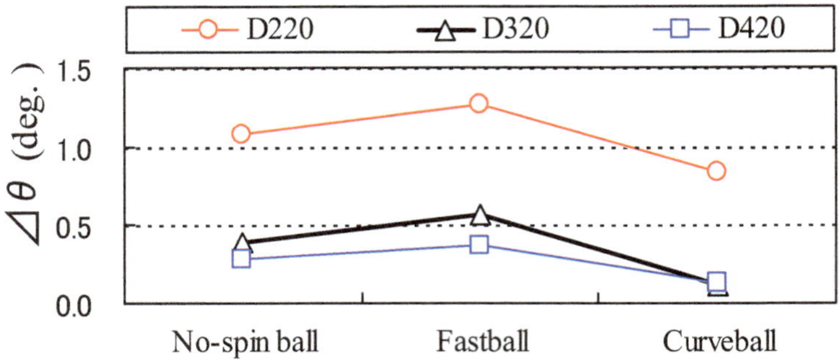

Figure 17. Comparison of several rollers on $\Delta\theta$.

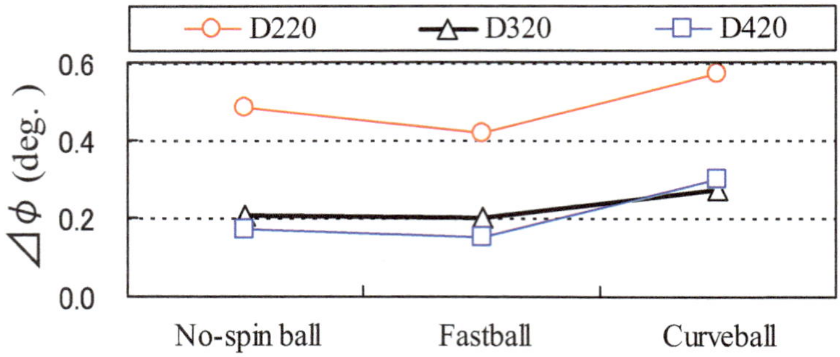

Figure 18. Comparison of several rollers on $\Delta\phi$.

Figures 17 and 18 compare $\Delta\theta$ and $\Delta\phi$ for each model when pitched respectively for each pitch type. In the case of the fastball pitch, in comparison with other pitch types, it is understood that its value in all models is the largest. In the case of the compared models, the value of $\Delta\theta$ in D220 is bigger for all pitch types, and it is understood that D320 and D420 are smaller. Conversely, the value of $\Delta\phi$ in the curveball tends to become slightly larger than other pitch types without relation to the model. $\Delta\phi$ is smaller, so the roller diameter becomes bigger. Additionally, it is understood that there is not any difference in D320 and D420.

From the results of the throw analysis where the roller diameter D was changed, the spin performance and the throw accuracy are poor in D220 and the outer diameter is small. It is understood that the performance of D420, with the largest outer diameter, was the best.

In general, a roller with a big diameter is used, a motor capable of producing a large output becomes necessary. In this case, the pitching machine in itself becomes large-scale, and its gross weight increases. However, if such problems can be solved, in terms of throw accuracy, it is advantageous to make the diameter of the roller larger.

When these things are considered generally, there are a few differences regarding the throw performance of the D420 and D320. The roller diameter of the three-roller type pitching machine is regarded as the measurement that used ϕ320mm present are practical.

6. Analysis of models that changed radius of roller distance, *r*

6.1. Analysis models and conditions

A roller-type pitching machine is a machine in which a ball is picked up by the roller, and is thrown. The performance of roller-type pitching machines changes based on the radius r of each roller. Thus, the roller distance of the present machine (D320, radius r =25.1 mm) was as shown in Fig. 19. The roller distance of the models was analyzed which may have been narrow or broad. Also, the throw analysis was the same as in the previous section.

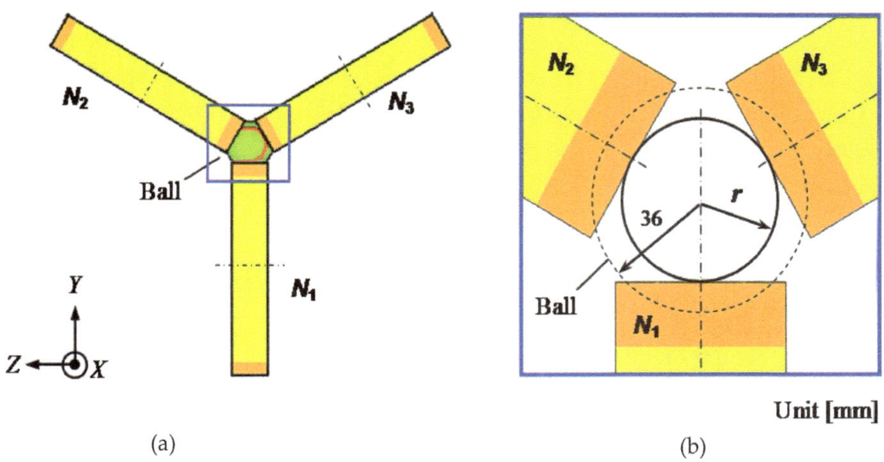

Figure 19. Radius of each roller distance, r a) Ball and three rollers. b) Close-up view.

6.2. Results and discussion

One example of the analysis results, the velocity of two- and four-seam fastballs of each model, is shown in Fig. 20. In this figure, the ball speed is understood to not be influenced by the seam posture in any of the cases. Additionally, the speed of the pitched ball is fixed at about 25m/s within the range of 23.1mm < *r* < 27.1mm. The speed rises with increases in *r*, and the greatest speed was approximately *r* = 31.1mm. The ball speed suddenly decreases when *r* > 31.1mm. It is thought that the ball is not held strongly enough between the rollers when *r* > 31.1mm. Figure 21 shows the relationship between $\Delta\theta$, $\Delta\phi$ and *r* from the analysis results. It is understood that $\Delta\theta$ is at its minimum value at 0.19 degrees when *r* =33.1mm, and $\Delta\phi$ is moves towards its minimum value at 0.10 degrees when *r* = 27.1mm.

From these results, when the speed of a pitched ball thrown from a three-roller type machine is considered, it is thought that the radius of a roller distance *r* in the range of 23.1mm < *r* < 33.1mm is practical. If *r* is limited to this range, the throw accuracy becomes highest at approximately *r* = 27.1mm when $\Delta\theta$ and $\Delta\phi$ are both at their minimum values.

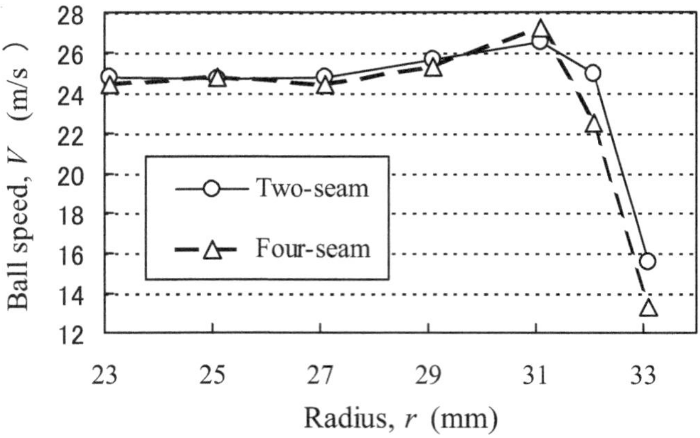

Figure 20. Relationship between ball speed and of each roller distance, *r*.

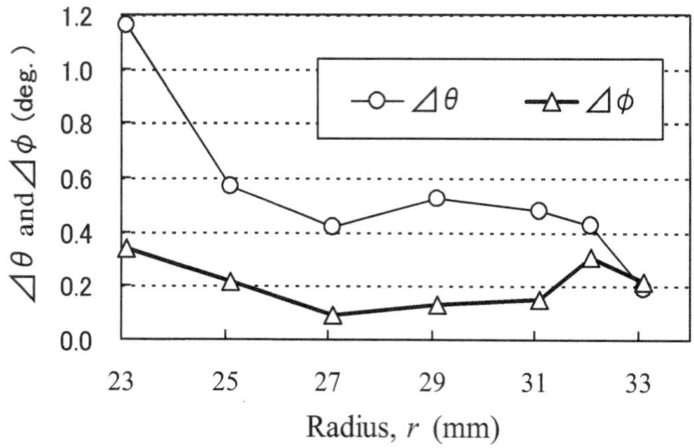

Figure 21. Relationship between $\Delta\theta$, $\Delta\phi$ and of each roller distance, *r*.

In theory, the control accuracy of the pitched ball by a pitching machine in this case is entered in the vertical and horizontal errors at 140mm and 50mm, respectively. Therefore, the measurements needed to create a new type of pitching machine have been achieved. (Mizota et al., 1995, Himeno, et al., 1999, Nathan, 2008).

7. Throw analysis changed roller shape

7.1. Analysis models and conditions

In order to throw a two- or four-seam ball, it is necessary to prepare the posture of the ball artificially. Thus, the roller developed by us shows that the throw accuracy is not affected by a seam and its posture (the robustness roller). We let the shape change in the roller section. Above, the roller section used was a flat type roller (flat roller) of a rectangle. For the concave type roller the central part of which where the ball touched became hollow (see Fig. 22), while the convex type roller which swelled (see Fig. 23) were devised. The radius of curvature R=100mm, and the distance from the center to the outside surface of the roller (the radius of each roller distance) was r=25.1mm in both models.

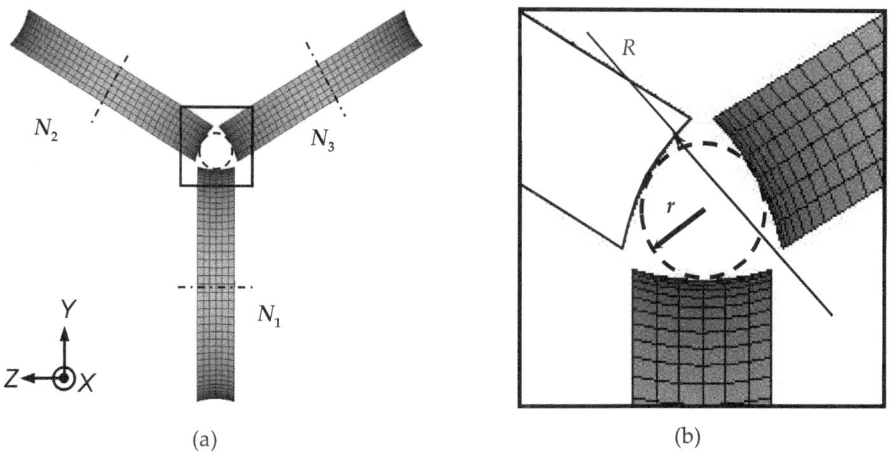

Figure 22. Concave type roller. a) roller shape. b) Close-up view.

7.2. Analytical results and discussion

The analytical results of a pitched fastball are shown in Figs. 24, 25 and 26. Figure 24 shows a time series of the two-seam ball's velocity. The velocity-time curve was almost the same for all roller types. The ball's velocity suddenly rises when the ball begins to come in contact with the roller, and is pitched at an almost constant speed after release. However, while not shown here, this is almost the same as other pitch types and different seam postures. The speed of a pitched ball was understood not to be influenced by the roller shape.

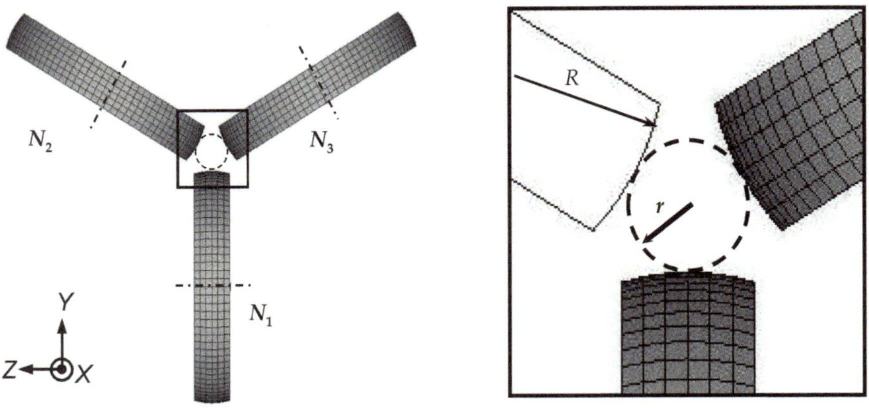

(a) (b)

Figure 23. Convex type roller. a) roller shape. b) Close-up view.

On the other hand, the difference of the vertical angle $\Delta\theta$ (see Fig. 25) and horizontal angle $\Delta\phi$ (see Fig. 26) with several rollers in the fastball pitch changed based on seam posture and roller shape dramatically. In both figures, it is understood that the variation in the range of the convex type roller by seam posture was smaller than the other roller types. Here, the value of difference of both seams (two- and four-seam) was small, meaning that the throw accuracy is higher.

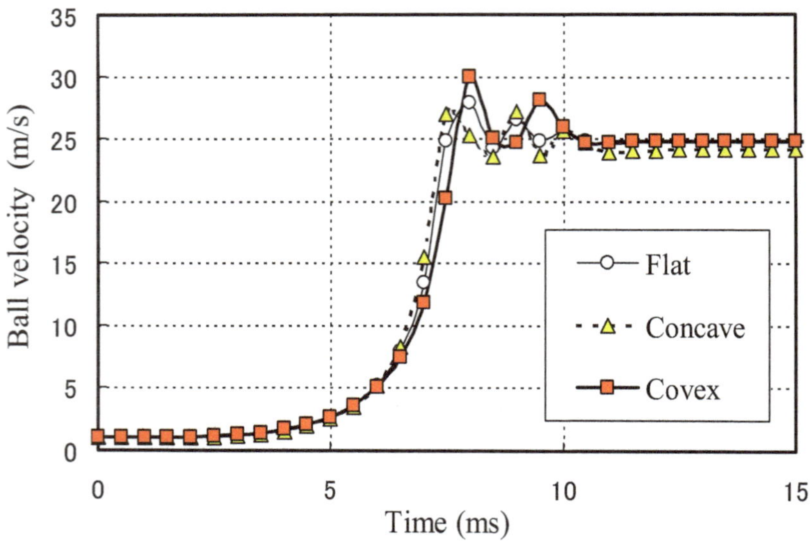

Figure 24. Time series of the ball velocity with several rollers for two-seam fastball.

In particular, the value of difference of horizontal angle $\Delta\phi$ becomes a factor of a pitch which hits the batter in an actual baseball game. Therefore, the value of $\Delta\phi$ of a commercial pitching machine is an important factor in the throw accuracy. The throw accuracy of the convex type roller is higher than the flat and concave type rollers.

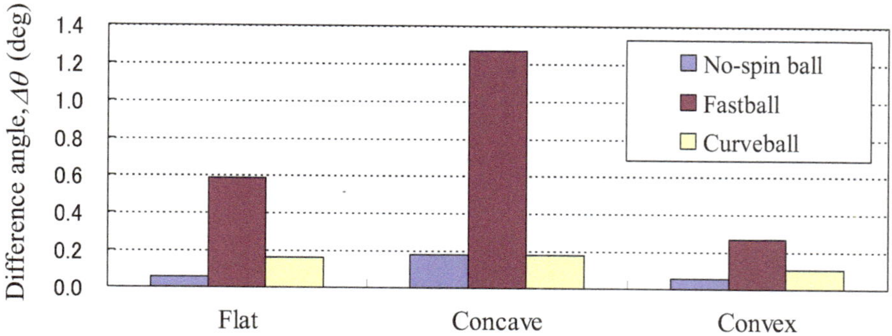

Figure 25. Difference of vertical angle $\Delta\theta$ with several rollers for three pitch types.

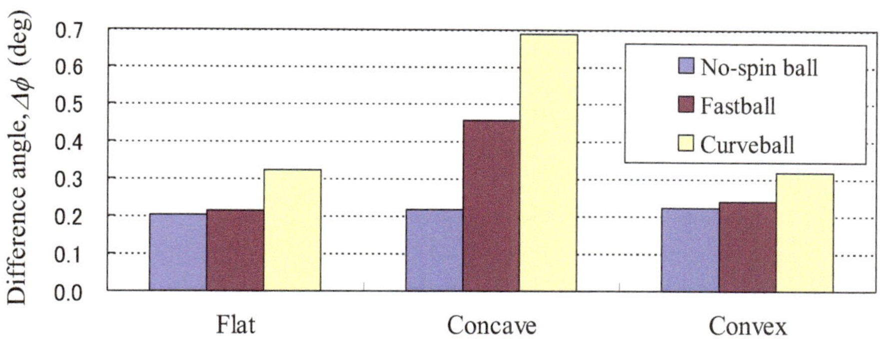

Figure 26. Difference of horizontal angle $\Delta\phi$ with several rollers for three pitch types.

8. Optimum design of roller

8.1. Optimization problems of roller

From the results in the previous section, when the roller shape is changed into convex from a flat design, the throw accuracy may become higher than the existing the flat type roller. In this section, the dynamic finite element analysis (FEA) is used, and the optimum design of the roller improving the throw accuracy is tried. Young's modulus E of the roller, the radius of curvature R and the radius of each roller distance r are the three design variables in Fig. 23. There is the optimization problems in designing rollers that have a different projection angle formed by differences in the seam posture becoming the smallest.

8.2. Response surface method and optimization

In order to obtain the optimum design, as evaluation of the roller optimization corresponding to the design variables, a difference of the projection horizontal angle ($\Delta\phi$) according to the two- and four-seam postures was decided upon. Objective function f_ϕ uses a difference of an argument regarded as most important to the throw accuracy. f_ϕ is minimized so that the effect of the seam becomes small, as in the equation given below:

$$Minimize \quad f_\phi = \Delta\phi + \eta \tag{5}$$

where η is a penalty coefficient. In the case of minimal spin rate S (min-1), not all pitch types are thrown. Thus, this consideration was removed from the optimum solution. It is prescribed in the next condition:

$$\left. \begin{array}{ll} S_2 \geq 400 \; and \; S_4 \geq 400 & \rightarrow \eta = 0 \\ S_2 < 400 \; or \; S_4 < 400 & \rightarrow \eta = 1 \end{array} \right\} \tag{6}$$

where 2 and 4 indicate the two-seam and four-seam, respectively. Eq. (5) is derived by using Response Surface Methodology (Myers et al., 1995, Khuri et al., 1996). The response surface is built by using a response surface tool: RSMaker for Excel (Todoroki, 2010). The response surface dispersion is shown in the standard three design variables (R, r, E) in Table 3. All analyses for 3×3×5 (45 ways) in Table 3 were executed. Additionally, the pitch type decided on was the curveball, because the change of the projection horizontal angle ϕ is the biggest over the other pitch types from the analytical results in the previous section.

Radius of curvature R [mm]	Center distance r [mm]	Young's modulus E [MPa]
36	22.1	10
		20
72	25.1	50
		100
100	30.1	500

Table 3. Design variables (1st analysis).

Radius of curvature R [mm]	Center distance r [mm]	Young's modulus E [MPa]
100	24.1	30
	24.6	40
	25.1	50
	25.6	60
	26.1	70
		80

Table 4. Design variables (re-analysis).

From the results, many peaks are recognized by the provided response values, and it was difficult to provide similar accuracy in all design domains. Thus, the response surface by the interpolation calculation was made, and the optimum point candidate was predicted. One example of the response surface made in interpolation calculation using the graph software ORIGIN is shown in Fig. 27 (E=50MPa). It is predicted that the optimum point for r=25mm and R=100mm. It is observed once again in the surrounding region that the zoomed response surface is made, and the optimum point is found in the small region. In the re-analysis, R was fixed at 100mm, the design variables were the two r and E (cf. Table 4). The reason for this is because sensitivity for the objective function was sensitive both to r and E in the range of R>72mm.

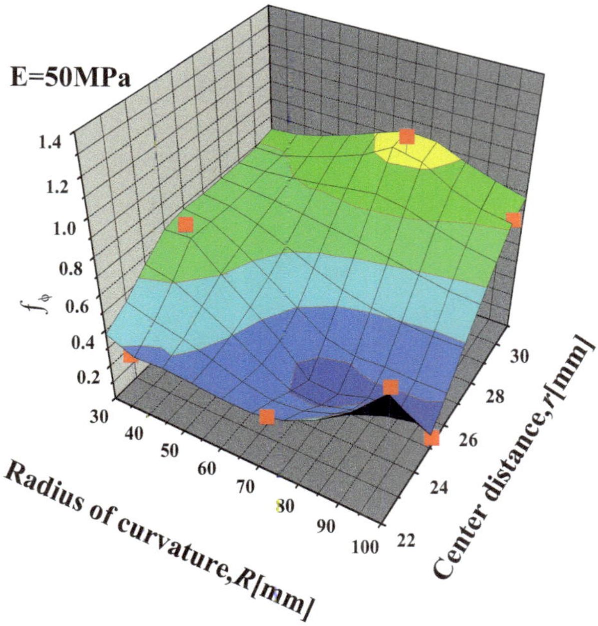

Figure 27. Response surface of interpolation calculation for curveball (E =50MPa).

The enlarged response surface demanded by the third multinomial expression approximation and the optimum point (seal ○) are shown in Fig. 28. In this way, the optimum condition of the convexity roller (r =25.4mm, E =52.4MPa) was pursued. Additionally, the re-analysis was executed in these conditions once again, and it was confirmed that these values are the optimum values. From Fig. 28, it is understood that the optimum point neighborhood is a gentle curved surface and range.

Next, when the pitch type was changed mid-pitch from a fastball to a curveball, roller geometry was optimized such as for the curveball. The zoomed response surface for the fastball is shown in Fig. 29 (seal ○ is the optimum point in the curveball). The changed values of both design variables are small in comparison with the curveball. Also, the optimum point was almost the same as the value pursued in the curveball.

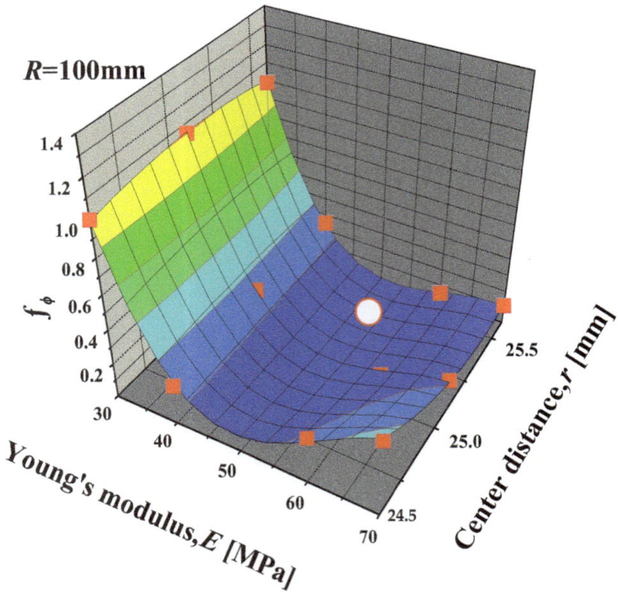

Figure 28. Zoomed response surface and optimum solution for curveball (*R* =100mm)

We think that the optimum value provided by the curveball is close to the optimum value of other pitch types.

8.3. Throw experiment using optimized convex type roller and throw accuracy evaluation

In the previous section, the optimum condition of the convex type roller was decided. The proprieties of these analysis results are examined. The convex type roller pursuing the optimum condition has been produced, and the pitching machine using its rollers was tested to throw the ball. The convex type roller produced the shape geometry that is the radius of curvature *R*=100mm and rubber as a soft material (Young's modulus, *E* =52MPa). This is called the soft convex type roller. In Fig. 23, it was a set of three of these rollers, and it was adjusted to become the center distance *r*=25.4mm.

The state of the throw experiment filmed the behavior of the ball just after release at intervals of 2000 fps using the high-speed video camera. For comparison of the roller shape, the produced soft convex type roller was filmed. The throw condition was experimented on in the same way as the analysis shown in Table 2, in order to compare it with the FEA.

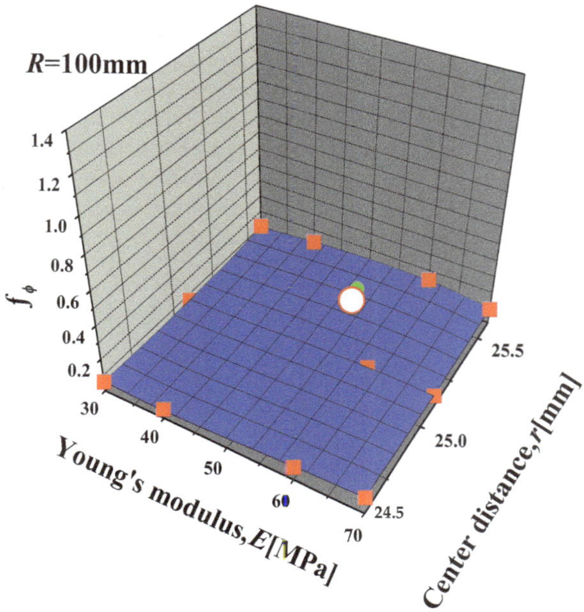

Figure 29. Zoomed response surface and optimum solution for fastball (*R* =100mm)

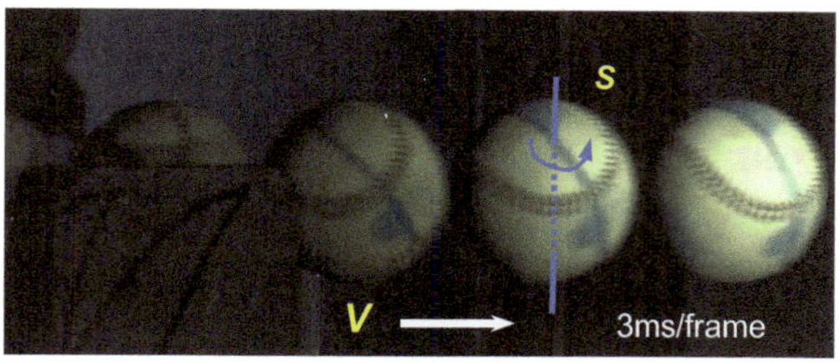

Figure 30. Flight behavior of ball after release using high-speed video camera for four-seam curveball by convex roller (*V*=25.3 m/s, *S*=820min-1).

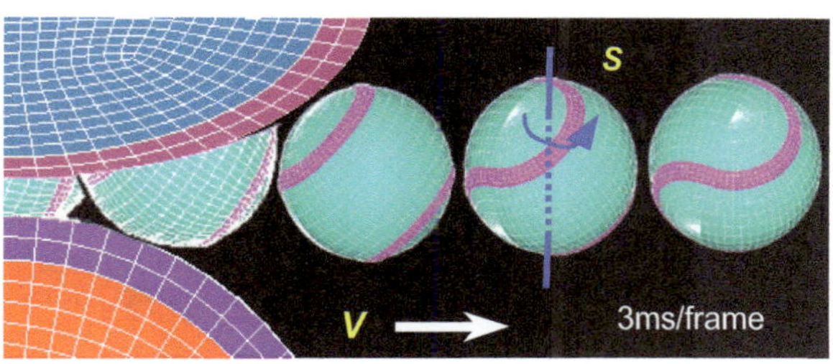

Figure 31. Flight behavior of ball after release by FEA results for four-seam curveball by convex roller (*V*=24.9 m/s, *S*=810min-1).

One example of the throw experiment results with the soft convex type roller, a flight image of the pitch of a four-seam curveball, is shown in Fig. 30. Otherwise, a state of the flight of the FEA that pitches a ball in the same condition in Fig. 31 is shown. From both figures, the spin axis of the ball is the axis which is inclined to the Z-axis (side of a paper plane) from the vertical direction. It is understood that the ball is pitched spinning around its axis. The speed of a pitched ball V and number of spins S were calculated from both images. The experimental and analytical values were almost the same. Additionally, similar results were acquired in the experiments and the simulation (FEA) of other pitch types and seam postures. These results show good agreement between the experiments and the FEA.

Figure 32 compares the roller geometry of the experiment and analytical values of the balls' speed V for the two pitch types (fastball and curveball) using the four-seam ball. It is understood that the balls' speed in the convex type roller increases a little more than in the flat type roller without a relation to pitch types. Its velocity increases by about 3%. These results are the same as the former FEA results (cf. Fig. 24).

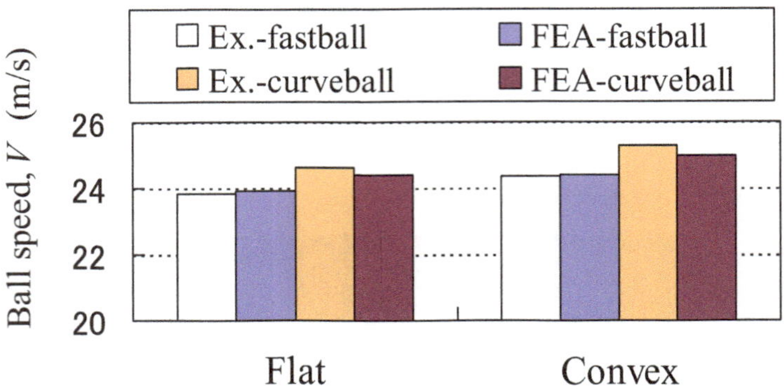

Figure 32. Comparison of roller geometry on ball speed V for four-seam ball after pitching.

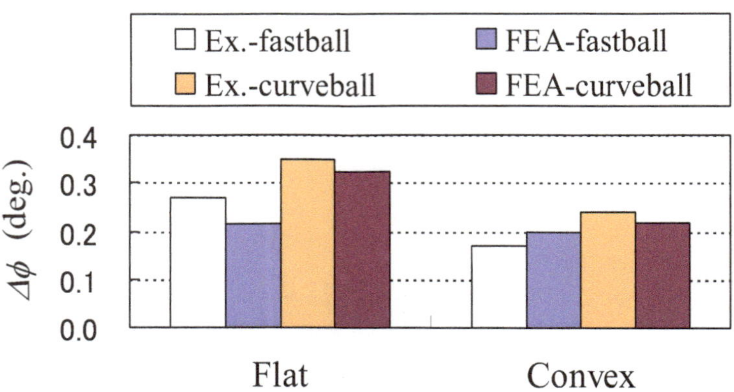

Figure 33. Comparison of roller geometry on the difference of horizontal angle $\Delta\phi$

Additionally, the experiment and analytical values were compared for roller geometry. The experimental and analytical values were almost the same. It was identified that the reproduction of the throw of the ball was experimented on several times.

On the other hand, the difference of the projection horizontal angle ($\Delta\phi$), which is one of the important performance considerations of a pitching machine, was considered in order to evaluate the throw accuracy of the machine by the seam posture quantitatively. If the value of $\Delta\phi$ is small, the throw accuracy of the pitching machine is higher with respect to the seam position and posture. The fastball and curveball were thrown using the two- and four-seam balls, respectively, and the projection horizontal angle ϕ was measured for each pitch type. The throw experiments were tested with two kinds of rollers of the flat and soft convex types.

Figure 33 shows the difference of $\Delta\phi$ values which were calculated in Eq. (4) from the values measured in these experiments. The values of the FEA in the throw condition are the same as the experiment shown in the figure. The values of the experiments were compared with the roller shape. The convex type roller showed smaller values than the flat type regardless of pitch type. The convex type roller was affected by the improvement in throw accuracy.

Finally, the analysis accuracy was described. In Fig. 33, each ball type is shown, and the $\Delta\phi$ of the analysis and experiment are compared, respectively. This included some errors which the experiment values (less than 0.07 degrees) considered. It is thought that the analysis accuracy is high enough. From these results, the validity of the computational model and its results in this study were confirmed.

9. Conclusion

In this chapter, the throw simulation using the finite element analysis (FEA) and experiments analyzed the improvement in the throw accuracy of three-roller type pitching machines for the purpose of studying the effect of the seam on the behavior and throw accuracy of the ball. The analytical models of the changes in the outer diameter, the distances and the form of the roller were made. The roller based on the convex shape was suggested, and the materials and the shapes of its roller were optimized. Additionally, throwing the ball using its optimized roller was experimented on, and the propriety of the analysis results was inspected. The conclusions provided are as follows:

1. In the roller type baseball pitching machine, the seam of a ball does not influence the ball's speed or spin rate after release, while the effects of the seam on the projection angles $\theta$and ϕ vary.
2. The throw accuracy of the three-roller type pitching machines was improved through the utilization of a larger roller diameter.
3. In the case of the flat roller, the radius of the three-roller distance is 27.1mm, the seam of a ball increase, and the throw accuracy of the pitching machine is improved.
4. The soft convex type roller (R=100mm and E=52.4MPa), which is the optimized roller, does not have an affect on the seam of a ball, and the throw accuracy and robust are high.

Author details

Shinobu Sakai*
School of Mechanical Engineering, Kanazawa University, Japan

Hitoshi Nakayama
Development Division, Sunaga Kaihatsu Co., Ltd., Japan

Acknowledgement

Great assistance was given to this study by Mr. Yuichiro KITAGAWA and Hirotoshi KAKUDA, who were seniors at this university while it was being conducted. For their help in making the finite element models and test of the machine, the authors wish to express great thanks. Additionally, concerning the three-roller type pitching machine described. Here, including the structure and control method, Japanese patent applications (No. 2001-45941 and No. 2007-301892) have already been filed at the Japan Patent Office.

This work was supported by Grants-in-Aid for Scientific Research (Grant No. 24560255) from the Ministry of Education, Culture, Sports, Science & Technology (MEXT) in Japan. The authors would like to thank the support from MEXT.

10. References

Adair, R.K. (1994). The Physics of Baseball, *Harper Perennial*, New York

Alaways, L.W., Mish, S.P. & Hubbard, M. (2001). Identification of release conditions and aerodynamic forces in pitched-baseball trajectories, *Journal of Applied Biomechanics*, Vol. 17, No.1, (2001), pp. 63-76

Alaways, L.W. & Hubbard, M. (2001). Experimental determination of baseball spin and lift, *Journal of Sports Sciences*, Vol. 19, No.5, (2001), pp. 349-358

ANSYS/LS-DYNA Theoretical Manual (2002), SAS IP Inc.

Chu, W.T., Wang, C.W. & Wu, J.L. (2006). Extraction of baseball trajectory and physics-based validation for single-view baseball video sequences, *IEEE International Conference on Multimedia and Expo*, ICME 2006, Art. No. 4036974, (2006), pp. 1813-1816

Frohlich, C. (1984). Aerodynamic drag crisis and its possible effect on the flight of baseballs, *American Journal of Physics*, Vol. 52, No.4, (1984), pp. 325-334

Hendee, S.P., Greenwald, R.M. & Crisco, J.J. (1998). Static and dynamic properties of various baseballs, *Journal of Applied Biomechanics*, Vol.14, No.4, (1998), pp. 390–400

Himeno, R., Sato, S. & Matsumoto, H. (1999). CFD study of the effect of seam on the flow around a baseball ball, *RIKEN Review, Focused on High Performance Computing in RIKEN*, No.25, (1999), pp. 124-127

Himeno, R. (2001). Computational Study of Influences of a Seam Line of a Ball for Baseball on Flows, *Journal of Visualization*, Vol. 4, No.2, (2001), pp. 197-207

* Corresponding Author

Jinji, T. & Sakurai, S. (2006). Direction of spin axis and spin rate of the pitched baseball, *Sports Biomechanics*, Vol. 5, No.2, (2006), pp. 197-214

Khuri, A.I. & Cornell, J.A. (1996). Response Surfaces; Design and Analysis, *Marcel Dekker*, New York, (1996)

Mehta, R. D. (1985). Aerodynamics of sports balls, *Annual Review of Fluid Mechanics*, Vol.17, (1985), pp. 151-189

Mish, S.P. & Hubbard, M. (2001). Design of a Full Degree-of-freedom Baseball Pitching Machine, *Journal of Sports Engineering*, Vol. 4, No.3, (2001), pp.123-133

Mizota, T., Kuba, H. & Okajima, A. (1995). Erratic Behavior of Knuckle Ball (1) Quasi-steady Flutter Analysis and Experiment, *Journal of Wind Engineering*, Vol. 62, No.62, (1995), pp. 3-13

Mizota, T., Kuba, H. & Okajima, A. (1995). Erratic Behavior of Knuckle Ball (2) Wake Field and Aerodynamic Forces, *Journal of Wind Engineering*, Vol. 62, No.62, (1995), pp. 15-22

Myers, R.H. & Montgomery, D.C. (1995). Response Surface Methodology: *Process and Product Optimization Using Designed Experiments*, John Wiley & Sons. Inc., (1995)

Nathan, A.M. (2008). The effect of spin on the flight of a baseball, *American Journal of Physics*, Vol. 76, No.2, (2008), pp. 119-124

Nicholls, R.L., Miller, K. & Elliott, B.C. (2003). Bat kinematics in baseball: implications for ball exit velocity and player safety, *Journal of Applied Biomechanics*, No.19, (2003), pp. 283-294

Nicholls, R.L., Miller, K. & Elliott, B.C. (2004). Modeling Deformation Behavior of the Baseball, *Journal of Applied Biomechanics*, Vol. 21, No.1, (2004), pp. 9-15

Nicholls, R.L., Miller, K. & Elliott, B.C. (2006). Numerical analysis of maximal bat performance in baseball, *Journal of Biomechanics*, Vol. 39, No. 6, (2006), pp. 1001-1009

Oda, J., Sakai, S., Yonemura, S., Kawata, K., Horikawa, S. & Yamamoto, H. (2003). Development research of intelligent pitching machine using neural network, *Transactions of the Japan Society of Mechanical Engineers*, Series C, Vol.69, No.678, (2003), pp.135-140, (in Japanese)

Sakai, S., Oda, J., Yonemura, S., Kawata, K., Horikawa, S. & Yamamoto, H. (2007). Research on the Development of Baseball Pitching Machine Controlling Pitch Type Ball Using Neural Network, *Journal of System Design and Dynamics*, Vol. 1, No.4, (2007), pp. 682-690

Sakai, S., Oda, J., Kawata, K. & Kitagawa, Y. (2007). Study on Throw Accuracy for Baseball Pitching Machine with Roller, *Transactions of the Japan Society of Mechanical Engineers*, Series C, Vol. 73, No.735, (2007), pp. 90-95, (in Japanese)

Sakai, S., Oda, J., Yonemura, S. & Sakamoto, J. (2008). Study on Impact Loading and Humerus Injury for Baseball, *Journal of Computational Science and Technology*, Vol. 2, No.4, (2008), pp. 609-619

Takahashi, M., Fujii, M. & Yagi, N. (2008). Automatic pitch type recognition from baseball broadcast videos, *10th IEEE International Symposium on Multimedia*, ISM 2008 , Art. No. 4741142, (2008), pp. 15-22

Todoroki, A., "RSMaker for Excel", 20. 05. 2010, Available from

http://todoroki.arrow.jp/ssoft/RSMkaisetsu.pdf
Watts, R.G. & Sawyer, E. (1975). Aerodynamics of Knuckleball, *American Journal of Physics*, Vol.43, No.11, (1975), pp. 961-963

Section 5

Applications of FEA in "Machining and Product Design"

Chapter 15

Finite Element Analysis of Machining Thin-Wall Parts: Error Prediction and Stability Analysis

YongAn Huang, Xiaoming Zhang and Youlun Xiong

Additional information is available at the end of the chapter

1. Introduction

Thin-wall workpieces are usually low rigidity and complex shapes, which results in great challenges in machining. The machining accuracy, both geometric accuracy and surface integrity, plays a significant role in achieving overall product's functional performance. Machining is a key process since it is easy to result in deflection and chatter[1, 2]. In stable machining, surface dimensional error due to the workpiece deflection affects machining precision[1, 3]. In unstable machining, the chatter becomes a critical problem for high surface quality. Currently, the existing methods for predicting the cutting forces, deflections and stability can be divided into experimenting and modeling. The direct trial-and-error approach is often expensive and time consuming. Producing the right profile in such parts increasingly depends on specialized CAD/CAE/CAM packages for defining appropriate cutting strategies and tool paths[4-7]. However, most of the existing techniques and models are based on idealised geometries and do not take into account factors such as variable cutting force, part/tool deflection, machining stability[8].

In many cases, the parameters of milling system are uncertain derived from the measurement errors, system nonlinear behavior, use or not use of coolants and other environmental noise. In addition, the stability boundaries are highly sensitive to the milling system parameter uncertainties. Therefore it is questionable that the usefulness of stability Lobes for estimating the milling stability obtained by applying deterministic parameters. Nominal machining parameters from the deterministic machining parameters optimization cannot guarantee the stability of milling process and lead to an actual maximization of material removal rate(MRR) and minimization of surface location error (SLE[1]). In practice,

[1] Surface location error is defined as the error in the placement of the milling cutter teeth when the surface is generated.

the cutting forces that statically and dynamically excite the tool and part, reducing the validity of the CAD/CAE/CAM output and leading to additional machining errors. For example, it induces chatter and large deformation in thin-wall workpiece, and very high cutting temperature which causes excessive thermal expansion specifically under dry machining conditions. Finite element analysis (FEA) has been widely adopted in numerical simulation of the machining process. FEA-based simulation, considering physical factors, such as material properties, tool geometry etc., is required to accurately predict the deflection and stability. Normally, the aim of cutting parameters optimization is to improve the part quality, maximize MRR and the constraint conditions are to keep the spindle speed below a predifined one and require the cutting process stable.

The cutting force and the part deflection are usually solved by an iterative simulation algorithm. Wang et al[9] studied force-induced errors, and developed a quasi-static error compensation method. Law et al.[10] calculated the cutting force from the measured milling torque, and integrated both the force and deflection models to develop a compensation methodology. However, currently the force modeling research has been mainly focused on theoretical rigid force models or mechanistic force models [11]. Budak and Altintas[12] considered the peripheral milling of a very flexible cantilever plate that incorporate a mechanistic force model and finite element methods. Feng and Menq [13, 14] developed a cutting force model taking into account the engaged cut geometry, the undeformed chip thickness distribution and the effect of the cutter axis offset. Budak et al.[15] proposed an oblique cutting mechanics model, in which the oblique cutting force are obtained from orthogonal cutting force. Ratchev et al.[16] proposed a flexible force model to study the tool deflection based on an extended perfect plastic layer model. They considered the end-milling cutter as a cantilever with the force acting at the cutter tip centre position[11]. However, the dynamic effect and the generated heat during machining were not considered, and the tool and the workpiece were assumed to deform to their static equilibrium position at any milling instant[1, 17]. Spence et al.[18] indicated that most existing machining simulation techniques were geometric and ignored the physical aspects of the process. Therefore, for making an appropriate choice of the cutting parameters and the operation sequence, deep understanding of the induced cutting deformation and the heat transfer is necessary. In the recent past, various techniques have been developed for studying the force- and temperature-induced aspects of machining, but separately. There is still a lack of a comprehensive milling simulation model which, taking into account the effects of the tool-path, and the cutting variables, simulates the thermomechanical aspects of machining[19]. Their applicability to model force in machining of thin-wall workpiece is limited due to the variability of material properties, cutting force, non-linear dependency on tool immersion angle and chip thickness[11]. Increased attention is being focused on the development of a computationally efficient milling process model, well capable to perform thermomechanical analysis of the metal cutting process.

Although there are many mechanisms of instability or chatter as mentioned in Wiercigroch and Budak[20], instability due to regeneration of surface waviness is by far the primary cause of instability. The stability analysis of the milling system can be performed only by applying approximated numerical methods since there is no closed solution to time-delay differential equation of milling dynamics. Alternatively it can be carried out by means of time-domain simulations[21], in the frequency domain[22] or by applying methods based on delay differential equation theory, such as the semi-discretization method[23] and the time finite element approach (TFEA)[24]. Budak and Tekeli[25] proposed a method to determine the optimal combination of depths of cut, so that chatter free material removal rate is maximized, with application on a pocketing example and significant reduction in the machining time is obtained. Altintas and Merdol[26] developed a generalized process simulation and optimization strategy for increasing MRR while avoiding machining errors and considering the chatter, and spindle's torque or power limits. However, the works of milling stability and cutting parameters optimization were addressed little to take into account the parameters uncertainties. Duncan et al.[27] used the Monte Carlo method to determine the associated uncertainties in the stability limit at each spindle speed. However the estimated stability intervals are too large to supply a useful advice to the parameters selection in an actual milling process. Kuidi[28] studied the robust optimization of SLE and MRR in milling process with uncertainties. The deterministic optimization formulation was modified to account for the axial depth uncertainty. But the uncertainties of Lobe diagram and SLE, which determine the milling stability and part quality were not taken into account. And the detailed optimization procedures are absent. Totis[29] used a new probabilistic algorithm for a robust analysis of stability in milling process, which performs the stability analysis on an uncertain model. The main objection to the general use of probabilistic analysis techniques is that non-deterministic properties cannot always be exactly represented using the probabilistic concept. Indeed, probabilistic methods cannot deliver reliable results unless sufficient experimental data are available to validate the assumptions.

In this chapter, we describe the force and temperature-induced error prediction and stabilty analysis of milling process by using the finite element method. Firstly a thermo-mechanical analysis is established to predict the force- and temperature-induced deflection. Then, TFEA is adopted as the deterministic model to obtain SLE and milling stability Lobe diagram. The uncertainties of modal shape parameters of spindle-tool system are investigated and sensitivity analysis is used to evaluate the upper and lower bounds of SLE and stability Lobe diagram. Finally, the formulation of robust spindle speed optimization is given to minimize the maximal SLE in a milling process and constraint condition is to maximize spindle speed and keep the machining parameters below the lower bound of Lobe diagram. With two optimization results, derived from robust and deterministic optimization formulations, experimental verifications are given.

2. Flexible thermo-mechanical model for error prediction

The surface error is induced by the deflection of tool and workpiece. Most of the existing techniques are based on idealized geometric profile and do not take into account tool/workpiece deflection, which results in a significant deviation between the planned and machined workpiece profiles[11]. This chapter focuses on deflection of thin-wall workpieces induced by cutting force and temperature. Reliable quantitative predictions of the cutting force and temperature components are essential for determining geometric errors of machined workpieces. Force predictions are required for arriving at constrained optimization strategies in computer-aided process planning[15].

2.1. Flexible force model for thin-wall workpiece

The research in the area of cutting covers a very wide range, since there are many independent influencing factors, including the cutting parameters, material properties, properties of the machine–tool–workpiece system, and tool geometry. A general model for the determination of cutting forces in ball-end milling operations was presented in reference[30]. The mathematical model represents the relations between the cutter and the workpiece, the change of chip thickness and the milling cutter rotation angle. The cutting forces are divided into several parts which depends on the number of cutting edges, cutting edge length and milling cutter rotation angle. Figure 1 shows the schematic diagram of a cutter of a ball-end milling and its configuration parameters.

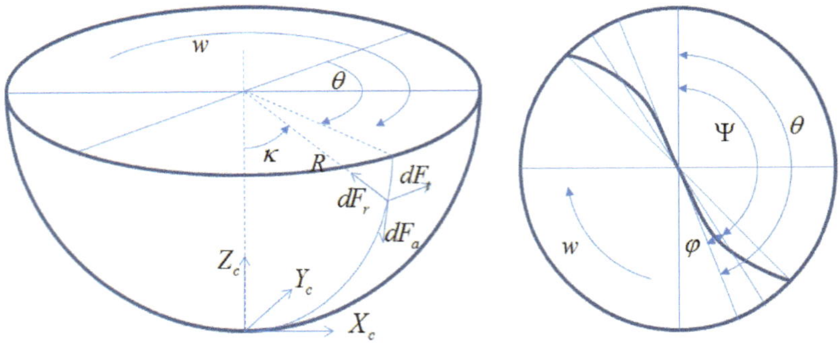

Figure 1. Constant lead spiral cutting edge for ball-end milling

The following is the main idea how to calculate it [31, 32]. The lag between the tip of the flute at $z = 0$ and at axial location z is[30]

$$\varphi = \frac{z}{R_0} \tan i_0 \tag{1}$$

where, R_0 is the radius of the hemispherical part, i_0 is the helix angle of the cutting edge. A point on the flute j at height z is referenced by its angular position in the global coordinate system,

$$\Psi_j(z) = \theta - \varphi + (j-1)\frac{2\pi}{N_f} \tag{2}$$

where N_f is the number of flutes, θ is the tool rotation angle.

The angle position in the direction of z-axis from the center of the hemispherical part to the point on the cutting edge can be written as

$$\begin{cases} f(z)=\sqrt{2R_0(R_0+z)} \\ \varphi(z)=z\tan i_0/R_0 \\ \kappa(z)=\arcsin\left(\sqrt{1-(1-z/R_0)^2}\right) \end{cases} \Rightarrow \begin{cases} x=R(\varphi)\sin(\varphi) \\ y=R(\varphi)\cos(\varphi) \\ z=R_0\varphi\cot i_0 \end{cases} \tag{3}$$

The tangential, radial and binormal components are calculated as[15, 32, 33],

$$\begin{cases} dF_t(\theta,z)=K_{te}dS+K_{tc}t_n(\Psi,\theta,\kappa)db \\ dF_r(\theta,z)=K_{re}dS+K_{rc}t_n(\Psi,\theta,\kappa)db \\ dF_a(\theta,z)=K_{ae}dS+K_{ac}t_n(\Psi,\theta,\kappa)db \end{cases} \tag{4}$$

where, $t_n(\Psi,\theta,\kappa)=f_{zb}\sin\Psi\sin\kappa$ is the uncut chip thickness normal to the cutting edge, and varies with the position of the cutting point, K_{tc}, K_{rc}, K_{ac} (N/mm^2) are the shear specific coefficients, K_{te}, K_{re}, K_{ae} (N/mm) are the edge specific coefficients. $dS=\sqrt{(R'(\varphi))^2+R^2(\varphi)+R_0^2\cot^2 i_0}\,d\varphi$ (mm) is the length of each discrete elements of the cutting edge, f_{zb} is the feeding per tooth, $\kappa=\arcsin(R(\Psi)/R_b)$, and db (mm) is the differential length of cutting edge. In many mechanistic models for the milling process, the milling force coefficients K_{tc}, K_{rc}, K_{ac}, K_{te}, K_{re}, K_{ae} are established from specially devised milling tests (e.g. average cutting force coefficient model) followed with linear regression analysis. Usually there are two methods to predict the parameters which are mechanistic evaluation and prediction from an oblique cutting model[15].

The cutting forces in Eq.(4) are modeled in terms of two fundamental phenomena, an edge force component due to rubbing or ploughing at the cutting edge, and a cutting component due to shearing at the shear zone and friction at the rake face[15]. The cutting force model including explicitly the ploughing component can obtain more precise prediction accuracy. Once the tangential $F_t(\theta,z)$, radial $F_r(\theta,z)$, and axial $F_a(\theta,z)$ cutting force at the tooth-workpiece contact point are determined, the resultant forces in Cartesian coordinates are obtained by introducing the transformation matrix[32],

$$\{dF_{x,y,z}\}=[T]\{dF_{r,t,a}\} \tag{5}$$

where, $\{dF_{x,y,z}\}=\begin{bmatrix} dF_x \\ dF_y \\ dF_z \end{bmatrix}$, $[T]=\begin{bmatrix} -\sin\kappa\sin\Psi & -\cos\Psi & -\cos\kappa\sin\Psi \\ -\sin\kappa\cos\Psi & \sin\Psi & -\cos\kappa\cos\Psi \\ \cos\kappa & 0 & -\sin\kappa \end{bmatrix}$, $\{dF_{r,t,a}\}=\begin{bmatrix} dF_r \\ dF_t \\ dF_a \end{bmatrix}$.

Then one can get [32]

$$\begin{cases} F_{xj}(\theta)=\int_{z_1}^{z_2}\left(-\sin\kappa_j\sin\Psi_j dF_{rj}-\cos\Psi_j dF_{tj}-\cos\kappa_j\sin\Psi_j dF_{aj}\right)dz \\ F_{yj}(\theta)=\int_{z_1}^{z_2}\left(-\sin\kappa_j\cos\Psi_j dF_{rj}+\sin\Psi_j dF_{tj}-\cos\kappa_j\cos\Psi_j dF_{aj}\right)dz \\ F_{zj}(\theta)=\int_{z_1}^{z_2}\left(\cos\kappa_j dF_{rj}-\sin\kappa_j dF_{aj}\right)dz \end{cases} \tag{6}$$

Since the cutting force coefficients (K_{tc}, K_{rc}, K_{ac}) may be dependent on the local chip thickness, the integrations given above should be calculated digitally by evaluating the contribution of each discrete cutting edge element at dz intervals. There are many models to calculate the coefficients, such as bi-linear force model, exponential chip thickness model, high-order force model, and semi-mechanistic model. Budak et al[15] presented the cutting transformation model.

$$K_{tc}=\frac{\tau}{\sin\phi_n}\frac{\cos(\beta_n-\alpha_n)+\tan\eta_c\sin\beta_n\tan\beta}{\sqrt{\cos^2(\phi_n+\beta_n-\alpha_n)+\tan^2\eta_c\sin^2\beta_n}} \tag{7}$$

$$K_{rc}=\frac{\tau}{\sin\phi_n\cos\beta}\frac{\sin(\beta_n-\alpha_n)}{\sqrt{\cos^2(\phi_n+\beta_n-\alpha_n)+\tan^2\eta_c\sin^2\beta_n}} \tag{8}$$

$$K_{ac}=\frac{\tau}{\sin\phi_n}\frac{\cos(\beta_n-\alpha_n)-\tan\eta_c\sin\beta_n}{\sqrt{\cos^2(\phi_n+\beta_n-\alpha_n)+\tan^2\eta_c\sin^2\beta_n}} \tag{9}$$

where, τ is shear stress at the shear plane, ϕ_n and β_n are normal shear and friction angles in oblique cutting respectively, α_n is normal rake angles, η_c is chip flow angle in the rake face, β is friction angle at the rake face. For example, when Cutter radius is 3mm, Feed is 0.02mm/rev, Depth of cutting is 0.02m, the cutting force can be calculated as Figure 2.

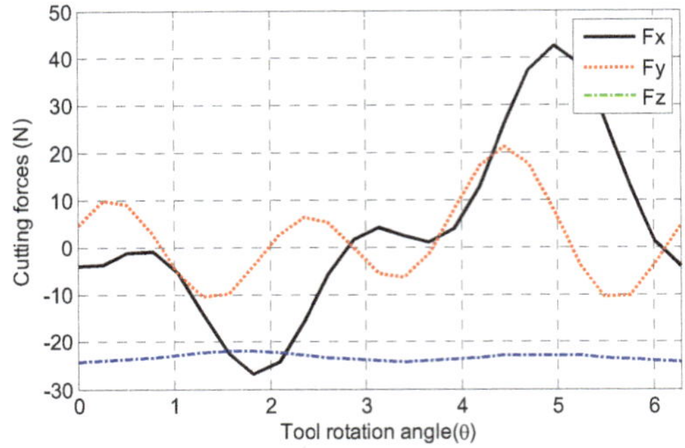

Figure 2. Predicted cutting forces for slot cutting tests

Most of reported papers in the area of cutting force- induced error belongs to those caused by large deformation of thin-wall workpieces under load[34, 35]. Peripheral milling of flexible components is rather complicated due to periodically varying cutting-forces. These forces statically and dynamically excite the tool and workpieces, which leads to significant and often unpredictable deflections. Static deflections produce dimensional profile errors, and dynamic displacements lead to poor surface finish quality and machining stability problems in addition to dimensional profile errors[16, 36]. In the known condition of cutting force, the deflection of the thin-wall workpiece can be calculated through finite element analysis in real time.

There are two kinds of force model, theoretical rigid force model and adaptive theoretical flexible force model. High complexity is associated with modeling of cutting forces in machining due to the variable tool/workpiece deflection and changing tool immersion angle. To address this complex dependency an interactive approach integrating an extended perfect plastic layer force model is adopted to link force prediction with workpiece deflection modeling[8]. The predicted profile of the workpiece is utilized to identify the "real" material volume that is removed during machining, instead of the "ideal" one defined by the "static" NC simulation packages [11]. In milling a thin-wall workpiece, the differential cutting force on the engaged infinitesimal tool cutting edge varies with the cutting depth that is effected by workpiece deflection[17]. The force is calculated by taking into account the changes of the immersion angle, ϕ, of the engaged teeth. As soon as the deflection, u_y and the coordinate (x, y, z) of point, a, are known (Fig. 4), the instant immersion angle ϕ in milling after deflection can be calculated using[1]

$$\Psi(t,z) = \cos^{-1}\frac{R-(h_r(z,t)-U_y(z,t))}{R} \tag{10}$$

where R is the cutter radius, and h_r is the designed milling depth in the workpiece thickness direction, while u_y is the deflection in the corresponding point predicted through FEA. Ratchev et al[3, 11] proposed a flexible force model for machining dimensional form error prediction of thin-wall components. Here, the thermal deflection is introduced into the the flexible force model to consider the temperature effects.

FEA softwares are used to calculate the deflection caused by the cutting force at each sampling point through the following equation,

$$[K]\{U\} = \{F\} \tag{11}$$

where $[K]$ is the stiffness matrix of the workpiece, $\{U\} = [U_x, U_y, U_z]^T$ and $\{F\} = [F_x, F_y, F_z]^T$ are nodal displacements of workpiece and the external cutting force acting on the tool-workpiece transient contact surface, respectively. As the boundary conditions are specified, the nodal displacements can be obtained through solving the above equations.

2.2. Thermo-mechanical analysis for thin-wall workpiece

The errors are usually caused by excessive deformation at the interface due to the cutting force and temperature, so they have to be considered simultaneously. With the recent developments in machining automation, various cutting temperature measurement techniques, including tool-work thermocouple, embedded thermocouple, and infrared pyrometer, emerged[37-41]. The aforesaid experimental techniques have been widely applied in machining due to its simplicity. Lin[42] and Kwon[43] studied the transient interfacial temperature and heat dissipation in the workpiece during a slot milling process. Fang and Zeng[44] utilized FEM to develop a 3D model of the oblique cutting process for the prediction of temperature distribution in the workpiece, tool and chip. Temperature distribution in the workpiece was estimated for a simple single pass slot milling operations in only a few reported works[45, 46]. However, these studies ignore the structural analysis for predicting part deformations under coupled thermomechanical loading conditions.

The milling techniques still face to a severe problems of inducing very high cutting temperatures causing excessive thermal expansion of the workpiece, especially in dry machining. The cutting parameters, namely, cutting speed and feed rate, have the greatest influence on the cutting temperature. For analyzing the phenomenon of heat dissipation into the workpiece and its influence on part deformation, a 3D model of the transient milling process was developed based on commercial FEM program, such as Abaqus, Ansys and Comsol. These systems allow: (1) creating 3D FEM models of the fixture–workpiece configurations, (2) applying appropriate materials for the workpiece, (3) applying appropriate machining boundary conditions, and (4) performing transient thermal and structural analysis where the transient temperature distribution profiles are applied along with the cutting forces to predict part deflections.

If the thermal conductivity (K) and heat capacity (ρC_p) of the work material is higher, then the generated heat is more readily conducted into workpieces and causes thermal expansions which produce severe irregularities. The cutting force more easily induces large deformation in the workpiece. In return, the deformation resulted from the cutting force and the generated heat will change the cutting parameters. However, most models are based on idealized geometries and do not take into account the factors, such as variable cutting force, thermal load, part/tool deflection, etc[1]. Before calculating the temperature distribution, cutting temperature at the interface need to be predicted. For speed-, and feed-dependent boundary conditions in machining, the thermal source problem in machining thin-wall workpiece is very difficult to be solved analytically. Therefore, the temperature prediction at the interface is usually achieved by non-linear empirical modeling approaches. The average interface temperature, as measured by the tool-work thermocouple, is[47]

$$T_{avg,interface}(°\text{C}) = 1700V^{0.5}d^{0.2}f^{0.4} \tag{12}$$

where V is cutting speed (m/s), d is depth of cut (mm), and f is feed (mm/rev). The above empirical model of the interface temperature is developed for turning of 4140 Steel alloy with tungsten carbide tools. The relation of cutting temperature and feed is plotted in Fig.3 where the temperature at the interface is very high.

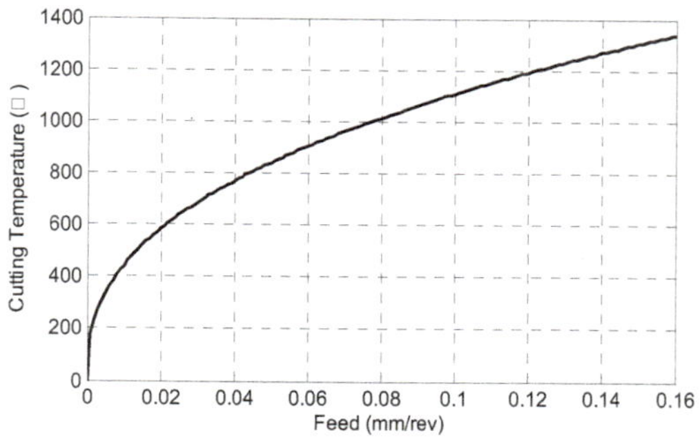

Figure 3. Cutting temperature versus cutting feed, depth of cut 0.763mm, cutting speed 3m/s

The temperature variations are related to the heat source movement, heat source intensity, and thermal resistance coefficient. The fundamental generalized problem to be solved analytically is the heat conduction in a thin infinite plate with a convective and radial boundary condition on the face. The time-dependent heat transfer process is governed by the following differential equation[48],

$$\frac{\partial^2 T}{\partial r^2}+\frac{1}{r}\frac{\partial T}{\partial r}-\eta T+\frac{g(r,t)}{k}=\frac{1}{\alpha}\frac{\partial T}{\partial t} \tag{13}$$

where, $\eta = h/kw$, h is the convection coefficient of heat transfer, k is the thermal conductivity of the material, w is the plate thickness, α is the thermal diffusivity of the material, and $g(r,t)$ is the internal heat generation rate per unit volume, the variable r is the radial distance from the heat source.

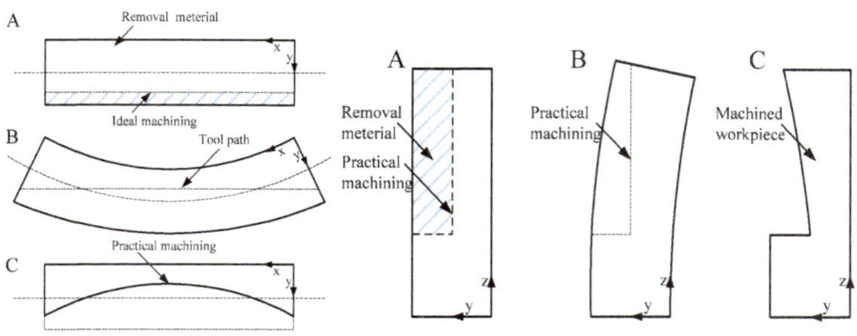

Figure 4. Temperature-effected machining: (a) Top view;(b) Side view

The dynamic non-uniform temperature distribution roots in time/position-dependent thermal deformations. Eq.(13) is to be solved by finite element method here. Ref. [48] proposed another computational methods. There are two basic thermal error modes, namely, thermal expansion and thermal bending. Figure 4 is the schematic diagram of machining results due to temperature effects. Figure 4a-4b show the deflections from the top and side views respectively. It can be observed from Figure 4 that the machining topography will be very complex when the temperature effects are considered. In return, the thermal expansion and thermal bending affect the cutting force. It is a complex, back and forth cycle, which has to be simulated by interactive algorithm.

2.3. Interactive algorithm for thermo-mechanical analysis

The cutting forces depend on the chip thickness which is a function of the tool immersion angle. The tool immersion angle is a function of the part deflection which itself depends on the cutting forces[47]. The deflection is determined by the cutting force, as well as temperature distribution. The cutting forces depend on the chip thickness which is a function of the tool immersion angle, and the machining temperature is a function of the cutting speed, depth of cut and the feed. An iterative procedure is used to determine the milling error. The predicted deflection from CAE is used to identify the practical material volume that is removed during machining instead of the ideal one defined by CAM. The thermo-mechanical analysis is outlined in Figure 5. While milling a thin-wall workpiece, as soon as the cutter is engaged, the workpiece deflects to a new position, and the cutting temperature also changes at the same time. In more detail, the tool-workpiece intersection line can be used as an example to explain the impact on the cutting force and deflection[16].

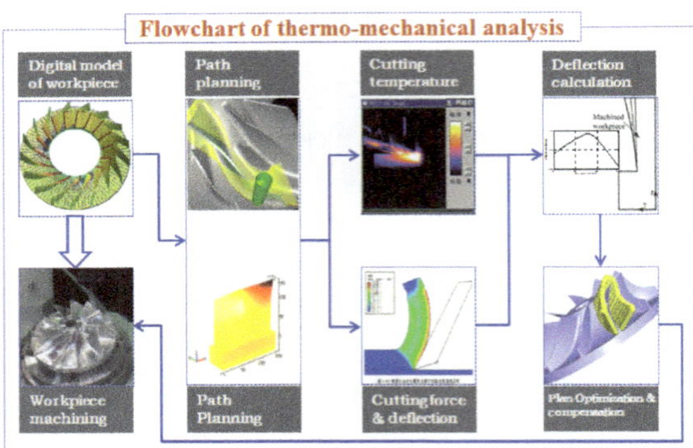

Figure 5. Flowchart of complex surface machining based on thermo-mechanical analysis

Although commercial FEA softwares are also used to simulate manufacturing processes, they cannot be solely used to simulate multi-step cutting processes of thin-wall parts. The main difficulty is that material removal and remeshing of part model are very complex for the multi-step processes, and all these FEA softwares do not integrate an appropriate theoretical force model for workpiece/tool deflection prediction, so varieties of models and software are involved[1]. There is a need to link the mainstream commercial FEA software with force prediction models, thermal prediction models and material removal models in order that data exchange among them can be achieved.

2.4. Error definition and thermo-mechanical deflection

The surface dimensional error is the normal deviation of the actual machined surface from the desired machined surface. For example, at point P in Figure 6, the corresponding surface dimensional error is $e_P = \delta_{t,P} + \delta_{f,P}$, where $\delta_{t,P}$ and $\delta_{f,P}$ are the normal projections of the temperature- and force-induced deflection corresponding to Point P, respectively. For the convenience of study, the distance between the initial surface to be machined and the desired machined surface is named as the nominal radial depth of cut symbolized by R_N. In actual machining, to ensure that the surface dimensional error does not violate the tolerance, R_A is often specified to be different from R_N ($R_A \neq R_N$)[3]. In this case, $e_P = \delta_{t,P} + \delta_{f,P}$ has to be adjusted to the calculation of surface dimensional error

$$e_P = \delta_{t,P} + \delta_{f,P} + R_N - R_A \tag{14}$$

Note that R_N and R_A are the nominal and specified radial depth of cut, respectively. For a certain surface generation line, the steps adopted to calculate the error distributions can be found in Ref.[3].

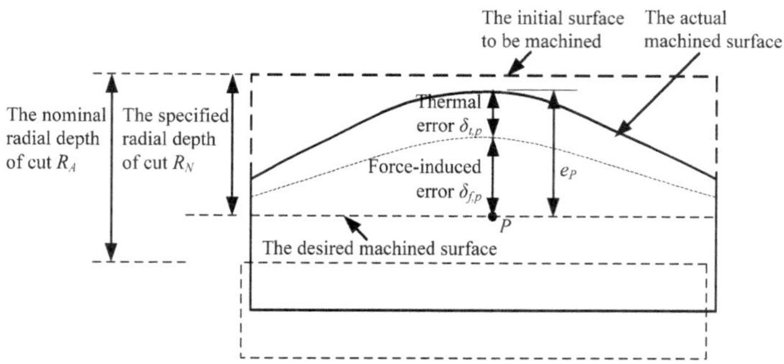

Figure 6. Definition of the surface dimensional error

2.5. Examples of thermomechanical models

For simplification the part is assumed to be a thin-wall rectangular workpiece. The required machined profile is a flat surface parallel to the plane OXY. During milling, the workpiece deflection in its thickness direction has a significant impact on forming the surface profile error. The contributions of the workpiece deflection in the feed direction and the tool axial direction can be ignored. Therefore, the investigations focus only on error prediction in y-axis direction. The simulations were based on clamped-free-free-free cantilever plates with dimension $150 \times 120 \times 5mm^3$ and *Aluminium alloy 6063 T83*. The quasi-static cutting force is treated as a moving-distributed load acting on the workpiece-tool contact zone in the milling process. To compute the workpiece/tool response to the cutting force, the continuous machining process was simulated by multi-step cutting processes[18]. Comsol is employed to estimate the force- and temperature-induced deflection.

The input to finite element model is the chip-related cutting force and temperature, and a set of parameters describing material properties, boundary conditions and other constraints. In order to simplify the complex simulation, we employ an assumption that the instantaneous stiff variation due to material removal will be not taken into account. To compute the workpiece response, the continuous machining process was simulated by multi-step cutting processes. During machining, the tool moves along the machining surface. Here, we took a different approach that uses a moving coordinate system fixed at the tool axis. After making the coordinate transformation, the heat transfer problem becomes a stationary convection-conduction problem in COMSOL. With the workpiece undergoing deformation, the cutting force is also changing considerably. These changes is taken into account by computing the cutter on a moving mesh attached to the workpiece.

2.5.1. *Machining based on a designed tool path*

The results based on the rigid force model are discussed. Supposed that the force is $0.23N / mm^{-2}$, the temperature is 1200℃, and the two loads act on the interface workpiece and cutter at neighborhood of $x = 75mm$ of the workpiece. It can be noted that the maximal deflections are 0.348mm and 0.384mm in Figure 7-10 in the positive direction and 1.5mm in the passive direction, respectively. Figure 9 is the force-induced and temperature-induced deflection of the top edge of the workpiece. It can be observed that the deflections at $x = 75mm$ in the two figures have different directions. The force-induced deflection is along the positive direction, however the temperature-induced deflection along the passive direction. In reported studies, the force-induced and the temperature-induced deflection are investigated dividually, which will leads large error in machining.

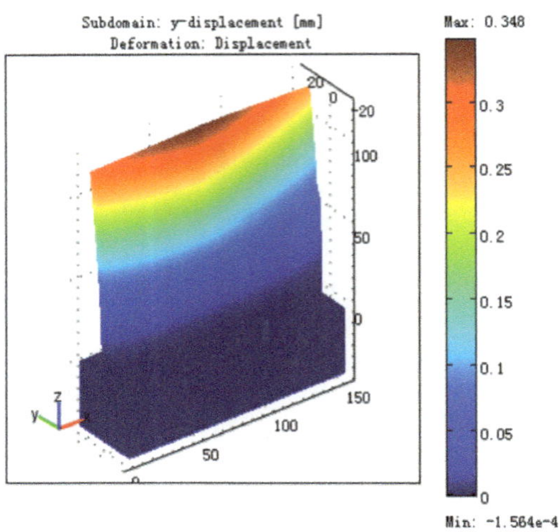

Figure 7. Deflection results of the rigid force model

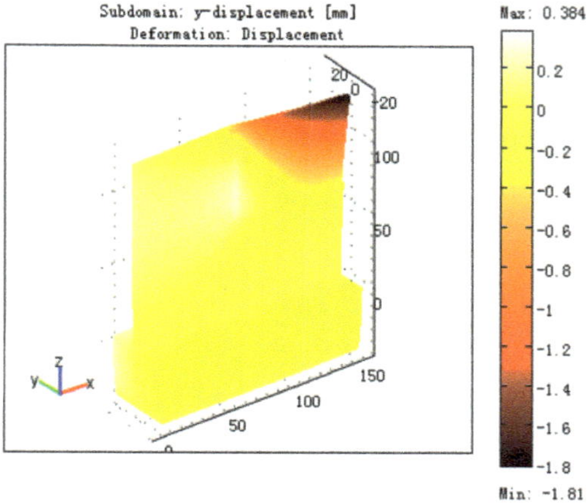

Figure 8. Thermal deflection results

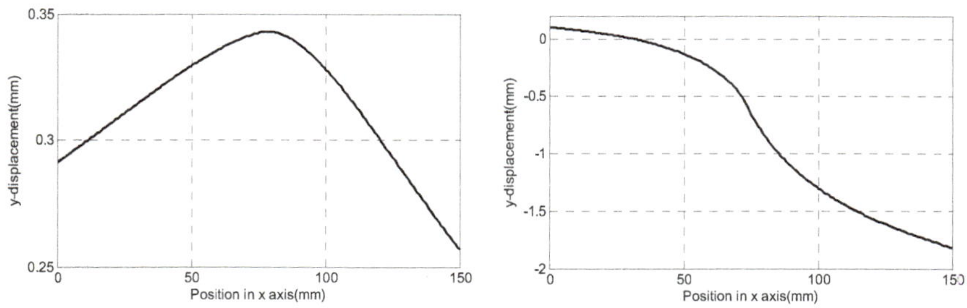

Figure 9. Deflection of the top edge of the workpiece due to force and temperature

2.5.2. *Machining based on flexible force model*

Figure 10 is the results based on the flexible force model. The maximal deflection is 0.278mm. Comparing with Figure 7, one can observe that the max stress and the max deflection in Figure 10 are all smaller. The simulation shows the deflection of workpiece affect the cutting force, and the cutting force also affect the deflection in return. We can get the following conclusion: (1) if the workpiece is machined based on CAM, there exists large errors, and (2) the prediction from the flexible force model is smaller than that from the rigid force model. Obviously, the flexible force model is closer to the real machining that the rigid force model. The temperature is not considered in both models.

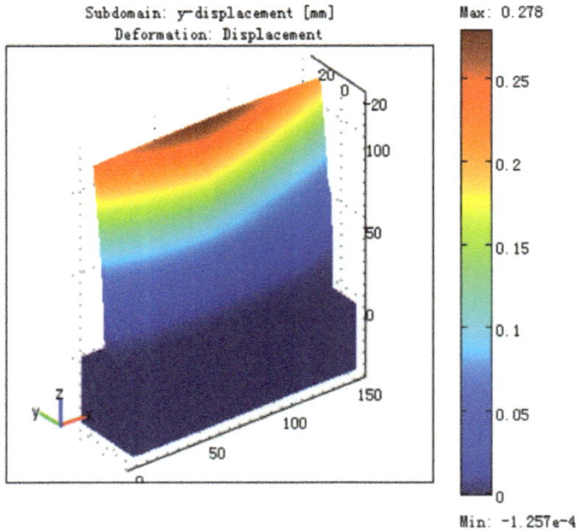

Figure 10. Stress and deflection results of the flexible force model

2.5.3. Machining based on thermo-mechanical analysis

The results plotted in Figure 11 and Figure 12 are based on the thermo-mechanical analysis. The maximal deflection in Figure 11 is 0.527mm. It is larger than the deflection in the rigid force model and the flexible force model, but smaller than the sum of 0.278mm of the flexible force model and 0.384mm of the temperature-induced deflection. So the thermo-mechanical model is not a simple combination of the flexible force model and the thermal deflection. The temperature has heavy effect on the deflection that the curve in Figure 12 is very similar with the curve in Figure 9. In practical error compensation, there will overcut or undercut if the force and the temperature studied separately. In order to reduce the machining error, it is important to reduce the cutting temperature during machining.

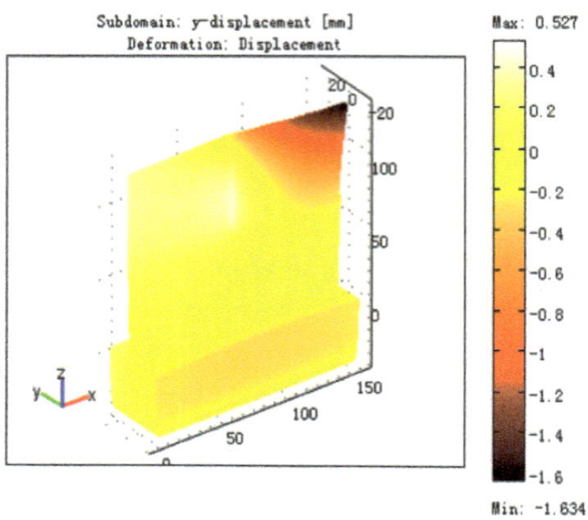

Figure 11. Deflection results of the thermo-mechanical analysis

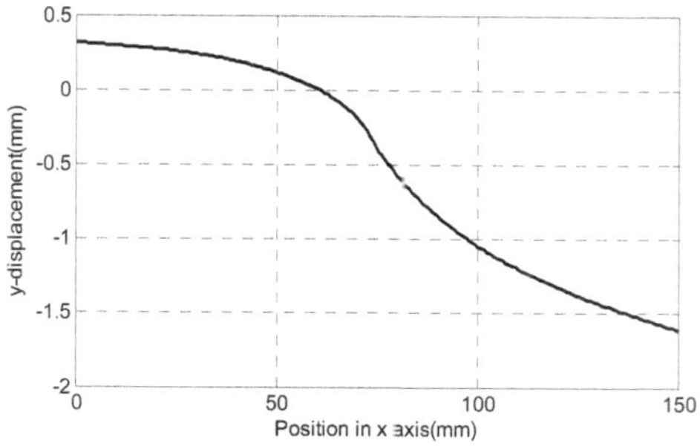

Figure 12. Deflection of the top edge of the workpiece due to thermo-mechanical action

3. Stability analysis of machining of thin-wall parts [49]

3.1. Milling dynamics

The standard two degree of freedom milling process is shown in Fig.1. The tool is assumed to be compliant relative to the rigid workpiece. The vibration is excited by the summation of cutting force. The governing equation of motion has the following form

$$\mathbf{M}\ddot{\mathbf{x}}(t) + \mathbf{C}\dot{\mathbf{x}}(t) + \mathbf{K}\mathbf{x}(t) = \mathbf{K}_c(t)b(\mathbf{x}(t) - \mathbf{x}(t-\tau)) + \mathbf{f}_0(t)b \tag{15}$$

where, $\mathbf{M} = \begin{bmatrix} m_x & 0 \\ 0 & m_y \end{bmatrix}$, $\mathbf{C} = \begin{bmatrix} c_x & 0 \\ 0 & c_y \end{bmatrix}$, $\mathbf{K} = \begin{bmatrix} k_x & 0 \\ 0 & k_y \end{bmatrix}$ and $\mathbf{F}$ are the modal mass, damping, stiffness, and cutting force matrices, respectively. The terms $m_{x,y}$, $c_{x,y}$, $k_{x,y}$ and $F_{x,y}$ are the corresponding components in the flexible directions of the system. b is the axial depth of cut. $\tau = 60/N\Omega$ is the tooth passing period in seconds, in which N is the number of teeth on the cutting tool and Ω the spindle speed in rpm. $\mathbf{x}(t) = \begin{bmatrix} x(t) & y(t) \end{bmatrix}^T$ is the dynamic response vector and $\mathbf{x}(t) - \mathbf{x}(t-\tau)$ the dynamic chip thickness, as shown in Figure 13.

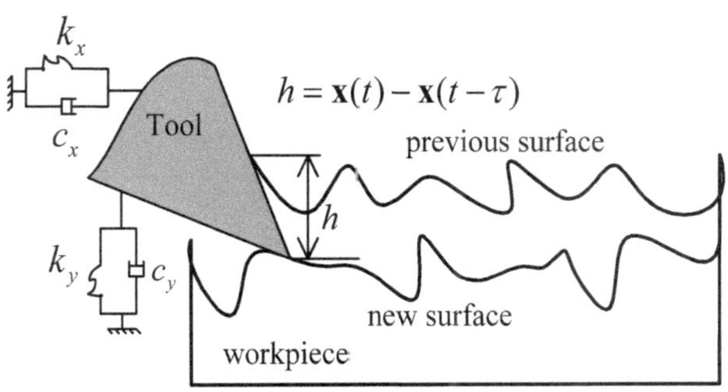

Figure 13. Dynamic chip thickness

$\mathbf{K}_c$ is a matrix given as follows, which represents the component of cutting forces that depend on the position vector

$$\mathbf{K}_c(t) = \sum_{j=1}^{N} g[\varphi_j(t)] \begin{bmatrix} -K_t sc - K_n s^2 & -K_t c^2 - K_n sc \\ K_t s^2 - K_n sc & K_t sc - K_n c^2 \end{bmatrix} \tag{16}$$

where K_t and K_n are the tangential and normal cutting force coefficients components, respectively. $s = \sin\varphi_j(t)$, $c = \cos\varphi_j(t)$ and the function $g[\varphi_j(t)]$ acts as a switching function, which is equal to 1 if the jth tooth is active and 0 if it is not cutting.

$$g[\varphi_j(t)] = \begin{cases} 1 & \varphi_e < \varphi_j(t) < \varphi_a \\ 0 & \text{otherwise} \end{cases} \tag{17}$$

where φ_e and φ_a are the angles where the jth tooth enters and exits the cut, respectively. For down-milling operation, $\varphi_a = \pi$, for up-milling, $\varphi_e = 0$. Note that the entry and exit angles may vary due to heavy vibrations of the tool. This effect is neglected here, and the angles φ_e and φ_a are approximated by constant values as it is usually done in the literature. $\mathbf{f}_o(t)$ is the stationary cutting force vector (f_z is the feed per tooth):

$$\mathbf{f}_o(t) = f_z \begin{bmatrix} K_t sc + K_n s^2 \\ -K_t s^2 + K_n sc \end{bmatrix} \tag{18}$$

3.2. Deterministic model for predicting milling stability and calculating SLE

Here the TFEA is introduced for brevity as a basis for the further uncertain analysis. The initial work of applying TFEA to the delay equations can be found in Ref. [5]. The main idea of TFEA is that the dynamic behavior of the milling process is governed by TFEA as a discrete linear map which relates the vibration response while the tool tooth is engaged in the cut, which depends on previous tooth passages and therefore includes the time delay τ, to free vibration while the tooth is not engaged in the cut. The dynamic map is expressed as

$$\begin{Bmatrix} q \\ \dot{q} \end{Bmatrix}^n = \mathbf{A} \begin{Bmatrix} q \\ \dot{q} \end{Bmatrix}^{n-1} + \mathbf{B} \tag{19}$$

where $\mathbf{A}$ is the state transition matrix, the size of which depends on the number of time finite elements and polynomial order representing one time period. $\mathbf{B}$ is a vector that depends on the process parameters. q and $\dot{q}$ are the sets of x and y positions and velocities for all nodal times in one tooth passage, respectively. Stability of the milling process is determined from the eigenvalues of $\mathbf{A}$, i.e. $\lambda(\mathbf{A})$. The maximum magnitude of the map eigenvalues is described by

$$\lambda_{\max}(\mathbf{A}) = \max_k |\lambda_k| \tag{20}$$

where λ_k denotes the kth eigenvalue of the dynamic map matrix $\mathbf{A}$, which is a function of the cutting conditions. Unstable conditions exist if

$$\lambda_{\max}(\mathbf{A}) > 1 \tag{21}$$

The stability boundary is defined by the boundary curve of axial depth b and spindle speed Ω. A combination of b and Ω values below the stability boundary, $b_{\lim}$, gives stable cutting conditions, whereas a combination above the stability boundary leads to an unstable cut. The stability boundary corresponds to the cutting conditions at which

$$\lambda_{\max}(\mathbf{A}(b_{\lim},\Omega)) = 1 \tag{22}$$

When the milling process is stable, SLE can be obtained from fixed points of the dynamic map as given in Eqs. (21):

$$\begin{Bmatrix} q \\ \dot{q} \end{Bmatrix}^{n} = \begin{Bmatrix} q \\ \dot{q} \end{Bmatrix}^{n-1} = \begin{Bmatrix} q \\ \dot{q} \end{Bmatrix}^{*} \tag{23}$$

Substituting Eqs.(23) into Eqs.(19) gives the solution of fixed point map:

$$\begin{Bmatrix} q \\ \dot{q} \end{Bmatrix}^{*} = (\mathbf{I} - \mathbf{A})^{-1}\mathbf{B} \tag{24}$$

The solution of fixed point displacement can be obtained and used to specify SLE as a function of cutting parameters.

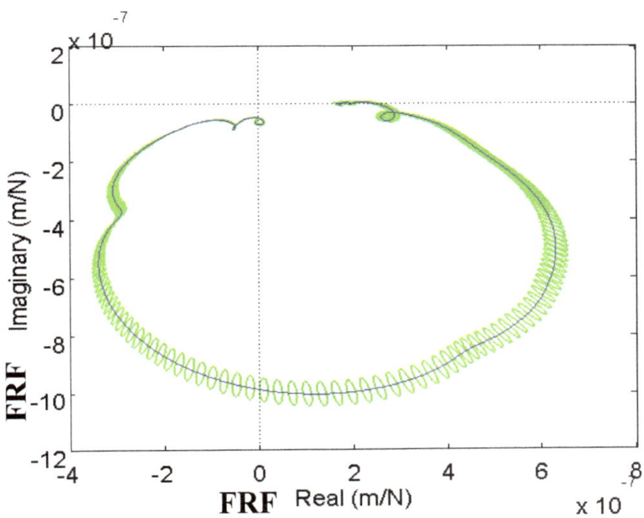

Figure 14. Nyquist plot of FRF of spindle-tool system with uncertainty regions[51]

3.3. Upper and lower bounds of Lobe diagram and SLE

Since the uncertainties occur in the experimental modal analysis, the measured values for the structural parameters given in Eq.(15) are different from the actual ones. Much experiments had been done so that the errors can be observed, see Ref.[50, 51]. A detailed study was carried out in the work[51] to discuss the modal test. As shown in Figure 14, Nyquist plot of FRF of the spindle-tool system with uncertainty regions was given.

In this section, we address the uncertainty problems in model parameters, which are modeled as the interval parameters as follows:

$$\mathbf{M}^I\ddot{\mathbf{x}}(t)+\mathbf{C}^I\dot{\mathbf{x}}(t)+\mathbf{K}^I\mathbf{x}(t)=\mathbf{K}_c(t)b(\mathbf{x}(t)-\mathbf{x}(t-\tau))+\mathbf{f}_0(t)b \tag{25}$$

where parameters $\mathbf{M}$, $\mathbf{C}$ and $\mathbf{K}$ are assumed to be bounded in the intervals and can be stated as:

$$\mathbf{M}\in\mathbf{M}^I \quad \mathbf{C}\in\mathbf{C}^I \quad \mathbf{K}\in\mathbf{K}^I \tag{26}$$

The problem can be formulated as given the uncertainties as shown in Eqs.(26) how to estimate the stability of the milling system and SLE (Eqs.(25)). Noted the linear transform matrix $\mathbf{A}$ given in Eq.(19), the right eigenvalue problem can be stated as

$$\mathbf{Au}=\lambda\mathbf{u} \tag{27}$$

where λ is the eigenvalue and $\mathbf{u}$ the corresponding eigenvector. The accompanying left eignenvalue problem is

$$\mathbf{A}^T\mathbf{v}=\lambda\mathbf{v} \tag{28}$$

Combined Eqs.(27) and (28), we can obtain

$$\mathbf{v}^T\mathbf{Qu}=\lambda\mathbf{v}^T\mathbf{u} \tag{29}$$

For a parameter z, i.e., a spindle-tool modal parameter or a part geometrical one, the sensitivity of system eigenvalue with respect to z, $\partial\lambda/\partial z$ can be expressed as follows

$$\frac{\partial\lambda}{\partial z}=\frac{\mathbf{v}^T}{\mathbf{v}^T\mathbf{u}}\frac{\partial\mathbf{A}}{\partial z}\mathbf{u} \tag{30}$$

where $\partial\mathbf{A}/\partial z$, the derivative of the matrix $\mathbf{A}$ is obtained by the central difference method. So the sensitivity expression in Eq.(30) can be treated as a semi-analytical one[52]. It should be noted that

$$|\lambda_{\max}|^2=\lambda_{\max}\lambda_{\max}^{'} \tag{31}$$

where $\lambda_{\max}^{'}$ is the complex conjugate of $\lambda_{\max}$, then

$$\frac{\partial|\lambda_{\max}|}{\partial z}=\frac{\lambda_{\max}\frac{\partial\lambda_{\max}^{'}}{\partial z}+\lambda_{\max}^{'}\frac{\partial\lambda_{\max}}{\partial z}}{2|\lambda_{\max}|} \tag{32}$$

Substituting Eq.(30) into Eq.(32), we can get the explicit expression for $\partial|\lambda_{\max}|/\partial z$. More details of sensitivity of stability boundary, especially the error analysis can be referred with the reference[52]. When there is a perturbation for the milling system parameter z, i.e., $z \mapsto z+\delta z$, the increment of $\lambda_{\max}$ is

$$\delta\lambda=\frac{\partial|\lambda_{\max}|}{\partial z}\delta z \tag{33}$$

Denote $\lambda_{\max}^{U}$ and $\lambda_{\max}^{L}$ the upper and lower bounds of the eigenvalues of milling systems, respectively. They read

$$\begin{aligned}\lambda_{\max}^{U}&=\max\{|\lambda_{\max}+\delta\lambda|,\ |\lambda_{\max}-\delta\lambda|\}\\ \lambda_{\min}^{L}&=\min\{|\lambda_{\max}+\delta\lambda|,\ |\lambda_{\max}-\delta\lambda|\}\end{aligned} \tag{34}$$

Then the upper and lower bounds of the milling stability Lobes, c^{U} and c^{L} can be obtained as follows

$$\begin{cases} c^{U} & \text{when } \lambda_{\max}^{L}(b_{\lim},\Omega)=1\\ c^{L} & \text{when } \lambda_{\max}^{U}(b_{\lim},\Omega)=1\end{cases} \tag{35}$$

The central difference method used for the sensitivity analysis of system eigenvalues can be also applied to the sensitivity analysis of SLE. Thus we get the upper and lower bound of SLE, i.e. f_{SLE}^{U} and f_{SLE}^{L}. The detailed discussions are omitted here.

3.4. Robust machining parameters optimization

In general, a deterministic cutting parameters optimization problem can be stated as follows:

$$\begin{aligned}&\min\ \{f_{SLE}(b,\Omega),\ -\Omega\}\\ &\text{s.t.}\ \begin{cases}\lambda_{\max}(\mathbf{A}(b,\Omega))\le 1\\ \Omega\le\Omega_{0}\end{cases}\end{aligned} \tag{36}$$

where the optimization objects are to minimize SLE and maximize the spindle speed, and constraint conditions are to keep the milling process stable and the spindle speed is less than the predefined one, which is set by the spindle system and tool suppliers. Since there are uncertainties between the real structure and the model, the predicted performance including the milling stability and SLE will not be guaranteed. When the uncertainties exist, as given in Section 5, the optimized machining parameters, i.e. axis depth b and spindle speed Ω

obtained from the deterministic optimization formulation (36) cannot guarantee the stability of the milling process, and minimization of SLE and maximization of spindle speed cannot be achieved. As a comparison, we give a formulation of robust machining parameters optimization as follows:

$$\begin{aligned} &\min \ \{\max f_{SLE}(b,\Omega), -\Omega\} \\ &\text{s.t.} \begin{cases} \lambda_{\max}^{U}(\mathbf{A}(b,\Omega)) \le 1 \\ \Omega \le \Omega_0 \end{cases} \end{aligned} \tag{37}$$

where the optimization object is to minimize the maximal SLE and maximize the spindle speed, and constraint conditions are to keep the machining parameters b and Ω below the lower bound of Lobe diagram given by $\lambda_{\max}^{U}(\mathbf{A}(b,\Omega)) \le 1$ and the spindle speed is less than the predefined one. The formulation of robust machining parameters optimization ensures the machining process stable and can lead to minimization of SLE under uncertainties.

3.5. Implementation of robust optimization formulation

The robust optimization formulation presented in Eqs.(37) can be transformed into the sequential optimization ones, that is:

$$\begin{aligned} &\min \ \max f_{SLE}(b,\Omega) \\ &\text{s.t.} \begin{cases} \lambda_{\max}^{U}(\mathbf{A}(b,\Omega)) \le 1 \\ \Omega \le \Omega_i \quad i = 1,2,\cdots k \end{cases} \\ &\text{for series of } \Omega_i \le \Omega_0 \end{aligned} \tag{38}$$

The optimization problem given in Eq.(38) is solved by the augmented Lagrangian function method. The basic idea of this method is that it transforms the nonlinear optimization problem into an unconstrained optimization one by introducing a penalty function, named augmented Lagrangian function[53]. The augmented Lagrangian function $L_A(x,\lambda;\mu)$ achieves these goals by including an explicit estimate of the Lagrange multipliers λ in the objective. The augmented Lagrangian function can be defined as

$$L_A(\mathbf{x},\mu;\lambda) = f_{SLE}(\mathbf{x}) - [\lambda_1 c_1 + \lambda_2 c_2] + \frac{1}{2\mu}(c_1^2 + c_2^2) \tag{39}$$

where $\mathbf{x} = \left[b,\Omega\right]^T, c_1 = l(\mathbf{x}), c_2 = \Omega - \Omega_i$, $l(\mathbf{x})$ is a cubic polynomial curve to interpolate the obtained Lobe diagram. It can be easily to obtain the differentiation of the augmented Lagrangian function with respect to the machining parameters, $\nabla_{\mathbf{x}} L_A(\mathbf{x},\mu;\lambda)$. Then we can obtain the following algorithmic framework.

Algorithm: the augmented Lagrangian function method.

Step 0: Given initial points $\mathbf{x}_0$ and λ_0, $\mu_0 > 0$, tolerance $\varepsilon \ge 0$;

Step 1: Find an approximate minimizer $\mathbf{x}_k$ of $L_A(\mathbf{x},\mu_k;\lambda_k)$, i.e. $\mathbf{x}_k = \arg\ \min L_A(\mathbf{x},\mu_k;\lambda_k)$;

Step 2: If $\left\|\nabla_{\mathbf{x}} L_A(\mathbf{x}, \mu_k; \lambda_k)\right\| > \varepsilon$, set $k = k+1$ and update Lagrange multipliers $\lambda_{k+1} = \lambda_k - c_i(x_k)/\mu_k \quad i = 1,2$ to obtain λ_{k+1} and go to **Step 1**; else exit and report the minimizer $\mathbf{x}^*$.

3.6. Experimental verification

In this section, experiments of cutting parameters optimization are carried out to verify the numerical optimization modeling results. First the modal shape parameters of the spindle-tool system are identified, with up and lower bounds, and then the optimization modeling results are presented by using the system parameters as input parameters. As a comparison, the deterministic optimization results are also presented. With the optimized machining parameters obtained from the robust optimization formulation and deterministic one, cutting experiments are implemented to check if the optimized cutting parameters are really "optimized", which means there is no chatter arising with the optimized spindle speed.

3.6.1. System parameters identification

The system parameters as input variables to the optimization formulations are the structural dynamics parameters, i.e. modal shape parameters of the spindle-tool system. An impact hammer test is implemented to obtain the structural dynamics parameters of spindle-tool system. To obtain a reasonable interval, normally the repeated modal hamper impact experiments (see Fig.3) are necessary. Based on these experimental results, the modal parameters are identified by a Rational Fraction Polynomial (RFP) method[54], as given in Figure 16 and Figure 17. The mean values with upper and lower bounds for the modal parameters are given as follows (Table 1), which are derived from the repeated modal hamper impact experiments. M_x and M_y (kg) are the modal mass in x and y directions, respectively. ξ_x and ξ_y (kg/s) are the modal damping coefficients in x and y directions, respectively. K_x and K_y (MN/m) are the modal stiffness, in x and y directions, respectively.

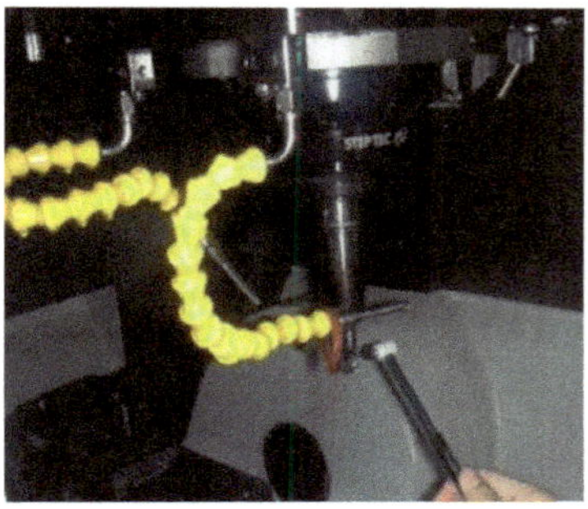

Figure 15. Experiment setup for modal shapes of spindle-tool system

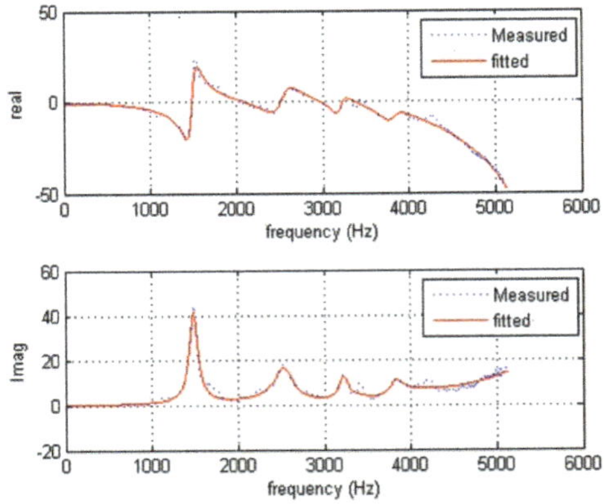

Figure 16. Fitting of mode parameters of tool system in X-axis

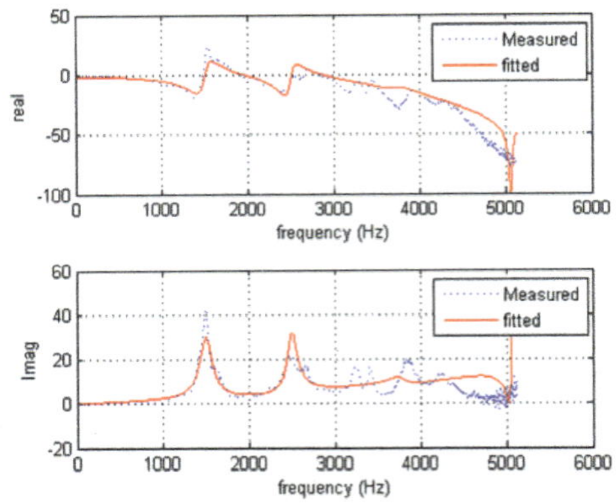

Figure 17. Fitting of mode parameters of tool system in Y-axis

M_x	M_y	ξ_x	ξ_y	K_x	K_y
0.7769E-2	0.7709E-2	4.0385%	6.0179%	0.6723E6	0.6868E6
0.7473E-2	0.6499E-2	4.1938%	5.3940%	0.6437E6	0.5664E6

Table 1. Modal parameters of spindle-tool system

The tool for the simulation and experiments is a 10mm cylindrical cutter with 4-tooth and the cutter parameters are shown in Table 2.

Radius (mm)	Number of tooth	Helix angle (degree)	Edge length (mm)	Total length (mm)	Material
10	4	30	8.5	33	Hard alloy

Table 2. Cutter parameters

3.6.2. Modeling results

In the experiment of machining of an impeller blade(single blade), the upper limit of the spindle speed is 20000rpm. The cutting depth is 0.560mm, which is derived from the geometry of the semi-finished and finished workpieces and the given tool path[55].

In table 1, we have presented the upper and lower bounds of the modal shapes parameters of spindle-tool system with mean values. The deterministic Lobe diagram is calculated by the TFEA[24]. Using the proposed method given in Section 4, we can obtain the upper and lower bounds of the boundary curve of Lobe diagram and SLE, as given in the upper part of fig.8. After the lower bound of the Lobe diagram and upper bound of SLE have been obtained, they are selected as the constraint conditions in the robust optimization formulations given as:

$$\begin{aligned}&\min\ \max f_{SLE}(b,\Omega)\\&\text{s.t.}\ \begin{cases}\lambda^{U}_{\max}(\mathbf{A}(b,\Omega))\le 1\\ \Omega\le\Omega_i\ \ i=1,2,\cdots k\end{cases}\ \text{for series of } \Omega_i\le 20000\end{aligned}\tag{40}$$

As a comparison, the mean values of the modal shapes parameters and SLE are adopted in the deterministic ones, stated as:

$$\begin{aligned}&\min\ f_{SLE}(b,\Omega)\\&\text{s.t.}\ \begin{cases}\lambda_{\max}(\mathbf{A}(b,\Omega))\le 1\\ \Omega\le\Omega_i\ \ i=1,2,\cdots k\end{cases}\ \text{for series of } \Omega_i\le 20000\end{aligned}\tag{41}$$

We solve the sequential optimization formulations as shown in Eqs. (40)-(41) by setting the spindle speed $\Omega_i = 20000-100*(i-1),\ i=1,2,\cdots 51$. That means we are interested in the spindle speed interval from 15000rpm to 20000rpm, and in this interval the optimization formulations are solved for 51 times. Augmented Lagrangian function method is adopted to solve the above two constrained optimization problems and the detailed procedures are given in Section 6. The modeling results are shown in Table 3.

	SLE	Spindle speed
Deterministic optimization	[1.88E-5m, 2.08E-5m]	17900 rpm
Robust optimization	1.99E-5 m	19100 rpm

Table 3. Comparison of the optimization results

3.6.3. Experiment of machining stability

The cutting experiments of impeller blade milling are implemented in Mikron HSM 600U, a five-axis NC machining center. Experiment setup including the machining setup and Labview signal acquisition interface are shown in Figure 18. First, we prepare two semi-finished blades, using the same workpiece geometry, tool path as well as machining parameters to ensure that the two blades are almost the same. Then the two semi-finished blades are used for the following finish milling with different spindle speeds.

The optimized spindle speeds derived from the robust and deterministic optimization formulations are used respectively to check the chatter occurrence. When the first blade is machined with spindle speed 19100 rpm, we observe that the milling process is stable and the quality of resulted workpiece is quite good, as shown in Figure 19(R). Then the second blade is machined and the spindle speed is set at 17900rpm, and the noise level dramatically becomes very high, which indicates the energy of the vibration at the frequencies related to chatter increases, see lower part of Figure 20. And we can see that chatter deteriorates the quality of the workpiece, as shown in Figure 19(L).

Spindle speed (rpm)	Resonance (Hz)	Resonance source
17900	1480	natural frequency of structures
19100	1271	spindle revolution frequency

Table 4. Resonances of sound pressure signal from different spindle speeds

Further frequency response function (FRF) analysis of sound pressure signal indicates that when the spindle speed is at 17900 rpm, the resonance is about 1480Hz, which is approaching the first modal shape frequency of the spindle-tool system given by $\omega_0\sqrt{1-\xi^2}/2\pi$. This indicates the occurrence of chatter. And when the speed is 19100rpm, the resonance is about 1271Hz, which is the spindle revolution frequency and its higher harmonics $kn\Omega/60$, with $k\in\mathbb{Z}^+$ and n is the number of the tooth. We list the results in Table 4. From Table 4, we can see that the FRF analyses also suggest the chatter occurrence when the spindle speed is at 17900rpm and no chatter at 19100rpm.

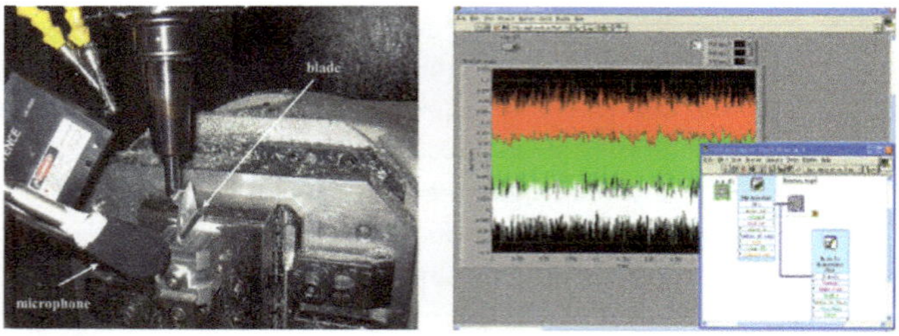

Figure 18. Experiment setup, (L)Machining setup (R)Labview signal acquisition interface

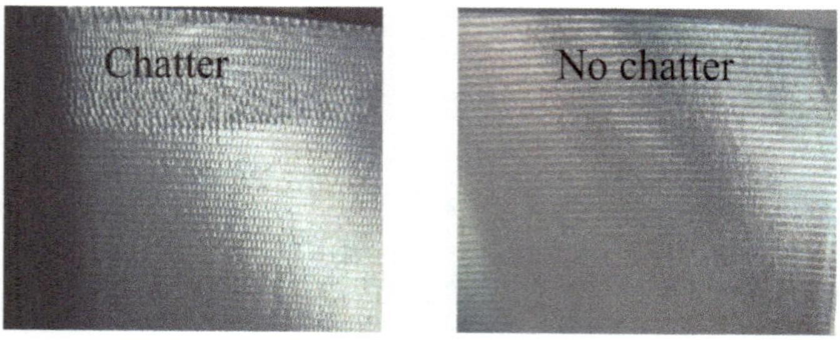

Figure 19. Comparison of finished blades using different spindle speeds, (L) from deterministic optimization (R) from robust optimization

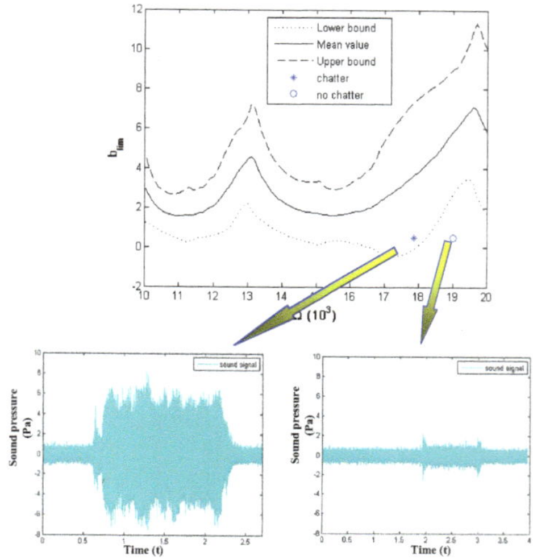

Figure 20. Signal of sound pressure with different spindle speeds. (U) Upper and lower bounds of Lobe diagram (L) Signal of sound pressure

4. Conclusions

In machining thin-wall workpieces, the temperature and force-induced deflection contribute significantly to the surface error. The proposed methodology is based on coupling effects between cutting forces and temperature and their induced deflection during machining. There is still a knowledge gap in identifying the impact of deflection on the process of metal removal, and hence there are not systematic approaches to modeling, prediction of the component errors due to thermo-mechanical deflection in thin-wall structures. And we develop a robust spindle speed optimization formulation. The quantitative analysis on how the uncertainties in milling process affect the milling stability and SLE are presented. Comparing with the traditional deterministic spindle speed optimization formulation, our model can take into account the uncertainties, i.e. modal shape parameters of spindle-tool system and the resulted optimized spindle speed ensures the milling stability.

Author details

YongAn Huang, Xiaoming Zhang and Youlun Xiong
State Key Laboratory of Digital Manufacturing Equipment and Technology,
Huazhong University of Science and Technology, Wuhan, China

Acknowledgement

The authors are grateful for supports from the National Natural Science Foundation of China (50835004,51005087).

5. References

[1] S. Ratchev, S. L. Liu, W. Huang and A. A. Becker (2007) Machining simulation and system integration combining FE analysis and cutting mechanics modelling. Int J Adv Manuf Tech. 35: 55-65.

[2] S. Herranz, F. J. Campa, L. N. L. d. Lacalle, A. Rivero, A. Lamikiz, E. Ukar, J. A. Sánchez and U. Bravo (2005) The milling of airframe components with low rigidity: A general approach to avoid static and dynamic problems. Proceedings of the Institution of Mechanical Engineers. Part B. Journal of engineering manufacture. 219: 789-801.

[3] M. Wan, W. H. Zhang, G. H. Qin and Z. P. Wang (2008) Strategies for error prediction and error control in peripheral milling of thin-walled workpiece. Int J Mach Tool Manu. 48: 1366-1374.

[4] S. Ratchev, S. Nikov and I. Moualek (2004) Material removal simulation of peripheral milling of thin wall low-rigidity structures using FEA. Advances in Engineering Software. 35: 481-491.

[5] S. Ratchev, S. Liu, W. Huang and A. A. Becker (2004) Milling error prediction and compensation in machining of low-rigidity parts. International Journal of Machine Tools and Manufacture. 44: 1629-1641.

[6] X. M. Zhang, L. M. Zhu, G. Zheng and H. Ding (2010) Tool path optimisation for flank milling ruled surface based on the distance function. International Journal of Production Research. 48: 4233-4251.

[7] H. Ding, Q. Z. Bi, L. M. Zhu and Y. L. Xiong (2010) Tool path generation and simulation of dynamic cutting process for five-axis NC machining. Chinese Science Bulletin. 55: 3408-3418.

[8] S. Ratchev, W. Huang, S. Liu and A. A. Becker (2004) Modelling and simulation environment for machining of low-rigidity components. J Mater Process Tech. 153–154: 67-73.

[9] S. M. Wang, Y. L. Liu and Y. A. Kang (2002) An efficient error compensation system for CNC multi-axis machines. Int J Mach Tool Manu. 42: 1235-1245.

[10] K. M. Y. Law and A. Geddam (2003) Error compensation in the end milling of pockets: a methodology. J Mater Process Tech. 139: 21-27.

[11] S. Ratchev, S. Liu and A. A. Becker (2005) Error compensation strategy in milling flexible thin-wall parts. J Mater Process Tech. 162: 673-681.

[12] E. Budak and Y. Altintas (1995) Modeling and Avoidance of Static Form Errors in Peripheral Milling of Plates. Int J Mach Tool Manu. 35: 459-476.

[13] H. Y. Feng and C. H. Menq (1994) The Prediction of Cutting Forces in the Ball-End Milling Process .1. Model Formulation and Model-Building Procedure. Int J Mach Tool Manu. 34: 697-710.

[14] H. Y. Feng and C. H. Menq (1994) The Prediction of Cutting Forces in the Ball-End Milling Process .2. Cut Geometry Analysis and Model Verification. Int J Mach Tool Manu. 34: 711-719.

[15] E. Budak, Y. Altintas and E. J. A. Armarego (1996) Prediction of milling force coefficients from orthogonal cutting data. J Manuf Sci E-T Asme. 118: 216-224.

[16] S. Ratchev, S. Liu, W. Huang and A. A. Becker (2004) A flexible force model for end milling of low-rigidity parts. J Mater Process Tech. 153: 134-138.

[17] Y. Huang, H. Liu, Z. Yin and Y. L. Xiong (2009) Complex Surface Machining: Thermo-mechanical Analysis for Error Prediction of Low-Rigidity Workpiece. Intelligent Robotics and Applications: 666-677.
[18] A. D. Spence, F. Abrari and M. A. Elbestawi (2000) Integrated solid modeller based solutions for machining. Comput Aided Design. 32: 553-568.
[19] J. K. Rai and P. Xirouchakis (2009) FEM-based prediction of workpiece transient temperature distribution and deformations during milling. Int J Adv Manuf Tech. 42: 429-449.
[20] M. Wiercigroch and E. Budak (2001) Sources of nonlinearities, chatter generation and suppression in metal cutting. Philos T Roy Soc A. 359: 663-693.
[21] M. L. Campomanes and Y. Altintas (2003) An improved time domain simulation for dynamic milling at small radial immersions. J Manuf Sci E-T Asme. 125: 416-422.
[22] Y. Altintas and E. Budak (1995) Analytical prediction of stability lobes in milling. Cirp Ann-Manuf Techn. 44: 357-362.
[23] T. Insperger and G. Stepan (2004) Updated semi-discretization method for periodic delay-differential equations with discrete delay. Int J Numer Meth Eng. 61: 117-141.
[24] B. P. Mann, PhD, Washington University, 2003.
[25] E. Budak and A. Tekeli (2005) Maximizing chatter free material removal rate in milling through optimal selection of axial and radial depth of cut pairs. Cirp Ann-Manuf Techn. 54: 353-356.
[26] Y. Altintas and S. D. Merdol (2007) Virtual high performance milling. Cirp Ann-Manuf Techn. 56: 81-84.
[27] G. S. Duncan, M. H. Kurdi, T. L. Schmitz and J. Snyder (2006) Uncertainty propagation for selected analytical milling stability limit analyses. Transactions of NAMRI/SME. 34: 17-24.
[28] M. H. Kurdi, Ph.D, University of Florida, 2005.
[29] G. Totis (2009) RCPM-A new method for robust chatter prediction in milling. Int J Mach Tool Manu. 49: 273-284.
[30] A. Milfelner, J. Kopac, F. Cus and U. Zuperl (2005) Genetic equation for the cutting force in ball-end milling. J Mater Process Tech. 164: 1554-1560.
[31] M. Milfelner and F. Cus (2003) Simulation of cutting forces in ball-end milling. Robot Cim-Int Manuf. 19: 99-106.
[32] P. Lee and Y. Altintas (1996) Prediction of ball-end milling forces from orthogonal cutting data. Int J Mach Tool Manu. 36: 1059-1072.
[33] A. Lamikiz, L. N. L. de Lacalle, J. A. Sanchez and M. A. Salgado (2004) Cutting force estimation in sculptured surface milling. Int J Mach Tool Manu. 44: 1511-1526.
[34] R. Ramesh, M. A. Mannan and A. N. Poo (2000) Error compensation in machine tools - a review Part I: geometric, cutting-force induced and fixture-dependent errors. Int J Mach Tool Manu. 40: 1235-1256.
[35] R. Ramesh, M. A. Mannan and A. N. Poo (2000) Error compensation in machine tools - a review Part II: thermal errors. Int J Mach Tool Manu. 40: 1257-1284.
[36] S. Ratchev, E. Govender, S. Nikov, K. Phuah and G. Tsiklos (2003) Force and deflection modelling in milling of low-rigidity complex parts. J Mater Process Tech. 143: 796-801.
[37] M. B. da Silva and J. Wallbank (1999) Cutting temperature: prediction and measurement methods - a review. J Mater Process Tech. 88: 195-202.

[38] M. R. Miller, G. Mulholland and C. Anderson (2003) Experimental cutting tool temperature distributions. J Manuf Sci E-T Asme. 125: 667-673.
[39] Y. K. Potdar and A. T. Zehnder (2003) Measurements and Simulations of Temperature and Deformation Fields in Transient Metal Cutting. Journal of Manufacturing Science and Engineering. 125: 645-655.
[40] N. A. Abukhshim, P. T. Mativenga and M. A. Sheikh (2006) Heat generation and temperature prediction in metal cutting: A review and implications for high speed machining. Int J Mach Tool Manu. 46: 782-800.
[41] M. A. Davies, T. Ueda, R. M'Saoubi, B. Mullany and A. L. Cooke (2007) On the measurement of temperature in material removal processes. Cirp Ann-Manuf Techn. 56: 581-604.
[42] J. Lin (1995) Inverse estimation of the tool-work interface temperature in end milling. International Journal of Machine Tools and Manufacture. 35: 751-760.
[43] P. Kwon, T. Schiemann and R. Kountanya (2001) An inverse estimation scheme to measure steady-state tool–chip interface temperatures using an infrared camera. International Journal of Machine Tools and Manufacture. 41: 1015-1030.
[44] G. Fang and P. Zeng (2005) Three-dimensional thermo-elastic-plastic coupled FEM simulations for metal oblique cutting processes. J Mater Process Tech. 168: 42-48.
[45] J. M. Lin (1995) Inverse Estimation of the Tool-Work Interface Temperature in End Milling. Int J Mach Tool Manu. 35: 751-760.
[46] M. Chen, F. H. Sun, H. L. Wang, R. W. Yuan, Z. H. Qu and S. Q. Zhang (2003) Experimental research on the dynamic characteristics of the cutting temperature in the process of high-speed milling. J Mater Process Tech. 138: 468-471.
[47] C. E. Leshock and Y. C. Shin (1997) Investigation on Cutting Temperature in Turning by a Tool-Work Thermocouple Technique. Journal of Manufacturing Science and Engineering. 119: 502-508.
[48] S. Fraser, M. H. Attia and M. O. M. Osman (1998) Modelling, identification and control of thermal deformation of machine tool structures, part 1: Concept of generalized modelling. J Manuf Sci E-T Asme. 120: 623-631.
[49] X. Zhang, L. Zhu, D. Zhang, H. Ding, Y. Xiong, Numerical robust optimization of spindle speed for milling process with uncertainties, International Journal of Machine Tools and Manufacture (2012), http://dx.doi.org/10.1016/j.ijmachtools.2012.05.002
[50] T. K. Hasselman and J. D. Chrostowski (1997) Effects of product and experimental variability on model verification of automobile structures. P Soc Photo-Opt Ins. 3089: 612-620.
[51] E. Balmes (1998) Predicted variability and differences between tests of a single structure. Imac - Proceedings of the 16th International Modal Analysis Conference, Vols 1 and 2. 3243: 558-564.
[52] M. H. Kurdi, R. T. Haftka, T. L. Schmitz and B. P. Mann (2008) A Robust Semi-Analytical Method for Calculating the Response Sensitivity of a Time Delay System. J Vib Acoust. 130.
[53] J. Nocedal and S. J. Wright (1999) Numerical optimization: Springer verlag.
[54] M. H. Richardson and D. L. Formenti, Parameter estimation from frequency response measurements using rational fraction polynomials, 1982.
[55] Y. Altintas, E. Shamoto, P. Lee and E. Budak (1999) Analytical prediction of stability lobes in ball end milling. J Manuf Sci E-T Asme. 121: 586-592.

Chapter 16

Concurrent Process-Product Design Optimization Using Coupled Nonlinear Finite-Element Simulations

Ali Najafi and Masoud Rais-Rohani

Additional information is available at the end of the chapter

1. Introduction

Finite-element (FE) simulations are often used to predict the response characteristics of a structural component under different boundary conditions and to help explore the design space for the optimum design while minimizing the need for physical testing. It has also been used to model various manufacturing processes, especially those involving the forming process (Cheng and Kikuchi 1985; Chung et al. 1998; Li et al., 2002; van den Boogaard et al., 2003).

Since such FE simulations and the accompanying structural design optimization studies rely on computer-based geometric model of the structural component and the tabulated material properties, the initial state (e.g., internal stresses and strains) of the component/material is most often ignored, which results in exclusion of the manufacturing process effects on the product (e.g., plastic strain, residual stress, thinning, springback, etc.). The selected manufacturing process and the choice of process parameters can also change the material microstructure (e.g., dislocation density, texture), thereby affecting the macroscale (e.g., stress-strain) behavior of the material and the structural component (Najafi et al., 2012; Oliveira et al., 2006).

A practical way to alleviate this shortcoming is to perform coupled process-performance simulations in a sequential manner whereby both changes in the material and component can be properly modeled and tracked from one stage to the next for a more accurate prediction of the structural performance measures (Noels et al., 2004). Consequently, process parameters can be evaluated based on both the process objectives and performance criteria. Coupling of the material, process, and performance models is an important step in modeling the actual physical behavior of the material and structure while facilitating the integrated material-process-product design (Olson, 1997; McDowell et al., 2007; Acar et al., 2009).

Coupled quasi-static analyses can be performed in some implicit finite element analysis (FEA) codes such as Abaqus/Standard where the boundary conditions can be changed and the material state can be tracked and passed from one solution step to the next by using the restart option, for example. The same capability also exists in some explicit FEA codes such as LS-DYNA and Abaqus/Explicit for coupled transient dynamic simulations. The increasing demand for coupled simulations has resulted in many commercial software codes to have an option to map some solutions as the initial state. However, when some combinations of dynamic and quasi-static analyses (i.e., explicit and implicit solvers) are required, a separate data flow management (DFM) program and strategy are required. The DFM procedure becomes more complex in the context of design optimization when the coupled analyses need to be performed numerous times for design space exploration in search of the optimum design. A particular case being considered in this chapter is the concurrent design of a coupled process-product system where both manufacturing process and product performance attributes are to be optimized by finding the optimum values of the manufacturing process on product design variables.

Coupled process-product (performance) simulation studies include those that have considered the manufacturing effects associated with forming and springback on crush/crash performance of tubular components (Oliveira et al., 2006; Kellicut et al., 1999; Kaufman et al., 1998; Grantab, 2006; Krusper, 2003). For example, Kellicut et al. (1999) considered springback, thinning, and other parameters such as plastic strain and residual stresses in bending-crush simulations of hydroformed tubes and showed that the plastic strain has the greatest effect on the crush behavior. Mayer (2004) and Williams et al. (2005) also performed integrated hydroforming-crush simulations, whereas Ryou et al. (2005) used ideal forming solution to extract the stress and strain responses from forming process and a hybrid membrane/shell method to pass the information to impact simulation. They improved the computation time while preserving the accuracy of the FE simulations. Simunovic and Aramayo (2002) studied the crash response of energy absorbing components of the ultra light steel auto body vehicle models and showed that by including the history effects the energy absorption properties can change even though the difference in the overall response was relatively modest.

Oliveira et al. (2006) and Williams et al. (2005) performed experimental and computational study of s-rail tubes and discovered that both the maximum and mean crush force values will change as a result of the manufacturing process effects. Bottcher and Frik (2003) did a similar study and showed that metal forming data is required in crash simulation of front rail panel of a vehicle model, especially in high strength dual phase steel due to its rapidly hardening characteristic. Krusper (2003) and Dagson (2001) performed analysis on a simple bar while considering the springback response of the material. Most of the studies cited only considered a material model with isotropic hardening while a few included the effect of kinematic hardening on the crush response. Recently, Williams et al. (2010) studied the effect of combined isotropic/kinematic hardening and strain rate sensitivity along with an anisotropic yield surface to study the crush behavior of hydroformed tubes. They showed that the combination of these parameters increases the capability of models to correctly predict the energy absorption performance of the crush tube. However, they did not investigate the capabilities of the model in coupled process-performance simulations.

This chapter presents the different steps in performing sequential coupled nonlinear FE simulations and their application in multi-objective process-product design optimization of thin-walled structures. The sheet stamping simulation includes both the deep drawing and the springback stages of the manufacturing process. Performance simulation considers the energy absorption response characteristics of the component under an impact (crash) loading condition. The coupled simulations involve both nonlinear explicit and implicit FEA. The developed computational framework is used for the analysis and design of a double-hat tube. A design sensitivity analysis is performed to investigate the effect of manufacturing process parameters and geometric attributes on the process and performance responses that are affected by the manufacturing process and geometric design parameters. The ensuing nonlinear design optimization problem is cast in a multi-objective formulation and solved for Pareto optimum design points using a multi-objective genetic algorithm.

2. Modeling of manufacturing effects

Constitutive models describe the stress-strain relationship for a given material and the influence of various factors such as temperature and strain rate. Plastic deformation requires variables that define the history of stress and temperature in the material. The history can be defined through functional analysis and mathematical theories known as the theory of material with memory (Lubliner, 2006). Manufacturing effects in most continuum-level material models are considered implicitly through the state variables defined in the model. For example, in the classical plasticity model, the total strain is written as an additive decomposition of elastic strain and plastic strain. Considering a piecewise linear isotropic hardening law derived from the stress-strain data, the equivalent plastic strain would represent, in a limited sense, the history or the manufacturing effect. Inclusion of manufacturing effects in the continuum plasticity models emerges by specifying the initial state for the numerical integration of the evolution equation of the state variable.

The constitutive relation of the rate dependent classical plasticity model with isotropic hardening can be represented through the following equations:

- Elastic stress-strain relationships in three-dimensional space

$$\dot{\boldsymbol{\sigma}} = \mathbb{C}\colon (\dot{\boldsymbol{\varepsilon}} - \dot{\boldsymbol{\varepsilon}}^{vp}) \tag{1}$$

- Yield surface (closure of elastic domain in the stress space)

$$f(\boldsymbol{\sigma}, \kappa) = |\boldsymbol{\sigma}| - (\sigma_y + H\kappa) \leq 0 \tag{2}$$

- Flow rule and hardening law

$$\dot{\boldsymbol{\varepsilon}}^{vp} = \dot{\lambda}\,\frac{\partial f(\boldsymbol{\sigma}, \kappa)}{\partial \boldsymbol{\sigma}}, \dot{\kappa} = \dot{\lambda} \tag{3}$$

$$\dot{\lambda}(\boldsymbol{\sigma},\kappa)=\begin{cases}\frac{1}{C}\left[\left(\frac{|\sigma|}{\sigma_y}\right)^{1/P}-1\right] & if\ f(\boldsymbol{\sigma},\kappa)\geq 0\\ 0\ if\ f(\boldsymbol{\sigma},\kappa)<0\end{cases} \tag{4}$$

where the bold symbols $\boldsymbol{\sigma}$ and $\boldsymbol{\varepsilon}$ represent the second-rank stress and stain tensors, respectively, and $\mathbb{C}$ is the forth-rank tensor of isotropic elasticity. The superscript vp designates the viscoplastic component. The $\dot{\blacksquare}$ terms indicate time derivatives. The yield stress is denoted by σ_y with parameters H and κ representing isotropic hardening modulus and plastic multiplier, respectively. Moreover, C and P in the hardening law equation are known as material constants associated with rate sensitivity, which are found based on experimental data on a specific material. As mentioned previously, the manufacturing effects emerge as the consequence of numerical integration of the model described by Eqs. (1) through (4).

The numerical integration scheme in the deviatoric space is defined based on updating the current state of stresses and state variables (with subscript n) by calculating their increment $\Delta\,(\blacksquare)$. The deviatoric part of total strain tensor $\boldsymbol{e}$ is updated as

$$\boldsymbol{e}_{n+1}=\boldsymbol{\varepsilon}_{n+1}-\frac{1}{3}tr\left(\boldsymbol{\varepsilon}_{n+1}\right)\boldsymbol{I} \tag{5}$$

where $\boldsymbol{I}$ is the identity tensor and $tr(\blacksquare)$ is the trace of the second rank tensor. The trial stress tensor $\boldsymbol{s}_{n+1}^{t}$ is calculated from the Hook's law given by

$$\boldsymbol{s}_{n+1}^{t}=2\mu\left(\boldsymbol{e}_{n+1}-\boldsymbol{e}_{n}^{p}\right) \tag{6}$$

where the $\boldsymbol{e}_{n}^{p}$ is the plastic strain at the nth increment and μ is the shear modulus. Then, the calculated stress magnitude value at n+1 increment $\|\boldsymbol{s}_{n+1}^{t}\|$ is evaluated with the yield surface f_{n+1}^{t} at the same increment. If $f_{n+1}^{t}=\|\boldsymbol{s}_{n+1}^{t}\|-\sqrt{\frac{2}{3}}\left[\sigma_y+H\kappa\right]\leq 0$ is met, the material is in the elastic region and the trial values are the admissible stress values and, therefore, the plastic multiplier $\dot{\lambda}$ or $\Delta\lambda_{n+1}$ is zero with

$$(\blacksquare)_{n+1}=(\blacksquare)_{n+1}^{t} \tag{7}$$

And if $f_{n+1}^{t}>0$, the trial values should be corrected due to an additional plastic strain and the yield surface is approximated through the first-order Taylor series expansion as

$$f_{n+1}^{t}\approx f_{n}^{t}+\frac{\partial f}{\partial\Delta\lambda}\Delta\lambda_{n+1} \tag{8}$$

where $\Delta\lambda_{n+1}$ is computed based on Newton-Raphson scheme (Wang and Budiansky, 1978; Simo and Hughes, 2000; Souza et al., 2008). Therefore, the next state of material (with subscript n+1) is derived by adding the increments $\Delta(\blacksquare)$ to the previous state of stress as

$$\kappa_{n+1}=\kappa_n+\sqrt{\frac{2}{3}}\,\Delta\lambda_{n+1} \tag{9}$$

$$\boldsymbol{\varepsilon}_{n+1}^{vp} = \boldsymbol{\varepsilon}_{n}^{vp} + \Delta\lambda_{n+1}\,\boldsymbol{n}_{n+1} \tag{10}$$

$$\Delta\boldsymbol{\sigma}_{n+1} = \bar{\lambda}\,tr(\Delta\boldsymbol{\varepsilon}_{n+1})\boldsymbol{I} + \Delta\boldsymbol{s}_{n+1}^{t} - 2\mu\,\Delta\lambda_{n+1}\,\boldsymbol{n}_{n+1} \tag{11}$$

$$\boldsymbol{\sigma}_{n+1} = \boldsymbol{\sigma}_{n} + \Delta\boldsymbol{\sigma}_{n+1} \tag{12}$$

where $tr(\blacksquare)$ represents the trace of the tensor, trial values are denoted with t superscript, and the normal to the yield surface is found as $\boldsymbol{n}_{n+1} = \boldsymbol{s}_{n+1}^{t}/\|\boldsymbol{s}_{n+1}^{t}\|$. By having $\kappa_n = \boldsymbol{\varepsilon}_{n}^{vp}$, the manufacturing effect can be considered through a non-zero state of $\boldsymbol{\sigma}_n$ and $\boldsymbol{\varepsilon}_{n}^{vp}$ at the beginning of the analysis as $\boldsymbol{\sigma}_1$ and $\boldsymbol{\varepsilon}_{1}^{vp}$.

Traditionally, the initial values for the first increment are considered to be zero assuming that the material deformation starts from a conceptual pristine state. However, incorporating proper initial values emanating from forming and springback simulations in the crush analysis couples the process and performance from the material standpoint.

3. Manufacturing process and product performance simulations

Coupled process and performance simulations are conducted sequentially by using Abaqus/Explicit for the deep drawing (loading) simulation, Abaqus/Standard for the springback (unloading) simulation under isothermal condition, followed by Abaqus/Explicit for the crush simulation. The illustrative example considered in this chapter is a double-hat, thin-walled tube that is modeled by joining two identical single hat sections.

3.1. Deep-drawing simulation

For forming or deep-drawing simulation, two sets of blank/holder/die/punch geometries are defined in the FE model with the model for one set shown in Fig. 1. A single die set is mirrored with respect to blank plane to simulate the forming of both pieces simultaneously. There are two basic steps in using the explicit FEA for the forming simulation of each single hat section depending on the specified boundary conditions.

The first step defines the gripping of the blank between the die set and the holders. While keeping the punch and dies fixed in their respective positions, the holding forces are increased linearly from zero to the specified value matching the corresponding manufacturing process control parameter. In this stage, the kinematic contact formulation is used because of the computational efficiency of the formulation (Rebel et al., 2002; Oden and Kikuchi, 1982; Oden and Pires, 1983; Bayram and Nied, 2000; Wriggers, 2006). Due to the simplicity of the geometry, die and holder surfaces are defined through standard analytical rigid surfaces. Contact surfaces are defined on both surfaces of the blank by considering the surface offset due to the blank thickness. Penalty formulation is used in tangential contact and a friction coefficient is defined as a manufacturing process parameter for an equivalent representation of both surface roughness and draw beads.

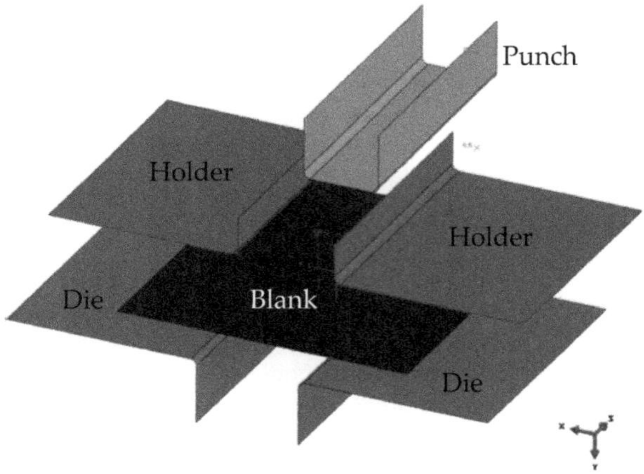

Figure 1. Sheet-forming FE model of a single hat section

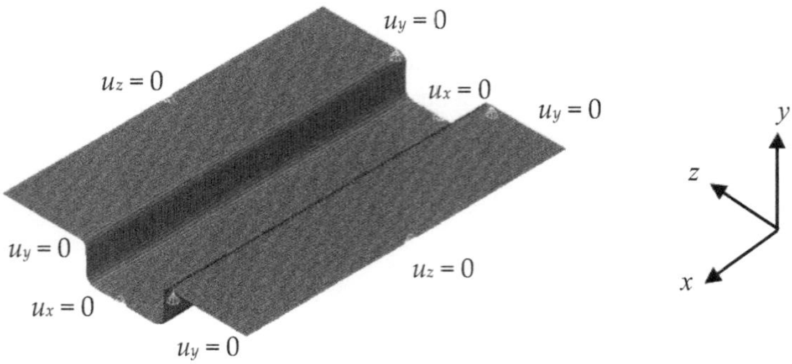

Figure 2. Formed hat section with the boundary conditions used for the subsequent springback simulation

The second step performs the deep drawing simulation by applying a constant velocity to the punch that is configured in the model to match the final geometry of the product (excluding the springback effect). Hence, the results of the previous step are directly transferred to this step where the boundary conditions on the fixed punch in the direction normal to the blank surface are removed and a constant velocity is applied to the punch to form the single hat section as shown in Fig. 2. The punch velocity is assumed to be constant for a linear displacement. This setup does not impose a uniform strain rate in all the elements. Thus, rate sensitivity of the material will not have a uniform effect on the structure. Dies remain clamped and the holders are fixed in all degrees of freedom except the direction parallel to the punch movement. In this direction, the constant holding force is applied to preserve the constant gripping force throughout the drawing process.

Rupture and thinning are two responses that can be calculated from the results of the deep-drawing simulation. Rupture is calculated by extracting the principal major and minor plastic strains in each element and taking the difference between the major strain and that extracted from the forming limit diagram (FLD) using the following equation

$$R = \begin{cases} \sum_{i=1}^{I} R_i^2 = \sum_{i=1}^{I} (\varepsilon_1^i - \phi(\varepsilon_2^i))^2 & \varepsilon_1^i > \phi(\varepsilon_2^i) \\ 0 & \varepsilon_1^i \leq \phi(\varepsilon_2^i) \end{cases} \quad (13)$$

where $\phi(\varepsilon_2)$ is the equation representing FLD curve and ε_1^i and ε_2^i are the principal major and minor strains at each integration point through the thickness, respectively, which are calculated in the deep-drawing simulation when the termination time reaches the limit. The parameter I represents the total number of integration points in the mesh. The FLD (Lee et al., 2008) that is used in this study is that for AZ31 magnesium alloy sheet and it is assumed to behave linearly in both compressive and tensile plastic strains as shown in Fig. 3.

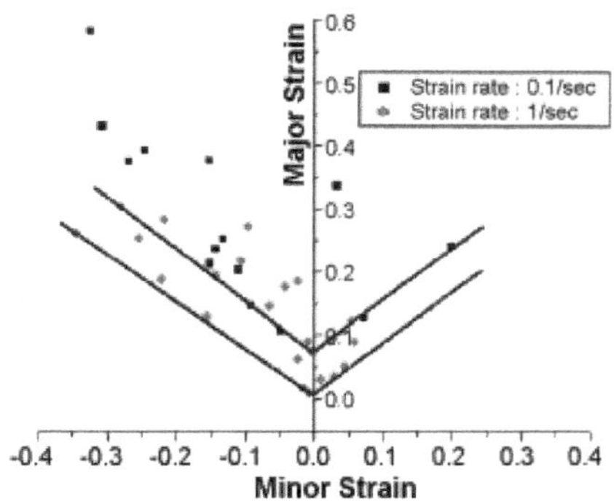

Figure 3. FLD for AZ31 at two different strain rates (Lee et al., 2008)

Thinning is measured by taking the difference between the final thickness of each element and its corresponding initial value and is calculated using a single metric, T defined as

$$T = \sum_{i=1}^{N} T_i^2 = \sum_{i=1}^{N} (\frac{t_i - t_o}{t_o})^2 \quad (14)$$

where t_o and t_i are the initial and final shell thicknesses, respectively, and N is the total number of elements in the mesh. Since the shell thickness in the blank is assumed to be constant for all the elements, t_o is the shell thickness assigned to the elements.

3.2. Springback simulation

The state variables and geometric information from the deep drawing simulation are transferred and treated as the initial state in the unloading stage (i.e., removal of all the tooling parts from the workpiece) for the springback analysis. Springback process is modeled as a quasi-static problem considering the stress distribution captured from deep drawing, dynamic effects, and the contact conditions. Additionally, all the rigid surfaces including punch, dies, and holders are removed from the FE model, which makes the model more suitable for implicit FEA considering the quasi-static nature of the springback phenomenon and the absence of highly nonlinear factors in the model. Convergence of the nonlinear implicit FEA are guaranteed by defining the boundary conditions in such a way that the two edges of the hat section are held fixed perpendicular to the actual normal surface. As shown in Fig. 2, the equilibrium condition is achieved by constraining the model in all the transverse directions. The boundary conditions defined in this stage are designed such that the effect of the force required to assemble a non-fitted double-hat section is already considered. The residual stresses and geometric attributes are updated during quasi-static analysis while the other computational state variables such as plastic strain remain unchanged. Similar to the deep-drawing simulation, the springback analysis is performed separately and simultaneously on the two identical single hat sections. There is no interaction between the two hat sections, however, in both the deep drawing and springback simulations. The two hat sections are then assembled in the next stage to produce a double-hat crush tube.

Springback is calculated by comparing the nodal coordinates obtained in the last step of the deep-drawing simulation with those in the last step of the springback simulation. A single springback metric, S representing the deviation of the nodal coordinates is calculated as

$$S = Max\left(\sqrt{(X_i - X_o)^2 + (Y_i - Y_o)^2 + (Z_i - Z_o)^2}\right) \quad \forall i = 1, NN \tag{15}$$

where X, Y, Z are the Cartesian coordinates with subscripts i and o representing the end of springback and deep-drawing stages, respectively with NN as the total number of nodes in the mesh.

An in-house FORTRAN code is used to automate the procedure to extract the rupture and thinning results from the Abaqus binary file, calculate the principal strains, and incorporate the equations mentioned above (without using Abaqus CAE) in the deep-drawing and springback simulations.

3.3. Joining and trimming

The two hat sections are joined longitudinally using tie contact formulation and trimmed by removing the outer flange elements as shown in Fig. 4. The tie contact constrains the master and slave surfaces similar to the multiple constraint points when the clearance between two surfaces is below the tolerance specified as input. If the surfaces are out of the prescribed tolerance, the interaction becomes a contact formulation. A preliminary study showed that switching the master and slave surfaces would not affect the crushing behavior of the tube. Once the distance between the two surfaces becomes more than the clearance tolerance, constraints are removed and contact formulation is activated similar to the conventional contact definition. It is worth noting that this kind of joining represents a fictitious weld seam along each joint line since the thermo-mechanical process involved in an actual welding process is not modeled here.

Figure 4. Joining and trimming of the two formed sections to generate the final double-hat tube geometry

3.4. Crush simulation

The explicit FEA for crush simulation is performed by holding one end of the tube fixed while applying an axial load through a moving rigid wall at the other end. The rigid wall moves with a constant speed to simulate constant loading rate defined with a prescribed displacement as shown in Fig. 5. The component geometry, residual stresses, and state variables at the start of crush stage are those accumulated through all the manufacturing stages discussed previously.

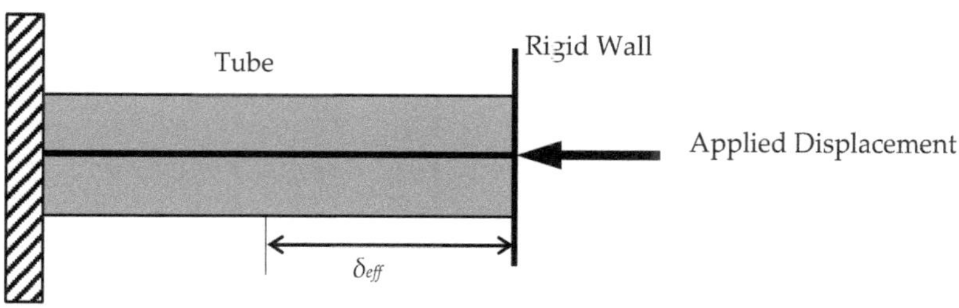

Figure 5. General setup for crush simulation

Six contact interaction sets among the elements are defined in the crush simulation, including interactions between the lower hat section and the rigid wall, the upper hat section and the rigid wall, interaction between the upper and lower hat sections, tie contact between the assembly edge of the upper and lower surfaces, and separate self contact interaction for the upper and lower hat sections. For all of the aforementioned contact interaction sets, penalty function formulation in both normal and tangential directions is used. Despite the computational cost, penalty function provides a proper flexibility for the explicit FEA to find a stable time step that is affected by severity of the contacts. Moreover, the maximum ratio of thickness-to-element length is used to overcome the difficulty of the fine mesh density that results in relatively thick shell elements.

The contact force history of the rigid wall during the crush simulation is used to calculate the maximum crush force, P_{max} while the mean crush force, P_m is found by dividing the area under the crush force-displacement curve by the effective crush distance. The mathematical equation for finding the mean crush force is expressed as

$$P_m = \frac{1}{\delta_{eff}} \int_0^t F(t)D(t)dt \tag{16}$$

where $F(t)$ is the instantaneous contact force normal to the rigid wall surface and $D(t)$ is the instantaneous cross-head axial displacement of the rigid wall. The final cross-head displacement of the rigid wall represents δ_{eff} and it is assumed to be 125 mm, or 50% of the tube length. The maximum crush force is the largest value of $F(t)$ found after applying the SAE filtering of 60 Hz to filter the noise in the force data. These results are extracted and filtered using a Python scripting application available in Abaqus and used as input to an in-house FORTRAN code to calculate the mean crush force values.

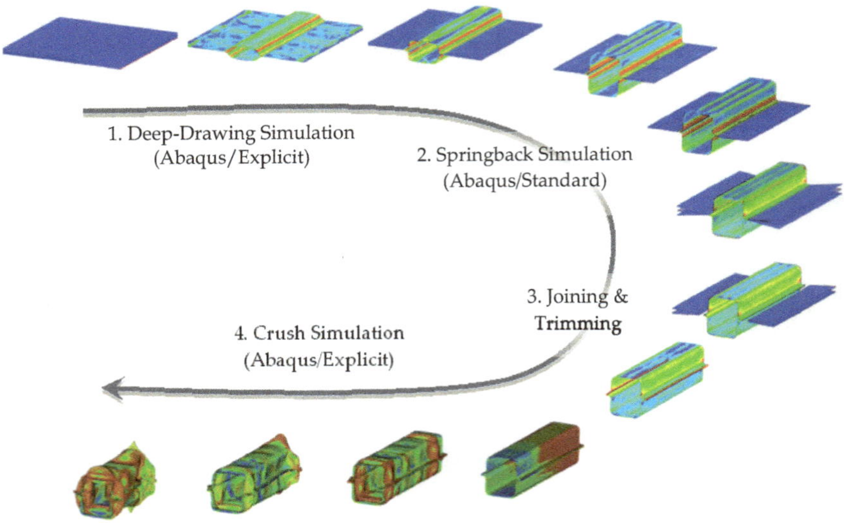

Figure 6. Sequence of coupled nonlinear FE simulations

Figure 6 shows the general four-step sequence of coupled simulations from the initial sheet forming to crush. The rupture and thinning responses are the outputs of step 1, springback is the result of step 2 while the mean and max crush force values are the outputs of step 4. Other than the joining of the two hat sections and removal of the excess tabs, no changes are induced in the tube model in step 3.

3.5. Material model

The material model uses piecewise linear isotropic hardening. The constant for the linear kinematic hardening is calculated based on the slope of a line connecting two adjacent points on the stress-strain curve. The material model uses von Misses yield surface and a one-dimensional stress-strain input is considered as equivalent von Misses stress versus effective plastic strain. Coupling scheme is utilized by transferring residual stresses and the equivalent plastic strains as the material state variables. The yield surface expands due to the isotropic hardening assumption in the model; therefore, the instantaneous yield point varies during the loading process. The yield point at the end of the forming simulation is captured by finding the plastic strain.

The AZ31 magnesium alloy sheet data is used for all the simulations. For modeling the stress-strain response at various strain rates, the stress-strain curves for two extreme rates (i.e., 1 s^{-1} and 4300 s^{-1}) are considered with those for the other rates found through interpolation. The elastic modulus, Poisson's ratio, and density are 45 GPa, 0.33, and 1.738 kg/m^3, respectively. The true stress-true strain curves for the two extreme strain rates are shown in Fig. 7. Adiabatic heating is not considered in any of the simulations.

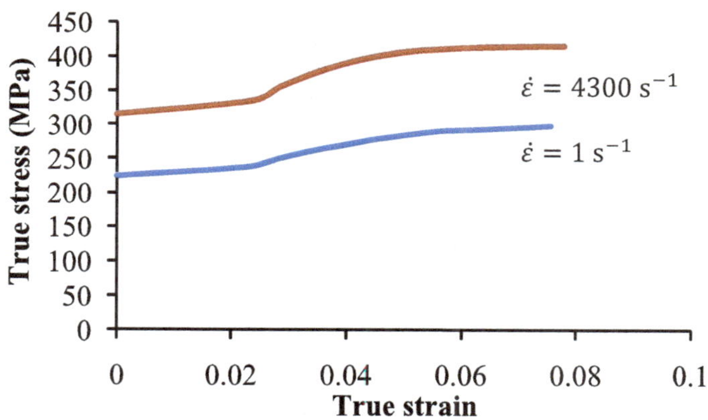

Figure 7. AZ31 magnesium alloy sheet stress-strain curves for two different strain rates

3.6. Effect of manufacturing process on product performance

To examine the role of the history effects on the axial crush response, two simulation cases are compared. In the first case, separate stand-alone performance simulation that does not include any history effects is performed, whereas in the second case, a sequential coupled process-performance simulation is performed that includes residual stresses, plastic strains, thinning, and springback information from process simulation (as manufacturing effects) together with a piecewise linear isotropic hardening material model in performance simulation. The 250-mm long tubes are modeled using the plane-stress shell element formulation. They are held fixed at one end and axially loaded with a flat rigid wall at the other end that is moving with a constant speed of 5 m/s. Both self-contact and surface-to-surface contact between rigid wall and tube are specified. A classical multi-linear kinematic hardening material model is used for this comparison.

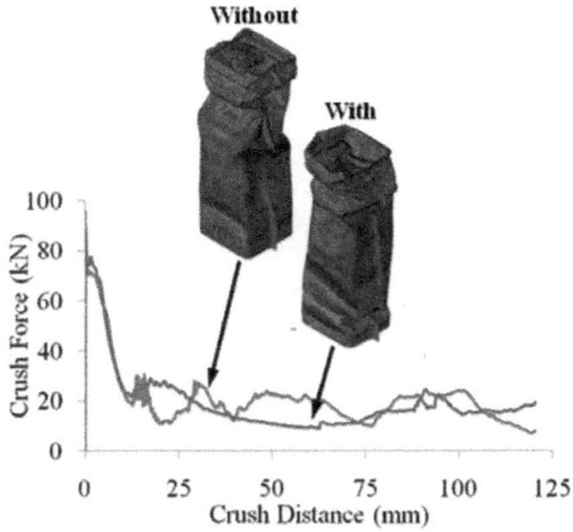

Figure 8. Crush behavior with and without consideration of manufacturing process effects

Figure 8 shows the two crush force-distance curves as well as the corresponding crush modes. The results clearly show that both the crush response and the collapse mode change due to the inclusion of the history effects. The maximum crush force increases by approximately 10% from 77 kN to 85 kN, whereas the mean crush force decreases from 20 kN to 18 kN when the history effects are considered.

4. Sensitivity analysis

A design sensitivity analysis is helpful in capturing the main effects of the individual process and product design variables on both the manufacturing and performance attributes (Najafi 2011). The product design variables are the tube cross-sectional dimensions (i.e., width, height of a single hat section, corner radius, and initial blank thickness) shown in Fig. 9, whereas the process design variables are the holding force, punch velocity and workpiece / die set friction coefficients.

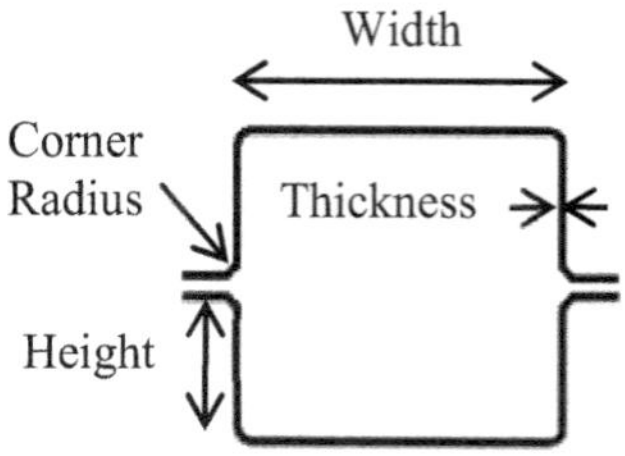

Figure 9. Description of geometric design variables for a 250-mm long tube

The friction coefficients for the holders, dies, and punch are assumed to be equal but can be treated as different design variables. The width design variable defines the punch width, the corner radius defines the die and holder corner radii, the thickness design variable is assigned directly to the shell elements that define the blank and the height design variable is controlled by the punch travel distance in the direction normal to the blank surface (this parameter also determines the simulation termination time as well as the prescribed punch velocity). Holding force defined as a manufacturing process parameter is the amount of maximum incremental force in the first step of deep drawing. The rate of holding force application is kept constant in all the simulations. Punch velocity is assumed to be constant in the direction perpendicular to the sheet metal; this parameter along with the height determines the deep-drawing simulation termination time. Friction coefficients are assigned to the contact tangential definition. Both kinematic and penalty tangential contact formulations produced the same response in the deep drawing simulation.

Design Variable	Width (mm)	Height (mm)	Corner Radius (mm)	Thicknes s (mm)	Holding Force (kN)	Punch Velocity (m/s)	Friction Coefficien t
Width	41.25	27.5	5.0	1.75	30	6.0	0.225
±15%	63.25	27.5	5.0	1.75	30	6.0	0.225
Height	55	20.625	5.0	1.75	30	6.0	0.225
±15%	55	31.625	5.0	1.75	30	6.0	0.225
Corner	55	27.5	3.75	1.75	30	6.0	0.225
Radius ±15%	55	27.5	5.75	1.75	30	6.0	0.225
Thickness	55	27.5	5.0	1.3125	30	6.0	0.225
±15%	55	27.5	5.0	2.0125	30	6.0	0.225
Holding	55	27.5	5.0	1.75	22.5	6.0	0.225
Force ±15%	55	27.5	5.0	1.75	34.5	6.0	0.225
Punch	55	27.5	5.0	1.75	30	4.5	0.225
Velocity ±15%	55	27.5	5.0	1.75	30	6.9	0.225
Friction	55	27.5	5.0	1.75	30	6.0	0.16875
Coefficient ±15%	55	27.5	5.0	1.75	30	6.0	0.25875
Upper Bound	70	35	7.5	2.5	50	10.0	0.35
Mean	55	27.5	5.0	1.75	30	6.0	0.225
Lower Bound	40	20	2.5	1.0	100	2.0	0.1

Table 1. The values assigned to design variables for sensitivity analysis

The sensitivity results are shown in Fig. 10. In each case, the sensitivity values are found by perturbing one design variable by +/-15% from its corresponding average value while holding the remaining design variables fixed at their respective average values shown in bold numbers in Table 1. Rupture is found to be most sensitive to the sheet thickness followed by the corner radius. In contrast, the friction coefficient, punch velocity, and holding force appear to have minimal effect. The rupture response was found to have a direct relationship with some parameters such as thickness and punch velocity and inverse relationship with others, corner radius being the most notable. The global measure of thinning is affected the most by changes in the corner radius, followed by blank thickness and height. In comparison, the manufacturing process parameters appeared to be less influential. Springback response is most sensitive to changes in blank thickness followed by the corner radius and friction coefficient. Both the maximum and mean values of the crush force increase as a result of increasing the blank or tube thickness. Generally, sensitivities to the geometric parameters seem to be greater than those of the manufacturing process parameters.

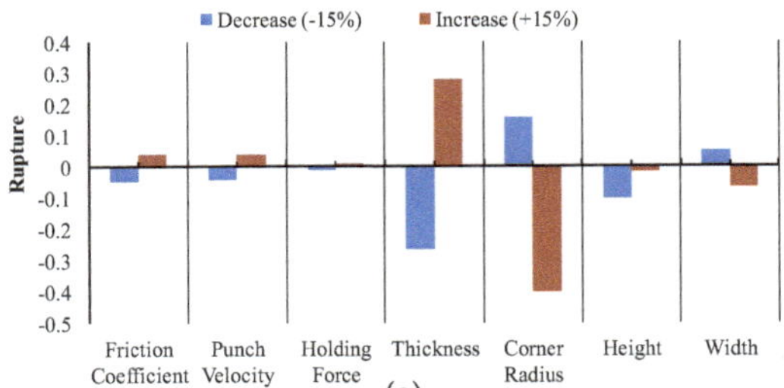

(a)

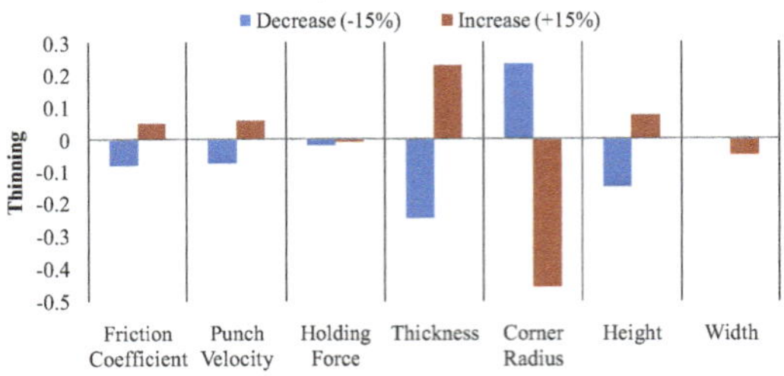

(b)

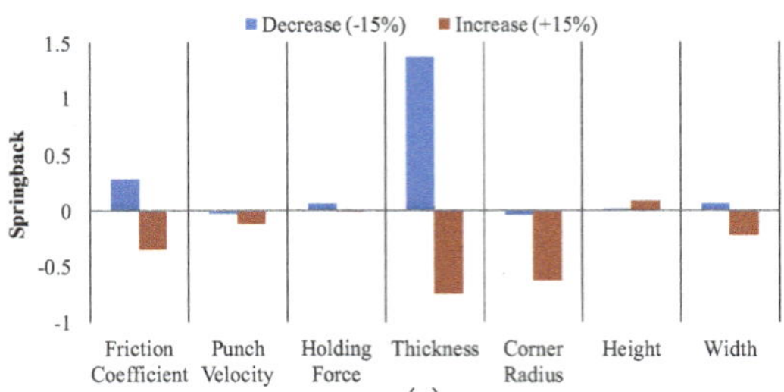

(c)

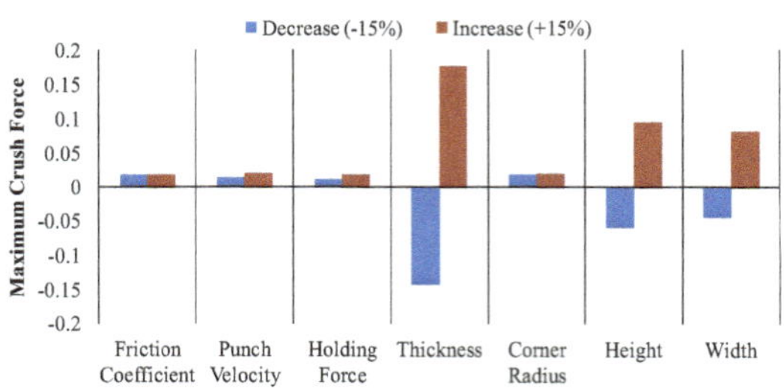

(d)

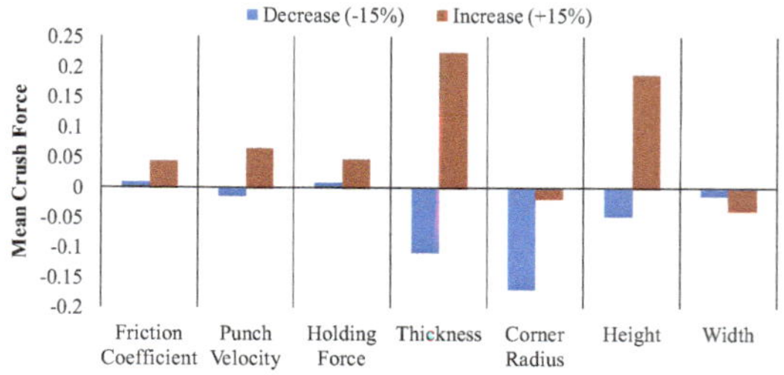

(e)

Figure 10. General sensitivity of rupture (a), thinning (b), springback (c), maximum crush force (d), and mean crush force (e) to the process and product design variables

5. Multi-objective design optimization

In a traditional process optimization problem, the manufacturing process objectives are optimized by varying the process control parameters (Sun et al., 2010; Wei and Yuying, 2008; Hu et al., 2008; Konak et al., 2006) whereas in product optimization, the geometry (e.g., shape, size) is altered to enhance the product performance. However, in a coupled process-performance optimization problem as considered here, both manufacturing- and performance-level attributes are optimized. When faced with competing objectives, the resulting multi-objective optimization problem becomes one of finding not just one but a collection of non-dominated design points that form the Pareto frontier. By specifying a particular target value for each objective, the multi-objective optimization problem is expressed as

$$\min_{\boldsymbol{x} = x_1, \ldots, x_7} \left\{ \begin{array}{c} (R(\boldsymbol{x}) - TR)^2, (T(\boldsymbol{x}) - TT)^2, (S(\boldsymbol{x}) - TS)^2, \\ (P_m(\boldsymbol{x}) - TPM)^2, (P_{max}(\boldsymbol{x}) - TPX)^2, (M(x_1, \ldots, x_4) - TM)^2 \end{array} \right\}$$

$$\begin{aligned} \text{s.t.} \quad & 40 \le x_1 \le 70 \text{ mm} \\ & 20 \le x_2 \le 35 \text{ mm} \\ & 2.5 \le x_3 \le 7.5 \text{ mm} \\ & 1.0 \le x_4 \le 2.5 \text{ mm} \\ & 10 \le x_5 \le 50 \text{ kN} \\ & 2 \le x_6 \le 10 \text{ m/s} \\ & 0.1 \le x_7 \le 0.35 \end{aligned} \tag{17}$$

where design variables x_1 to x_7 represent width, height, corner radius, thickness, holding force, punch velocity, and friction coefficient, respectively with each having both lower- and upper-bound side constraints. Also, $R(\boldsymbol{x})$ is rupture, $T(\boldsymbol{x})$ is thinning, $S(\boldsymbol{x})$ is springback, $P_m(\boldsymbol{x})$ is the mean crush force, $P_{max}(\boldsymbol{x})$ is the maximum crush force, and $M(\boldsymbol{x})$ is mass with *TR, TT, TS, TPM, TPX*, and *TM* as the corresponding target values, defined later in this section. The blank length is always equal to the tube length of 250 mm, whereas the blank width is selected to be twice as long as a single hat section's perimeter (developed width).

Considering the computational complexity and cost of coupled nonlinear FE simulations, reduced-order or surrogate models are used to approximate the responses defined in Eq. (17). Different metamodeling techniques have been developed and reported in the literature. Although they vary in terms of complexity and accuracy, they all rely on measured responses at the selected training points in the design space to find the unknown coefficients of a specific metamodel such that the approximation error is less than an acceptable threshold.

Radial basis functions (RBF) have been shown (Fang et al., 2005; Wang and Shan, 2007; Acar and Rais-Rohani, 2008; Parrish et al., 2012) to be suitable for approximating highly nonlinear responses using relatively small number of training points. In RBF formulation (Fang and Horstemeyer, 2006), the approximate response $\hat{f}(y)$ at a design point defined by the normalized design variable vector y is found as

$$\hat{f}(y) = \sum_{i=1}^{M} \lambda_i \phi(\|y - y_i\|) \tag{18}$$

where y_i is the vector of normalized design variables at the *ith* training point, $\|y - y_i\| = r_i$ is the Euclidian norm or distance from design point y to the ith training point, and M is the total number of functions included in the summation. The λ_i parameters are the unknown interpolation coefficients with ϕ representing the radially symmetric basis function that can take different forms. We considered both the multiquadric $\phi(r) = \sqrt{r^2 + c^2}$ and Gaussian $\phi(r) = \exp(-cr^2)$ basis functions, where c is a tuning parameter that can vary in the range of $0 < c \leq 1$ depending on the selected response.

In the coupled process-performance FE simulations, it is necessary to perform the deep-drawing, springback, and crush simulations in sequence. However, once a stand-alone surrogate model for each response is built, all the responses can be evaluated simultaneously, which provides considerable computational cost savings in the design optimization analysis. Using the Latin hypercube sampling (LHS) to produce a uniform distribution of design points, fifty training points were generated in the seven design-variable (7-dimensional) design space. Table 2 lists the training points and the corresponding values of the selected design variables. The maximum and minimum values for each design variable are shown in bold. Ten additional random design points are also generated as test points to measure the accuracy of the surrogate models.

Six responses are extracted for each set of simulations at each training point with the calculated response values listed in Table 3, where the value selected as the target for each response is shown in bold.

Two error metrics are considered. For the cross-validation normalized root-mean-square error (NRMSE) estimation (Lin et al., 1999), a metamodel is created using all except one training point and then the predicted response at the omitted point is compared to the corresponding true response value to measure the approximation error. This process is repeated for all the training points and the average is used as the overall error of the metamodel. NRMSE is calculated using

Point	x_1 (mm)	x_2 (mm)	x_3 (mm)	x_4 (mm)	x_5 (kN)	x_6 (m/s)	x_7	Point	x_1 (mm)	x_2 (mm)	x_3 (mm)	x_4 (mm)	x_5 (kN)	x_6 (m/s)	x_7
1	60.44	27.19	4.91	1.26	31.93	5.86	0.18	26	52.98	31.62	4.94	1.34	**21.35**	4.90	0.29
2	40.00	35.14	4.58	1.60	36.68	**7.76**	0.23	27	70.88	21.22	5.42	2.19	23.66	6.47	0.27
3	53.14	25.15	5.72	**2.27**	31.18	6.26	0.22	28	64.74	25.38	5.59	1.63	31.58	4.29	0.26
4	56.48	27.61	4.86	1.58	29.09	7.02	0.25	29	61.85	34.06	5.98	1.29	22.27	6.81	**0.16**
5	62.84	34.42	3.74	1.42	30.85	6.70	0.18	30	47.60	21.91	6.12	2.24	32.31	6.32	0.24
6	44.85	24.15	6.41	1.84	38.00	4.69	0.25	31	39.66	32.03	3.91	1.93	30.46	7.09	0.21
7	56.30	32.39	5.13	1.91	29.42	6.60	0.19	32	46.59	21.26	3.73	1.74	23.37	4.71	0.26
8	42.46	23.00	5.10	2.14	26.58	5.77	0.17	33	57.54	30.83	4.58	1.82	25.82	6.76	0.24
9	63.58	26.23	4.68	2.19	38.47	4.62	0.22	34	43.90	29.55	5.52	1.62	26.97	6.92	0.20
10	54.23	33.49	3.99	1.28	**38.96**	6.56	0.20	35	65.50	31.16	4.81	1.55	24.56	7.45	0.24
11	48.30	29.98	6.03	2.00	36.15	6.05	**0.29**	36	66.23	35.09	6.16	**1.23**	34.19	5.82	0.28
12	43.44	26.91	4.38	2.11	34.52	4.42	0.24	37	50.61	23.55	3.68	1.51	28.36	4.79	0.16
13	42.84	29.06	5.19	1.78	32.73	6.13	0.22	38	54.81	19.89	**6.50**	1.66	21.68	5.26	0.18
14	67.81	20.25	5.27	1.33	33.95	5.38	0.21	39	**38.59**	32.69	5.76	2.15	33.21	5.56	0.27
15	46.24	24.20	6.35	1.37	28.02	7.30	0.23	40	50.36	26.68	5.65	1.76	35.66	5.99	0.17
16	69.79	20.21	4.43	1.53	27.13	**4.25**	0.27	41	55.16	21.57	3.82	1.88	24.72	5.67	0.19
17	51.31	33.00	4.31	2.03	25.58	4.95	0.19	42	48.47	20.71	**3.52**	1.70	35.24	6.40	0.21
18	**70.76**	**19.49**	4.48	1.71	35.85	7.37	0.17	43	61.19	22.46	5.43	2.08	26.22	7.70	0.28
19	52.31	28.09	5.35	2.05	37.52	5.03	0.27	44	67.06	23.50	3.59	2.02	22.06	5.15	0.26
20	41.12	22.63	4.72	1.46	37.64	5.47	0.25	45	58.65	**35.70**	5.87	1.44	30.33	5.10	0.16
21	59.51	28.37	4.15	1.98	23.96	5.60	0.21	46	49.22	25.76	4.07	1.39	37.08	7.53	0.21
22	57.95	29.43	5.81	1.49	22.61	7.18	0.28	47	68.86	28.68	4.17	1.95	34.92	7.59	0.19
23	65.67	30.75	4.26	2.09	25.26	6.21	0.23	48	41.67	33.31	6.23	1.68	27.49	5.29	0.20
24	45.56	25.88	6.28	1.80	29.67	7.26	0.28	49	60.22	24.72	3.96	1.41	33.33	4.49	0.17
25	64.11	34.45	5.05	2.23	23.08	4.55	0.23	50	69.41	30.27	5.92	1.86	28.69	6.94	0.28

Table 2. Selected training points and design variable values in the process-product design space

$$NRMSE = \sqrt{\frac{1}{K}\sum_{i=1}^{K}(f_i - \hat{f}_i)^2 \Big/ (f_{max} - f_{min})} \tag{19}$$

where K is the number of training points, f_i is the actual response obtained from the coupled FE simulations (expect for mass), and $\hat{f}_i$ is the response predicted by the model that excludes the contribution of the ith point with f_{max} and f_{min} as the maximum and minimum values of the response, respectively.

The second error metric is obtained by fitting a metamodel based on all the training points and taking the absolute value of the difference between the approximate and the true response values at each of the selected test points with the average error of all the test points as the final indicator of metamodel accuracy.

The metamodels were tuned by selecting the parameter c and the RBF formulation that produced the least error for each response Table 4 shows the RBF tuning parameter and formulation used for each response and the corresponding NRMSE and average error

Point	R	T	S	P_{max} (kN)	P_m (kN)	M (kg)	Point	R	T	S	P_{max} (kN)	P_m (kN)	M (kg)
1	538.3	36.6	0.21	95.5	30.8	0.13	26	1003.5	88.4	1.83	101.0	34.7	0.14
2	1911.8	173.2	0.29	116.3	43.3	0.16	27	1672.9	109.0	0.31	160.9	62.1	0.22
3	1938.2	119.0	0.68	160.2	67.4	0.21	28	881.1	62.5	1.72	123.9	41.6	0.17
4	1370.0	116.8	0.79	116.8	43.1	0.16	29	455.3	32.0	2.56	109.6	33.8	0.15
5	1477.6	141.8	3.60	123.2	35.7	0.16	30	1521.7	81.5	1.16	144.6	63.7	0.18
6	870.2	45.3	2.10	114.8	54.0	0.15	31	3328.9	298.8	1.18	136.8	57.0	0.18
7	1697.4	139.7	0.34	154.0	60.0	0.20	32	2123.5	165.4	0.40	102.7	36.2	0.14
8	1774.8	107.1	0.92	132.0	54.3	0.17	33	2073.2	184.0	0.44	142.8	51.7	0.19
9	2098.0	161.2	0.50	167.0	66.3	0.22	34	1112.4	85.7	1.78	112.4	54.9	0.15
10	1253.6	117.3	3.54	100.6	30.1	0.14	35	1282.8	111.4	1.48	132.0	42.5	0.17
11	1663.9	123.8	1.36	145.5	54.7	0.19	36	633.2	59.6	1.63	108.4	33.9	0.15
12	2470.5	186.3	0.50	140.6	53.6	0.18	37	1427.3	112.1	1.03	96.6	28.8	0.13
13	1651.3	124.6	1.62	121.3	50.0	0.16	38	476.2	15.9	1.48	103.1	42.2	0.14
14	538.4	32.1	0.62	92.3	30.0	0.13	39	2113.2	154.7	0.85	153.6	64.9	0.20
15	573.7	34.9	1.16	86.4	31.6	0.12	40	1003.8	66.3	1.81	122.2	46.3	0.16
16	1115.1	78.7	0.59	108.5	37.0	0.15	41	2158.6	168.0	0.83	121.7	45.1	0.16
17	2118.4	184.5	1.77	158.6	50.3	0.21	42	2313.8	180.7	0.50	100.3	34.9	0.13
18	1288.4	92.9	0.32	120.6	39.6	0.17	43	1779.5	122.4	0.42	145.2	60.6	0.20
19	1971.0	141.4	0.77	148.5	57.9	0.20	44	3037.0	251.1	0.70	149.3	49.0	0.20
20	1148.4	81.1	1.34	84.0	29.9	0.11	45	528.6	36.1	1.04	123.7	45.6	0.17
21	2188.3	184.8	0.94	151.8	51.8	0.20	46	1497.3	133.9	1.21	93.0	30.5	0.12
22	938.5	76.0	1.68	115.4	43.1	0.15	47	2242.6	195.4	0.61	162.3	59.7	0.22
23	2675.3	228.7	1.16	175.1	54.0	0.23	48	818.8	57.2	2.95	124.5	47.1	0.16
24	1213.2	82.4	1.66	119.2	45.0	0.16	49	1017.6	78.1	0.66	100.6	30.9	0.14
25	1940.7	171.3	1.422	195.1	68.1	0.26	50	1265.0	103.4	0.81	158.0	53.4	0.21

Table 3. Results from sequential coupled process-performance simulations at the training points

estimates. It is seen that the average errors for the test points also validate the surrogate models created for the optimization problem. In order to enhance the accuracy of the metamodels for thinning and spring back responses, the actual responses are transformed using a logarithmic function as presented in Table 4.

Response	RBF Type[a]	Response Transformation	c	NRMSE (%)	Average Error (%)
Rupture, R	G	None	1.000	2.2	9.2
Thinning, T	G	$\ln(T)$	0.001	2.0	2.4
Springback, S	M	$\ln(1x10^7 S)$	0.500	1.8	3.5
Max Crush Force, P_{max}	M	None	0.500	4.5	2.0
Mean Crush Force, P_m	M	None	0.100	3.7	5.6
Mass, M	G	None	0.010	4.6	1.8

[a]G = Gaussian, M = Multiquadric

Table 4. Metamodel type, parameter, and approximation error for each response

6. Optimization results and discussion

The surrogate-based design optimization problem in Eq. (17) is solved using the multi-objective genetic algorithm (MOGA) toolbox in MATLAB (Fonseca and Fleming, 1993) with a randomly generated initial population size of 105 representing different combinations of design variable values within the specified bounds in Eq. (17). The subsequent generations are populated using the tournament selection algorithm with a crossover fraction of 80% using intermediate crossover function and the termination function tolerance of 1e-4. Stopping criterion is set at generation number 1400. The specific steps taken in the application of MOGA to this problem are as follows:

1. Design variables expressed in real number are converted into bit strings.
2. A random initial population is generated.
3. Using a fitness function, members of the population are examined by
    - assigning a rank to each solution based on non-dominated front (Sun et al., 2010).
    - assigning a fitness value based on Pareto ranking.
    - calculating the niche count of each solution.
    - calculating the shared fitness value of each solution.
    - normalizing the fitness values by using share fitness values.
4. Using a stochastic method to select parents for the next generation.
5. Performing crossover and mutation operations.
6. Establishing a new population.
7. Evaluating the population attributes.
8. Continuing steps 3 to 7 to evaluate all the objectives.
9. Selecting half of the individuals that have a higher rank than the rest.
10. Continuing the solution process until a stopping criterion is satisfied based on the average change in the spread of Pareto solution being less than the tolerance specified.

    The solution to the optimization problem in Eq. (17) for the Pareto optimum set converged after 139 GA generations and 14,701 function calls. The forty-five design points forming the Pareto frontier are listed in Table 5 in particular order. For ease of detection, the Pareto ID number and the corresponding best value for each objective are highlighted in the table. The results show that Pareto ID nos. 3, 4, 5, 6, 7, and 9 have the best values for rupture, springback, thinning, max crush force, mass, and mean crush force, respectively. A closer examination of the response values in Table 5 reveals that the response values at two or more design points may be very close to each other or nearly equal. This is because of the nonlinear distribution of the points on the Pareto frontier. For all the objectives, the preferred value is the smallest one except for the mean crush force. This is because the tube's energy absorption capacity improves by increasing the mean crush force value. However, it is preferred to reduce the max crush force for such components.

The range of variation over the Pareto frontier is different from one objective to another. Specifically, the relative variations from the best to the worst values are approximately 757%, 20%, 1354%, 133%, -59%, and 145% for rupture, thinning, springback, max crush force, mean crush force, and mass, respectively. The most significant range of variation in

Pareto ID	Design Variables							Responses					
	x_1 (mm)	x_2 (mm)	x_3 (mm)	x_4 (mm)	x_5 (kN)	x_6 (m/s)	x_7	R	T	S	P_{max} (kN)	P_m (kN)	M (kg)
1	43.35	27.93	4.86	1.28	25.06	6.95	0.20	848.0	17.28	2.25	86.0	36.2	0.11
2	48.98	30.77	6.36	1.55	22.50	6.24	0.23	624.2	17.04	2.32	112.5	46.2	0.15
3	40.44	31.85	5.55	1.31	26.93	5.12	0.18	**455.4**	16.65	3.16	95.9	38.8	0.12
4	58.36	33.16	5.35	2.09	32.40	6.85	0.23	2217.0	18.38	**0.24**	171.7	64.0	0.23
5	65.41	22.66	6.28	1.28	26.72	6.08	0.22	684.0	**15.93**	1.52	93.7	32.3	0.13
6	51.72	24.55	6.09	1.30	27.16	5.76	0.21	474.8	16.26	1.63	**84.0**	30.3	0.12
7	54.92	22.86	5.55	1.28	25.34	6.73	0.21	594.6	16.54	1.09	85.1	29.1	**0.11**
8	65.80	35.05	6.32	2.26	32.40	6.68	0.20	1543.5	18.16	1.21	195.4	71.4	0.27
9	63.30	35.61	6.44	2.26	32.41	6.68	0.20	1533.6	18.19	1.14	193.7	**71.6**	0.27
10	58.99	34.75	6.13	2.24	31.87	6.57	0.23	1924.1	18.27	0.61	186.9	69.5	0.25
11	40.05	31.91	6.33	1.65	28.18	5.00	0.18	569.7	16.78	3.39	119.3	47.8	0.15
12	43.97	31.87	3.88	2.25	29.99	6.46	0.27	3902.1	18.99	1.21	162.3	62.0	0.21
13	41.42	31.79	5.98	1.85	28.20	5.22	0.19	1143.4	17.41	2.98	133.9	53.4	0.17
14	44.79	31.86	3.74	2.13	29.97	6.42	0.27	3751.6	19.06	1.23	153.8	57.7	0.20
15	39.44	31.91	6.46	1.35	28.18	5.00	0.18	762.9	16.08	3.49	100.3	41.5	0.12
16	44.16	31.91	3.75	1.88	28.41	6.52	0.23	3130.2	18.91	1.88	134.9	51.3	0.18
17	40.52	31.89	4.45	1.99	28.33	5.08	0.28	2700.7	18.75	1.57	141.0	54.5	0.18
18	42.16	32.76	4.24	2.25	30.28	6.22	0.22	3606.3	18.83	1.76	164.0	63.7	0.21
19	55.93	33.29	5.12	2.25	31.21	6.57	0.24	2764.3	18.55	0.44	181.7	67.0	0.24
20	57.84	30.86	5.35	2.08	28.88	6.61	0.22	2046.6	18.21	0.51	166.7	63.6	0.22
21	40.67	31.85	5.88	1.63	27.44	5.15	0.19	811.9	17.13	3.09	117.3	46.9	0.15
22	41.36	31.88	6.13	1.68	28.18	5.16	0.18	724.3	17.01	3.20	121.7	48.4	0.16
23	48.48	31.92	3.91	2.16	30.45	6.46	0.27	3685.4	19.00	1.03	160.3	58.6	0.21
24	63.50	34.16	5.71	2.26	32.02	6.65	0.20	1983.8	18.29	0.78	192.6	70.3	0.26
25	40.06	31.96	6.34	1.51	29.18	5.57	0.22	503.7	16.94	2.88	108.1	44.2	0.14
26	57.74	33.56	5.22	2.17	32.15	6.61	0.23	2484.5	18.48	0.34	178.5	65.7	0.24
27	49.24	32.26	5.00	2.25	30.16	6.60	0.26	3062.1	18.60	0.80	170.3	64.7	0.23
28	40.60	32.47	3.78	2.19	28.72	6.34	0.26	3871.6	19.04	1.52	156.8	61.3	0.20
29	54.19	32.99	5.00	2.12	31.53	6.66	0.25	2706.8	18.57	0.42	168.1	62.1	0.22
30	42.42	31.88	5.24	2.02	29.14	6.09	0.19	2195.0	18.19	2.24	146.3	59.3	0.19
31	39.63	31.85	6.15	1.81	28.19	5.01	0.18	951.1	17.18	3.27	130.6	52.5	0.17
32	42.26	33.15	4.45	2.24	31.63	6.62	0.19	3423.0	18.75	1.93	164.0	64.4	0.21
33	41.73	27.49	3.92	1.81	28.97	5.97	0.28	2648.1	18.71	1.45	119.0	44.8	0.16
34	44.99	31.29	3.75	2.17	30.27	6.12	0.26	3741.3	19.03	1.33	155.9	58.2	0.21
35	43.41	32.29	5.04	1.74	29.59	5.88	0.21	1699.2	18.02	2.18	125.7	49.7	0.17
36	40.14	31.90	3.97	1.91	29.01	6.02	0.21	2982.7	18.75	2.33	135.1	53.6	0.17
37	41.30	31.81	5.54	1.93	28.85	5.41	0.19	1651.7	17.82	2.65	139.1	55.7	0.18
38	43.90	32.73	4.56	2.08	31.34	6.27	0.23	3026.6	18.65	1.47	152.4	58.8	0.20
39	60.43	34.90	6.28	2.26	32.08	6.59	0.19	1630.5	18.13	0.95	190.4	71.3	0.26
40	40.71	32.24	4.97	1.43	29.79	4.99	0.23	1029.4	17.67	2.55	101.7	38.7	0.13
41	42.69	34.76	5.63	2.10	29.45	5.17	0.21	2003.1	18.10	2.15	158.7	60.3	0.21
42	61.52	32.37	5.47	2.22	30.16	6.52	0.23	2230.0	18.30	0.53	185.4	67.8	0.25
43	56.76	32.84	5.36	2.25	32.40	6.61	0.21	2405.2	18.37	0.49	182.7	69.0	0.24
44	57.81	33.58	5.24	2.25	32.20	6.58	0.23	2609.8	18.50	0.41	184.9	68.0	0.25
45	52.62	34.49	4.55	2.26	31.66	6.56	0.27	3337.9	18.85	0.69	177.9	64.2	0.24

Table 5. Design points on the Pareto optimum frontier

the springback reveals the high sensitivity of this response, relative to the rest, to changes in the design variable values. In comparison, thinning seems to be affected the least by the changes in design. Such information helps in identifying the critical responses for both process and product design considerations.

Given the six-dimensional space of the process-product criteria space, the process and performance objectives are plotted separately and shown in Fig. 11. In addition, a sample subset of the Pareto set is selected with the individual tube geometries and crush deformation modes shown in Fig. 12. Among the six points shown, design points 6, 11, and 33 are shown among the best choices to minimize the mass and maximum crush force.

The results indicate that the Pareto set consists mostly of a tube design that is larger in total height than width with approximately 72% and 67% having larger thickness and longer corner radius than the respective average values, respectively. The general trend appears to be toward a tube design model with dissimilar width and height dimensions, which can be traced to two contributing factors: (1) while only a portion of width is work hardened, the entire height section undergoes plastic deformation during the forming process; and (2) the flanges (short tabs) in the double-hat geometry (see Fig. 9) influence how the different sides of the tube deform and contribute to the crush energy absorption. In most cases, it appears to be preferable for the holding force to be less than its average value in approximately 60% of the Pareto set. A nearly equal percentage prefers a lower friction coefficient while a slightly higher portion (roughly 67%) prefers a higher punch velocity than the respective average values.

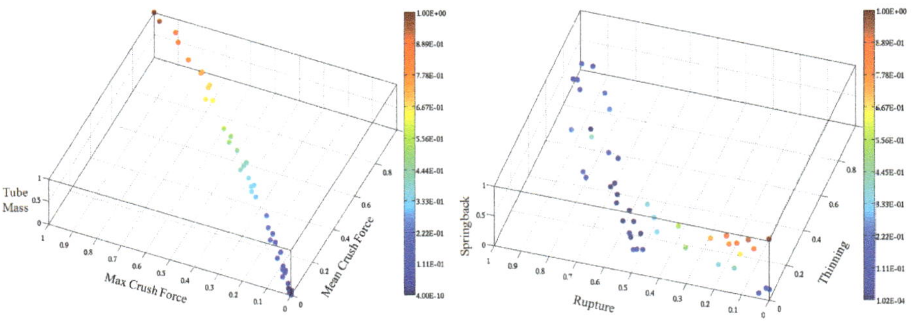

Figure 11. Distribution of Pareto optimal set in performance (a) and process (b) criteria subspaces

To measure the approximation error in the optimum process and performance objective values, a complete verification simulation was performed on seven samples with the relative errors shown in Table 6. Most error values are fairly low, less than 5%, with the highest reaching 14.7%, which is reasonable for the types of simulation involved and the approximation techniques used in solving the design optimization problem.

Figure 12. Selected design points from the Pareto set

Pareto ID	R (%)	T (%)	S (%)	P_{max} (%)	P_m (%)	M (%)
4	7.8	1.6	4.7	1.1	0.4	0.4
6	5.7	3.8	4.3	3.1	12.2	0.3
11	12.2	1.0	6.1	2.5	14.7	0.7
18	5.6	0.8	1.0	1.9	5.3	0.1
23	9.1	0.7	4.3	1.1	1.0	0.1
33	0.9	2.9	1.9	1.3	2.9	0.9

Table 6. Relative Error in Responses at Selected Pareto Points

7. Conclusion

A methodology for concurrent process-product design optimization using coupled nonlinear finite-element (FE) simulations was presented and applied to a sheet formed component made of AZ31 magnesium alloy. All FE simulations were performed using the Abaqus Explicit and Standard solvers. Surrogate models based on radial basis functions were developed for process and performance response approximations to facilitate the numerical multi-objective design optimization process.

The results of this investigation lead to the following conclusions:

- Material and component geometry variations can be modeled using a sequence of coupled nonlinear FE simulations with careful transfer of state variables and other information from one simulation stage to the next.

- Both the manufacturing and geometric design variables can have significant influence on the energy absorption behavior of the formed tube considered.
- All process responses (i.e., rupture, thinning, and springback) were greatly influenced by the initial blank thickness value and corner radius.
- The results of the multi-objective optimization problem highlighted different levels of conflict among the process and performance objectives considered. Moreover, the variation of objectives over the Pareto frontier indicated differing levels of sensitivity to changes in the design properties with springback being the most noteworthy.

Author details

Ali Najafi and Masoud Rais-Rohani
Mississippi State University, USA

Acknowledgement

This material is based on the work supported by the US Department of Energy under Award Number DE-EE0002323. Authors also acknowledge the collaboration provided through the SIMULIA Research & Development program under which licenses of Abaqus were provided.

8. References

Acar, E. & Rais-Rohani, M. (2008). Ensemble of Metamodels with Optimized Weight Factors. *Structural and Multidisciplinary Optimization*, Vol.37, No.3, pp.279-294

Acar, E.; Rais-Rohani, M., Najafi, A., Marin, E. B. & Bammann, D. (2009). Concurrent Design of Product-Material Systems Using Multilevel Optimization. *Proceedings of the 50th AIAA/ASME/ASCE/AHS/ASC Structures, Structural Dynamics, and Materials Conference*, Palm Springs, CA, USA, May 4-7, 2009

Bayram, Y.B. & Nied, H.F. (2000). Enriched Finite Element-Penalty Function Method for Modeling Interface Cracks With Contact. *Engineering Fracture Mechanics*, Vol.65, pp.541-557

Bottcher, C.S. & Frik, S. (2003). Consideration of Manufacturing Effects to Improve Crash Simulation Accuracy, 4th European LS-Dyna Users Conference, 2003

Cheng, J.H. & Kikuchi, N. (1985). An Analysis of Metal Forming Processes Using Large Deformation Elastic-Plastic Formulations. *Computer Methods in Applied Mechanics and Engineering*, Vol.49, pp.71-108

Chung, W.J.; Cho, J.W. & Belytschko, T. (1998). On the Dynamic Effects of Explicit FEM in Sheet Metal Forming Analysis. *Engineering Computations*, Vol.15, pp. 750–776

Dagson, N. (2001). Influence of the Forming Process on the Crash Response of a Roof Rail Component, Master Thesis, Department of Solid Mechanics, Linkping University, March 2001.

Fang, H. & Horstemeyer, M. F. (2006). Global Response Approximation with Radial Basis Functions. *Engineering Optimization*, Vol.38, No.4, pp.407–424

Fang, H.; Rais-Rohani, M., Liu, Z. & Horstmeyer, M.F. (2005). A Comparative Study of Metamodeling Methods for Multi-objective Crashworthiness Optimization. *Computers and Structures*, Vol.83, pp.2121-2136

Fonseca, M. & Fleming, P. J. (1993). Genetic Algorithms for Multiobjective Optimization: Formulation, Discussion And Generalization, *Proceedings of the 5th International Conference on Genetic Algorithms*, San Francisco, CA, USA: Morgan Kaufmann Publishers Inc., pp.416-423, 1993

Grantab, R. (2006). Interaction between Forming and Crashworthiness of Advanced High Strength Steel S-Rails, PhD dissertation, University of Waterloo, Canada, 2006

Hu, W.; Yao, L.G. & Hua, Z. Z. (2008). Optimization of Sheet Metal Forming Processes by Adaptive Response Surface Based on Intelligent Sampling Method. *Journal of Materials Processing Technology*, Vol.197, No.1-3, pp.77-88

Kaufman, M.; Gaines, D., Kundrick, K. & Liu, S. D. (1998). Integration of Chassis Frame Forming Analysis Into Performance Models to More Accurately Evaluate Crashworthiness. *SAE Proceedings*, paper no. 980551, 1998

Kellicut, A.; Cowell, B., Kavikondala, K., Dutton, T., Iregbu, S. & Sturt, R. (1999). Application of the Results of Forming Simulation in Crash Models. *Proceedings of Numisheet 99 Conference*, Besancon, France, pp. 509–514, 1999

Konak, A.; Coit, D.W. & Smith, A. E. (2006). Multi-Objective Optimization Using Genetic Algorithms: A Tutorial. *Reliability Engineering & System Safety*, Vol.91, No.9, Special Issue - Genetic Algorithms and Reliability, pp.992-1007

Krusper, A. (2003). Influences of the Forming Process on the Crash Performance - Finite Element Analysis, Master Thesis, Department of Structural Mechanics, Chalmers University of Technology, May 2003

Lee, S.; Kwon, Y.N., Kang, S.H., Kim, S.W. & Lee, J.H. (2008). Forming Limit of AZ31 Alloy Sheet and Strain Rate on Warm Sheet Metal Forming, *Journal of Materials Processing Technology*, Vol.201, No.1-3, pp.431-435

Li, K. P.; Carden, W. P. & Wagoner, R.H. (2002). Simulation of Springback, *International Journal of Mechanical Sciences*, Vol.44, pp.103-122

Lin, Y.; Krishnapur, K., Allen, J.K. & F. Mistree. (1999). Robust Design: Goal Formulations and a Comparison of Metamodeling Method, *Proceedings of DETC-99, 1999 ASME Design Engineering Technical Conferences*, Las Vegas, ND, September 12-15, 1999.

Lubliner J. (2006). *Plasticity Theory*, Revised Edition

Mayer, R. (2004). Theoretical Effects of Hydroforming on Crashworthiness of Straight Sections. *ASME Applied Mechanics Division*, Vol.255, 591-603

McDowell, D.L.; Choi, H. J., Panchal, J.H., Austin, R., Allen, J.K. & Mistree, F. (2007). Plasticity-Related Microstructure-Property Relations for Materials Design. *Key Engineering Materials*, Vol. 340-341, pp. 21-30

Najafi, A. (2011). Coupled Sequential Process-Performance Simulation and Multi-Attribute Optimization of Structural Components Considering Manufacturing Effects, PhD Dissertation, Computational Engineering, Mississippi State University, 2011.

Najafi, A.; Marin, E.B. & Rais-Rohani, M. (2012). Concurrent Multi-Scale Crush Simulations with a Crystal Plasticity Model. *Thin-Walled Structures*, Vol.53, pp. 176-187

Noels, L.; Stainier, L. & Ponthot, J.P. (2004). Combined Implicit/Explicit Time Integration Algorithms for the Numerical Simulation of Sheet Metal Forming. *Journal of Computational and Applied Mathematics*, Vol.168, pp. 331-339

Oden, J.T.; Kikuchi, N. & Song, Y.J. (1982). Penalty-Finite Element Methods for the Analysis of Stokesian Flows. *Computer Methods in Applied Mechanics and Engineering*, Vol.31, pp.297-326

Oden, J.T. & Pires, E.B. (1983). Numerical Analysis of Certain Contact Problems in Elasticity With Non-Classical Friction Laws. *Computers & Structures*, Vol.16, pp.481-485

Oliveira, D.; Worswick, M., Grantab R., Williams B. & Mayer R. (2006). Effect of Forming Process Variables on the Crashworthiness of Aluminium Alloy Tubes. *International Journal of Impact Engineering*, Vol.32, No. 5, pp. 826–846

Olson, G. B. (1997). Computational Design of Hierarchically Structured Materials. *Science*, Vol.277, No.5330, pp. 1237-42

Parrish, A.; Rais-Rohani, M. & Najafi, A. (2012). Crashworthiness Optimization of Vehicle Structures with Magnesium Alloy Parts, *International Journal of Crashworthiness*, Vol.17, Issue 3, pp. 259-281

Rebel, G.; Park, K. C. & Felippa, C.A. (2002). A Contact Formulation Based on Localized Lagrange Multipliers: Formulation and Application to Two Dimensional Problems. *International Journal for Numerical Methods in Engineering*, Vol.54, pp. 263-297

Ryou, H.; Chung K., Yoon J. & Han C. (2005). Incorporation of Sheet-Forming Effects in Crash Simulations Using Ideal Forming Theory and Hybrid Membrane and Shell Method. *Journal of Manufacturing Science and Engineering*, Vol.127, No.1, pp. 182–192

Simo, J. C. & Hughes, T. J. R. (2000). *Computational Inelasticity*, Springer, New York

Simunovic, S. & Aramayo G. (2002). Steel Processing Properties and Their Effect on Impact Deformation of Lightweight Structures. Computer Science and Mathematics Division, Computational Materials Science Group, American Iron and Steel Institute

Souza Neto, E. A.; Peric, D. & Owen, D.R.J. (2008). *Computational Methods for Plasticity: Theory and Applications*, John Wiley & Sons

Sun, G.; Li, G. Gong, Z., Cui, X., Yang, X. & Li, Q. (2010). Multiobjective Robust Optimization Method for Drawbead Design in Sheet Metal Forming. *Materials & Design*, Vol.31, Issue 4, pp.1917-1929

van den Boogaard, A.H.; Meinders, T. & Huétink, J. (2003). Efficient Implicit Finite Element Analysis of Sheet Forming Processes, *International Journal for Numerical Methods in Engineering*, Vol.56, No.8, pp. 1083–1107

Wang, N-M. & Budiansky, B. (1978). Analysis of Sheet Metal Stamping by Finite Element Method. *Journal of Applied Mechanics, Transaction ASME*, Vol.45, pp. 73–82

Wang, G. & Shan, S. (2007). Review of Metamodeling Techniques in Support of Engineering Design Optimization, *Journal of Mechanical Design*, Vol.129, No.4, pp.370–380

Wei, L. & Yuying, Y. (2008). Multi-Objective Optimization of Sheet Metal Forming Process Using Pareto-Based Genetic Algorithm. *Journal of Materials Processing Technology*, Vol.208, No.1-3, 21, pp. 499-506

Williams, B.W.; Simha, C.H.M., Abedrabbo, N. Mayer, R. & Worswick, M.J. (2010). Effect of Anisotropy, Kinematic Hardening, and Strain-Rate Sensitivity on the Predicted Axial Crush Response of Hydroformed Aluminum Alloy Tubes. *International Journal of Impact Engineering*, Vol.37, No.6, pp. 652-661

Williams, B.W.; Oliveira, D.A., Worswick, M.J. & Mayer, R. (2005) Crashworthiness of High and Low Pressure Hydroformed Straight Section Aluminum Tubes. *Proceedings of SAE World Congress*, paper No. 2005-01-0095, 2005

Wriggers, P. (2006). *Computational Contact Mechanics*, Springer; 2nd edition

Permissions

The contributors of this book come from diverse backgrounds, making this book a truly international effort. This book will bring forth new frontiers with its revolutionizing research information and detailed analysis of the nascent developments around the world.

We would like to thank Dr. Farzad Ebrahimi, for lending his expertise to make the book truly unique. He has played a crucial role in the development of this book. Without his invaluable contribution this book wouldn't have been possible. He has made vital efforts to compile up to date information on the varied aspects of this subject to make this book a valuable addition to the collection of many professionals and students.

This book was conceptualized with the vision of imparting up-to-date information and advanced data in this field. To ensure the same, a matchless editorial board was set up. Every individual on the board went through rigorous rounds of assessment to prove their worth. After which they invested a large part of their time researching and compiling the most relevant data for our readers. Conferences and sessions were held from time to time between the editorial board and the contributing authors to present the data in the most comprehensible form. The editorial team has worked tirelessly to provide valuable and valid information to help people across the globe.

Every chapter published in this book has been scrutinized by our experts. Their significance has been extensively debated. The topics covered herein carry significant findings which will fuel the growth of the discipline. They may even be implemented as practical applications or may be referred to as a beginning point for another development. Chapters in this book were first published by InTech; hereby published with permission under the Creative Commons Attribution License or equivalent.

The editorial board has been involved in producing this book since its inception. They have spent rigorous hours researching and exploring the diverse topics which have resulted in the successful publishing of this book. They have passed on their knowledge of decades through this book. To expedite this challenging task, the publisher supported the team at every step. A small team of assistant editors was also appointed to further simplify the editing procedure and attain best results for the readers.

Our editorial team has been hand-picked from every corner of the world. Their multi-ethnicity adds dynamic inputs to the discussions which result in innovative outcomes. These outcomes are then further discussed with the researchers and contributors who give their valuable feedback and opinion regarding the same. The feedback is then collaborated with the researches and they are edited in a comprehensive manner to aid the understanding of the subject.

Apart from the editorial board, the designing team has also invested a significant amount of their time in understanding the subject and creating the most relevant covers. They scrutinized every image to scout for the most suitable representation of the subject and create an appropriate cover for the book.

The publishing team has been involved in this book since its early stages. They were actively engaged in every process, be it collecting the data, connecting with the contributors or procuring relevant information. The team has been an ardent support to the editorial, designing and production team. Their endless efforts to recruit the best for this project, has resulted in the accomplishment of this book. They are a veteran in the field of academics and their pool of knowledge is as vast as their experience in printing. Their expertise and guidance has proved useful at every step. Their uncompromising quality standards have made this book an exceptional effort. Their encouragement from time to time has been an inspiration for everyone.

The publisher and the editorial board hope that this book will prove to be a valuable piece of knowledge for researchers, students, practitioners and scholars across the globe.

List of Contributors

Jamal Samir, Jamal Echaabi and Mohamed Hattabi
Applied Research Team on Polymers, Department of Mechanical Engineering, ENSEM, Hassan II University, Casa Blanca, Oasis, Casablanca, Morocco

Antonis P. Papadakis
Department of Electrical Engineering, Frederick University, Nicosia, Cyprus

A. Ghorbanpour Arani, R. Kolahchi, A. A. Mosalaei Barzoki and A. Loghman
Department of Mechanical Engineering, Faculty of Engineering, University of Kashan, Kashan, Iran

F. Ebrahimi
Department of Mechanical Engineering, Faculty of Engineering, International University of Imam Khomeini, Qazvin, Iran

Samia Dridi
Department of Mechanics of Structures and Materials of the ISAE, Institute Superieur of the Aeronautique and the Space, Edouard Belin, Toulouse, France

Khosrow Ghavami
Department of Civil Engineering, Pontificia Universidade Católica (PUC-Rio), Rio de Janeiro, Brazil

Mohammad Reza Khedmati
Faculty of Marine Technology, Amirkabir University of Technology, Tehran, Iran

B.H. Wu
Department of Mechanical and Power Engineering, Nanjing University of Technology, Nanjing, People's Republic of China

Mrityunjay R. Doddamani and Satyabodh M Kulkarni
Mechanical Engineering, National Institute of Technology Karnataka, Surathkal, India

Giovanni Leonardi and Michele Buonsanti
Faculty of Engineering, University of Reggio Calabria, Department of Mechanics and Materials - MECMAT, Reggio Calabria, Italy

Koji Hasegawa
Division of Information and Electronic Engineering, Graduate School of Engineering, Muroran Institute of Technology, Muroran, Japan

Takao Shimada
Department of Electrical and Electronic Engineering, Tsuyama National College of Technology, Tsuyama, Japan

Ahmad Ivan Karayan and Homero Castaneda
Department of Chemical and Biomolecular Engineering, The University of Akron, USA

Deni Ferdian, Sri Harjanto, Dwi Marta Nurjaya and Ahmad Ashari
Department of Metallurgy and Materials Engineering, University of Indonesia, Indonesia

Deni Ferdian, Sri Harjanto, Dwi Marta Nurjaya and Ahmad Ashari
Center for Materials Processing and Failure Analysis (CMPFA), University of Indonesia, Indonesia

Jairo A. Martins and István Kövesdy
Machines and Components Product Engineering, Metso Minerals, Sorocaba, Brazil

Toshihiro Saito
Honda R&D Co., Ltd. Automobile R&D Center, Japan

P.H. Darji
Department of Mechanical Engineering, C. U. Shah College of Engineering & Technology, Surendranagar, India

D.P. Vakharia
Department of Mechanical Engineering, S.V. National Institute of Technology, Surat, India

Shinobu Sakai
School of Mechanical Engineering, Kanazawa University, Japan

Hitoshi Nakayama
Development Division, Sunaga Kaihatsu Co., Ltd., Japan

YongAn Huang, Xiaoming Zhang and Youlun Xiong
State Key Laboratory of Digital Manufacturing Equipment and Technology, Huazhong University of Science and Technology, Wuhan, China

Ali Najafi and Masoud Rais-Rohani
Mississippi State University, USA